AF837025

HISTORICAL DICTIONARY

The historical dictionaries present essential information on a broad range of subjects, including American and world history, art, business, cities, countries, cultures, customs, film, global conflicts, international relations, literature, music, philosophy, religion, sports, and theater. Written by experts, all contain highly informative introductory essays of the topic and detailed chronologies that, in some cases, cover vast historical time periods but still manage to heavily feature more recent events.

Brief A–Z entries describe the main people, events, politics, social issues, institutions, and policies that make the topic unique, and entries are cross-referenced for ease of browsing. Extensive bibliographies are divided into several general subject areas, providing excellent access points for students, researchers, and anyone wanting to know more. Additionally, maps, photographs, and appendixes of supplemental information aid high school and college students doing term papers or introductory research projects. In short, the historical dictionaries are the perfect starting point for anyone looking to research in these fields.

HISTORICAL DICTIONARIES OF INTELLIGENCE AND COUNTERINTELLIGENCE

Jon Woronoff, Series Editor

British Intelligence, by Nigel West, 2005.
United States Intelligence, by Michael A. Turner, 2006.
Israeli Intelligence, by Ephraim Kahana, 2006.
International Intelligence, by Nigel West, 2006.
Russian and Soviet Intelligence, by Robert W. Pringle, 2006.
Cold War Counterintelligence, by Nigel West, 2007.
World War II Intelligence, by Nigel West, 2008.
Sexspionage, by Nigel West, 2009.
Air Intelligence, by Glenmore S. Trenear-Harvey, 2009.
Middle Eastern Intelligence, by Ephraim Kahana and Muhammad Suwaed, 2009.
German Intelligence, by Jefferson Adams, 2009.
Ian Fleming's World of Intelligence: Fact and Fiction, by Nigel West, 2009.
Naval Intelligence, by Nigel West, 2010.
Atomic Espionage, by Glenmore S. Trenear-Harvey, 2011.
Chinese Intelligence, by I. C. Smith and Nigel West, 2012.
Signals Intelligence, by Nigel West, 2012.

Historical Dictionary of Signals Intelligence

Nigel West

The Scarecrow Press, Inc.
Lanham • Toronto • Plymouth, UK
2012

Published by Scarecrow Press, Inc.
A wholly owned subsidiary of The Rowman & Littlefield Publishing Group, Inc.
4501 Forbes Boulevard, Suite 200, Lanham, Maryland 20706
www.rowman.com

10 Thornbury Road, Plymouth PL6 7PP, United Kingdom

Copyright © 2012 by Nigel West

All rights reserved. No part of this book may be reproduced in any form or by any electronic or mechanical means, including information storage and retrieval systems, without written permission from the publisher, except by a reviewer who may quote passages in a review.

British Library Cataloguing in Publication Information Available

Library of Congress Cataloging-in-Publication Data

West, Nigel.
Historical dictionary of signals intelligence / Nigel West.
p. cm.
Includes bibliographical references and index.
ISBN 978-0-8108-7187-8 (cloth : alk. paper) — ISBN 978-0-8108-7391-9 (ebook : alk. paper)
1. Electronic surveillance—History—Dictionaries. I. Title.
TK7882.E2W47 2012
327.12—dc23

2012014257

∞™ The paper used in this publication meets the minimum requirements of American National Standard for Information Sciences—Permanence of Paper for Printed Library Materials, ANSI/NISO Z39.48-1992.

Printed in the United States of America

In some quarters the idea got about, and was
not discouraged, that the British were
served by particularly intelligent and
active spies in Germany.
Sir Alfred Ewing
Lecture to the Edinburgh Philosophical Institution
13 December 1927

Of the many forms of espionage adopted by Germany,
the use of wireless telegraphy was one that, in the hands
of competent operators, promised many profitable
results to the enemy.
Daily Chronicle
12 November 1914

Failure in security of one's own communication
systems obviously lost ships, battles, if not lost
wars, and loses diplomatic conferences as surely
as a stud poker player who loses to opponents
who read his hole cards.
Lieutenant Commander A. S. Kramur, USN
8 June 1942

Contents

Acknowledgments

The author owes a debt of gratitude to Kenneth MacFarlan, Josef Garlinski, Jim Bamford, David Kahn, Robin Denniston, Cecil Phillips, Meredith Gardner, Robert Louis Benson, Bob Lamphere, Geoffrey Sudbury, Hayden Peake, Mick Smith, Ralph Erskine, Peter Calvocoressi, Harry Hinsley, John Croft, Tony Sale, Fred Winterbotham, Ursula and Jurgen Kuczinsky, and Jon Woronoff.

Editor's Foreword

In one sense, signals intelligence is the same thing it has always been, gathering intelligence by intercepting signals between people verbally or electronically. In every other sense, it has never stopped changing, and the speed of change keeps on accelerating. This is because of the revolution in communications, passing from personal conversation or over the phone, to telegraph and radio as well as the more recent electronic means, with each of these new channels increasingly complex. As if that were not enough, if the communications are crucial to the state, every effort is made to keep them from falling into the wrong hands. This has led SIGINT to delve into encryption of one's own messages and decryption of those of the opponents. All this is obviously not so simple, and major tasks range from signal detection, direction-finding, and traffic analysis to electronic order of battle. It is therefore not surprising that SIGINT has, especially in times of war but also in periods of peace, given rise to countless organizations, some of them truly mammoth in size. So they are also part of this story. But the most fascinating features still come down to human beings who have played a notable role.

This *Historical Dictionary of Signals Intelligence* is the latest addition to a series on intelligence and counterintelligence, which has largely covered the specific situation in major countries but now takes a look at one of the key features. Although barely known by the general public, SIGINT lies at the core of the whole process, and without the best possible personnel and machinery, no country can get far. Thus, the introduction has to explain not only what it is and does but also why it is so crucial. That there are indeed a lot of organizations becomes evident on looking at the many entries in the dictionary section, which presents the more important ones while also passing in review their key personnel and some of their targets. Other entries focus on the operations in specific countries or during major conflicts. Still, nothing can take the place of persons, individuals who run the organizations and ferret out the spies, many of whom actually outsmart the organizations (and, in that case, only appear in the dictionary if they finally get caught). Then, of course, there are entries on the various tasks and functions of SIGINT. Not surprisingly, there is need for a long list of acronyms, as nobody has time to call these bodies by their full names. And also there is a precious bibliography, not all that big as not all that much is known about the field and some of this is censored or blocked by the government.

This particularly useful addition to the series was written by Nigel West, who has already produced many other volumes, including those on British intelligence, international intelligence, Cold War counterintelligence, World War II intelligence, and naval intelligence. But this is just part of his output, since he has been publishing books in the field since 1980. This volume is just his latest but doubtlessly not his last contribution to enhancing our knowledge of a still pretty much unfathomable field.

Jon Woronoff
Series Editor

Acronyms and Abbreviations

AFB	Allgemeinefunkspruchbuch (German radio codebook)
AFESC	Air Force Electronic Security Command
AFISRA	Air Force Intelligence, Surveillance and Reconnaissance Agency
AFSA	Armed Forces Security Agency
AFSS	Air Force Security Service
AGI	Auxiliary General Intelligence
AGTR	Auxiliary General Technical Research
AI	Air Ministry Intelligence Branch
Amtorg	Amerikanskaya Torgovaya Kompaniya (American Trading Company)
ASA	Army Security Agency
ASIO	Australian Security Intelligence Organisation
ATS	Auxiliary Territorial Service
AVI	Afdeling Verbindingsinlichtingen (Dutch Signals Intelligence Service)
BND	Bundesnachrichtendienst (German Federal Intelligence Service)
BP	Bletchley Park
BRUSA	British–United States of America Communications Agreement
BSC	British Security Coordination
BVD	Binnenlandse Veiligheidsdienst (Dutch National Security Service)
CANUS	Canada–United States of America Communications Agreement
CBME	Combined Bureau Middle East
CBNRC	Communications Branch, National Research Council
CDAA	Circularly disposed antenna array
CIA	Central Intelligence Agency
CIFE	Combined Intelligence Far East
CNA	Computer network attack
COMINT	Communications intelligence

COMSEC	Communications security
CPGB	Communist Party of Great Britain
CPSU	Communist Party of the Soviet Union
CPUSA	Communist Party of the United States of America
CSE	Communications Security Establishment
CSO	New Zealand Combined Signals Organisation
CSO	Composite Signals Organisation
DCI	Director of central intelligence
D/F	Direction finding
DGI	Dirección General de Inteligencia (Cuban intelligence service)
DGSE	Direction Générale de la Sécurité Extérieure
DIA	Defense Intelligence Agency
DIRNSA	Director, National Security Agency
DIS	Danish Defence Intelligence Service
DSB	Defence Signals Branch
DSB	Defence Signals Bureau
DSD	Defence Signals Directorate
DSD	Defence Signals Division
DST	Direction de la Surveillance du Territoire (French security service)
DWS	Diplomatic Wireless Service
ECM	Electronic countermeasure
ELINT	Electronic intelligence
FAECT	Federation of Architects, Engineers, Chemists, and Technicians
FAPSI	Federalnoie Agentsvo Pravitelsvennoi Sviazi i Informatsi (Russian Federal Agency for Government Communications and Information)
FBI	Federal Bureau of Investigation
FCC	Federal Communications Commission
FECB	Far East Combined Bureau
FFB	*Flottenfunkspruchbuch* (Fleet radio codebook)
FISA	Foreign Intelligence Surveillance Act
FRA	Försvarets Radioanstalt (Swedish National Defense Radio Institute)
FRUMEL	Fleet Radio Unit Melbourne
FRUPAC	Fleet Radio Unit Pacific
FSO	Federal Protective Service
G-2	Irish Intelligence Service
GC&CS	Government Code and Cypher School
GCHQ	Government Communications Headquarters

GCR	Groupement des Contrôles Radio-électriques (French signals intelligence organization)
GCSB	Government Communications Security Bureau
GPO	General Post Office
GRU	Glavnoe Razvedyvatel'noe Upravlenie (Soviet Military Intelligence Service)
HCUA	House Committee on Un-American Activities
HFDF	High-frequency direction finding
HoW	Home Office Warrant
HVA	Hauptverwaltung Aufklärung (East German foreign intelligence service)
HVB	*Handelsverkehrsbuch* (German merchant shipping transport code)
IDB	Inlichtingendienst Buitenland (Dutch intelligence service)
IED	Improvised explosive device
INTELSAT	International Telecommunications Satellite Organization
JIC	Joint Intelligence Committee
JTUM	Joint Telecommunications Unit Melbourne
KGB	Komitei Gosudarstevnnoi Bezopasnosti (Soviet intelligence service)
LPG	London Processing Group
LSIB	London Signal Intelligence Board
MASINT	Measurement and signature intelligence
MfS	Ministerium für Staatssicherheit (East German Ministry of State Security)
MI1(b)	War Office signals intelligence branch
MI1(e)	War Office zeppelin branch
MI5	British Security Service
MI6	British Secret Intelligence Service
MI8	War Office Signals Intelligence Branch
MI14	War Office enemy order-of-battle section
MID	Militaire Inlichtingendienst (Dutch military intelligence service)
NATO	North Atlantic Treaty Organization
NKVD	Narodni Kommisariat Vnutrennih Dei (Soviet intelligence service)
NORUSA	Norway–United States of America Communications Intelligence Agreement
NRO	National Reconnaissance Office
NRSC	National Remote Sensing Center
NSA	National Security Agency
NSG	Naval Security Group

NSO	National Signals Intelligence Organization
NVA	North Vietnamese Army
OGPU	Obyedinennoye Gosudarstvennoye Politicheskoye Upravlenye (Soviet State Political Directorate)
ONI	Office of Naval Intelligence
OSS	Office of Strategic Services
OTP	One-time pad
PCO	Political Security Organization (Yemeni security service)
PCR	Police de Contrôle Radio-électronique (French radio security service)
PRC	People's Republic of China
PTT	Post, telegraph, and telephone service
RAAF	Royal Australian Air Force
RAB	Radio Analysis Bureau
RADINT	Radar intelligence
RAF	Royal Air Force
RAGFOR	Radio Analysis Group Forward
RAN	Royal Australian Navy
RFP	Radio fingerprinting
RID	Radio Intelligence Division
RIS	Radio Intelligence Service
RRU	Radio Research Unit
RSS	Radio Security Service
SAM	Surface-to-air missile
SCU	Special communications unit
SHAEF	Supreme Headquarters Allied Expeditionary Force
SIGINT	Signals intelligence
SIM	Servizio Informazioni Militare (Italian military intelligence service)
SIS	Secret Intelligence Service
SIS	Signal Intelligence Service
SKM	*Signalbuch der Kaiserlichen Marine* (Signal book of the Imperial Navy)
SLO	Special liaison officer
SLU	Special liaison unit
SOSUS	Sound Surveillance System
SS	Schutzstaffel (Nazi protection squad)
SSTB	Special Security Technical Branch
STANCIB	State-Army-Navy Communication Intelligence Board
Stasi	Staatspolizei (East German security police)
SUKLO	Special United Kingdom liaison officer
SUSLO	Special United States liaison officer

SWG	Special Wireless Group
SWS	Special Wireless Section
T/A	Traffic Analysis
TELINT	Telemetry intelligence
TICOM	Target Intelligence Committee
TIVC	Technische Informatie Vermerkings Centrum (Dutch technical collection information center)
TSP	Terrorist Surveillance Program
UCN	Unidentified covername
UKUSA	United Kingdom–United States of America Communications Agreement
UN	United Nations
UNRRA	United Nations Relief and Rehabilitation Agency
VB	*Verkehrsbuch* (German transport codebook)
VC	Vietcong
VHF	Very high frequency
WES	Wireless Experimental Station
WKC	Wiskundig Centrum (Dutch signals intelligence service)
WOYG	War Office Y Group
W/T	Wireless telegraphy
Y	Signals interception
ZfCh	Zentralstelle für das Chiffrierwesen (German cipher coordination bureau)

Chronology

1878 Alexander Graham Bell builds the first telephone exchange in London with seven subscribers.

1897 Guglielmo Marconi transmits a wireless message across the Solent to the Isle of Wight.

1899 Wireless signals are exchanged between an Admiralty station on South Foreland and Wimereux on the French coast.

1900 The British Army establishes a wireless station at Aldershot. HMS *Forte*, *Thetis*, and *Magicienne* connected by radio to Royal Navy shore stations at Durban and Delagoa Bay.

1901 Section H intercepts cable traffic to and from South Africa during the Boer War.

1902 Alfred Ewing appointed the Admiralty's director of education.

1903 The first international wireless telegraphy conference is held in Berlin. Wireless sets are carried on British Army camels during the Somali expedition.

1904 The Wireless Telegraphy Act is passed to license radio transmissions.

1909 The German transmitter at Nauen is linked to Atakpamé in Togoland, Windhoek, Tabora, and the Caroline Islands.

1910 The Committee of Imperial Defence plans an Imperial Wireless Chain to link London to Egypt, Australia, South Africa, and Hong Kong. The Eiffel Tower communicates with three sites in French West Africa.

1911 The Marconi Company complains to the General Post Office (GPO) about the interception of transatlantic transmissions from Poldhu.

1913 The Royal Engineers' Telegraph School at Chatham is amalgamated with the Army Signals School at Aldershot.

1914 The *Magdeburg*'s *Signalbuch der Kaiserlichen Marine* is delivered to the Admiralty. German interception of Russian communications during the Battle of Tannenberg leads to the destruction of the Second Army.

1915 The German High Seas Fleet is attacked at the Dogger Bank. A German diplomatic codebook is seized from the consulate in Bushehr. MI1(b) is established as the War Office's first signals intelligence organization.

1916 The *Allgemeinefunkspruchbuch* is recovered from the wreckage of the zeppelin *L-32* in Essex. MI1(e) is established by the War Office. Analysis of zeppelin movements alerts the Admiralty to an imminent sortie by the German High Seas Fleet, resulting in the Battle of Jutland.

1917 Dillwyn Knox breaks the U-boat code. The Zimmermann telegram is decrypted by Nigel de Grey. Room 40 is absorbed into the Naval Intelligence Division (NID). Mata Hari is executed in Paris.

1918 Herbert Yardley is appointed to head the U.S. War Department's Cipher Bureau. NID divers recover cryptographic material from the wrecks of the *UB-33*, *UC-11*, and *UB-110*.

1919 Establishment of the Government Code and Cypher School (GC&CS).

1920 The Official Secrets Act requires cable companies to supply the British government with copies of all messages. Ferdinand Tuohy publishes *The Secret Corps*. The British government releases decrypted Soviet diplomatic telegrams to newspapers.

1921 GC&CS moves into Watergate House, the old Marconi headquarters on the Strand. The State Department's Black Chamber reads the Japanese diplomatic traffic during the Washington naval conference.

1922 GC&CS reads American cipher traffic during the Washington naval conference. The U.S. Office of Naval Intelligence (ONI) copies a Japanese Imperial Navy cipher during a burglary of the New York consulate.

1923 Arthur Scherbius markets his Enigma machine. Lord Curzon's note to Grigori Chicherin reveals the Soviet diplomatic cipher has been broken by the British.

1924 GC&CS receives copies of Soviet diplomatic telegrams from the Beijing Central Post Office. Nobby Clarke heads a naval section at GC&CS. The ONI creates OP-20-G.

1925 Felix Fetterlein breaks two Soviet codes.

1926 The War Office Y Group (WOYG) commences intercept operations at Fort Bridgewoods, Chatham. GC&CS moves from Queensgate into the third floor of Broadway Buildings, Victoria.

1927 Soviet ciphers are changed following the Arcos raid and publication of Soviet diplomatic cables. A British patent granted on the Enigma machine. Winston Churchill discloses cryptographic information in *The World Crisis*.

1928 GC&CS acquires a commercial Enigma machine. The Y Committee is established to coordinate interception operations.

1929 John Tiltman joins GC&CS to work on Soviet traffic. The U.S. State Department's Black Chamber in New York is closed down.

1930 The Soviet Black Sea Fleet's call-sign system is solved at Sarafand. GC&CS creates an army section.

1931 GC&CS begins to read Japanese naval attaché traffic from Tokyo. Gustave Bertrand receives Enigma manuals from ASCHÉ. Herbert Yardley publishes *The American Black Chamber*.

1932 NID 9 created to manage all the Admiralty's wireless intercept sites. Hector Bywater publishes *Their Secret Purposes*.

1933 The Polish Cipher Bureau begins to read some Wehrmacht Enigma traffic. Soviet diplomatic missions are opened in the United States.

1934 GC&CS begins to read the MASK traffic. The Admiralty establishes a Far East Combined Bureau in Hong Kong. The Air Ministry creates AI-4 at RAF Waddington. Walter Gill publishes *War, Wireless and Wangles*. The Federal Communications Act outlaws interception.

1935 GC&CS monitors Italian military traffic exchanged between Abyssinia and Rome. Hans Schimpf commits suicide in Berlin.

1936 The Kriegsmarine uses plugboard Enigma machines during naval exercises in the Mediterranean. An air section is established at GC&CS.

1937 The MASK traffic is terminated. GC&CS concentrates on Spanish Civil War wireless traffic. Wilfred Bodsworth breaks the Spanish naval attaché's Enigma traffic. The GPO opens intercept sites at Sandridge, Brora, and Cupar.

1938 Admiralty direction-finding stations monitor the movements of the *Deutschland*. Bletchley Park (BP) becomes operational.

1939 Alan Turing begins work on a *bombe*. WOYG completes a direction-finding (D/F) network based on Helston, Sutton Vallance, and Montrose. Luftwaffe BROWN traffic is intercepted at Fort Bridgewoods.

1940 American cryptographers visit Bletchley Park. WOYG constructs a D/F station at Chartres. ASCHÉ holds a final meeting in Lugano with Henri Navarre. Italian naval Enigma traffic and the Luftwaffe's RED key are broken at BP. The Luftwaffe mounts MOONLIGHT SONATA, an air-raid on Coventry. Loss of British ciphers on the *Automedon*.

1941 LOBSTER is broken at BP. Crete falls to an inferior airborne force despite advance notice of the impending attack from ULTRA. The Italian Servizio Informazioni Militare (SIM) steals the American Black Code.

1942 ISK becomes available at BP. The Swedes break some German Geheimschreiber traffic. A German Würzburg radar is seized at Bruneval. Indian troops capture an Enigma machine at Sidi Rezegh. ULTRA assists an 8th Army victory at El Alamein.

1943 Fritz Kolbe provides the Office of Strategic Services (OSS) with original German Foreign Office telegrams in Bern. Death of Dillwyn Knox. General George Strong concludes a collaboration agreement with BP. U.S. Army Security Agency cryptographers at Arlington Hall begin work on "the Russian problem."

1944 Colossus begins to break some Geheimschreiber traffic. The U.S. Army Signal Security Agency is established. Signals intelligence fails to indicate the presence of two SS panzer divisions at Arnhem. The German offensive in the Ardennes is launched under radio silence.

1945 Alger Hiss attends the Yalta Conference. Igor Gouzenko defects in Ottawa with 109 Glavnoe Razvedyvatel'noe Upravlenie (GRU) documents. The Army Security Agency (ASA) negotiates the SHAMROCK Agreement with U.S.-based international carriers.

1946 The BRUSA Agreement is signed in Washington, DC. The Communications Branch of the Canadian National Research Council established from the Examination Unit. The U.S. Communications Intelligence Board created. Gustave Bertrand heads the Groupement des Contrôles Radio-électriques (GCR).

1947 The Government Communications Headquarters (GCHQ) site at Sarafand is dismantled in anticipation of the British withdrawal from Palestine.

1948 The Federal Bureau of Investigation (FBI) begins a mole hunt for SIMA. MI5 discloses VENONA material to the Australian government. Duncan Lee gives evidence to the House Committee on Un-American Activities. The Berlin Airlift begins without Soviet interference. William Weisband resumes contact with the NKVD and warns of cryptographic work at Arlington Hall. BOURBON is terminated on "Black Friday."

1949 The Australian Security Intelligence Organisation created. Judith Coplon and Valentin Gubitchev are arrested in New York. The Armed Forces Security Agency (AFSA) is established. The USS *Cochino is* lost during a HOLYSTONE mission. Kim Philby is indoctrinated into VENONA.

1950 Canada and the United States sign the CANUS Agreement. A U.S. Navy Privateer is shot down over Latvia. The AFSA fails to warn of the surprise North Korean attack on Seoul. Max Elitcher confesses to the FBI and implicates Morton Sobell, who flees to Mexico, and Julius Rosenberg. David Greenglass is arrested. Alger Hiss is convicted of perjury.

1951 Defection of Donald Maclean to Moscow days after he had been identified as HOMER in the VENONA traffic.

1952 Frank Rowlett transfers to the Central Intelligence Agency (CIA) and indoctrinates the Counterintelligence Staff into VENONA. The National Security Agency (NSA) is established. The Bundesnachrichtendienst establishes Asterhig II. William Marshall is convicted of passing information about the Diplomatic Wireless Service to the Komitei Gosudarstevnnoi Bezopasnosti (KGB) in London.

1953 GCHQ moves from Eastcote to Benhall in Cheltenham. The UKUSA Agreement is signed in Washington, DC. Execution of Ethel and Julius Rosenberg. An RAF Avro Lincoln from 192 Squadron is shot down over East Germany.

1954 Vladimir and Evdokia Petrov defect in Australia. The NORUSA Agreement is signed with Norway. Joseph Petersen is arrested.

1955 Bob Lamphere resigns from the FBI. The Combined Signals Organisation is created in New Zealand. The Berlin tunnel becomes operational. The Austrian Fernmeldeaufklarungs is established at Konigwarte. Brian Patchett defects in Berlin.

1956 Frantisek Tisler makes contact with the FBI and supplies cipher material from the Czech embassy in Washington, DC. The Berlin tunnel is closed after 11 months and 11 days. The NSA moves to Fort George G. Meade in Maryland.

1957 The Naval GRU VENONA traffic is broken by the NSA. Jack Soble and Dr. Robert Soblen arrested. The U.S. embassy code room in Taipei is compromised during a riot. Martha Dodd flees to Prague. Field Station Berlin becomes operational on the Teufelsberg. The USS *Gudgeon* is forced to the surface off Vladivostok while on a HOLYSTONE mission.

1958 A SUN VALLEY C-130 is shot down over Armenia. The Czech cipher clerk Frantisek Tisler defects in the United States.

1959 The NSA's Bernon Mitchell and William Martin visit Cuba. The Swedish Försvarets Radioanstalt reveals the existence of SKIM to GCHQ.

1960 Jack Dunlap visits the Soviet embassy in Washington, DC, and volunteers to spy. After Cyprus independence, Great Britain retains the Sovereign Bases. David Greenglass is released from prison.

1961 Morris and Lona Cohen are arrested in London. Wartime intercepts, codenamed HASP, are passed to GCHQ by the Försvarets Radioanstalt (FRA) in Stockholm. The USS *Valdez* commences SIGINT collection patrols from Cape Town. Boris Morros dies. The ASA's James T. David becomes the first American soldier to die in Vietnam.

1962 The USS *Oxford* commences SIGINT collection patrols in Latin America. Dr. Robert Soblen commits suicide to avoid extradition to the United States. An NSA linguist, Victor Hamilton, defects to the Soviet Union. The CIA discovers Soviet ballistic missiles in Cuba.

1963 Elizabeth Bentley dies. Jack Dunlap commits suicide. The NSA and ASA agree to a division of labor in South Vietnam.

1964 Anthony Blunt confesses in return for immunity from prosecution and is identified as JOHNSON in the NKVD's 1945 London VENONA traffic. *The Invisible Government* refers to the NSA. The USS *Maddox* reports having been attacked in the Gulf of Tonkin.

1965 The USS *Banner* commences collection patrols from Yokosuka. Diego Garcia is leased by the British government to the Naval Security Group. The ASA's Ramasun Station at Udorn, Thailand, becomes operational. NSA communications clerk Robert Lipka begins to spy for the KGB.

1966 Harry Gold is released from prison. The first North Vietnamese MiG is shot down following a HAMMOCK warning. Mark Zborowsky commits suicide. Kitty Harris dies in Gorky. Pine Gap becomes operational in Australia. Professor J. B. S. Haldane dies.

1967 A Rockex cipher machine stolen when the British embassy in Beijing is ransacked. Israel's Unit 8200 provides advance warning of an imminent Arab offensive. The USS *Liberty* is attacked in the Mediterranean. Transatlantic Intelsat telephone channels are intercepted by NSA and GCHQ ground stations. A Polish cipher clerk, Adam Kaczmaryk, is arrested in Warsaw. The NSA is described in *The Espionage Establishment*. The *Daily Express* headlines a cable vetting scandal. John Walker offers to spy for the Soviets. Robert Lipka leaves the NSA.

1968 The USS *Pueblo* is seized by North Korea. The NSA's HEAVY GREEN site on Phou Pha Thi is captured by North Vietnamese troops. The first CANYON satellite is launched. Chief technician, Douglas Britten, is arrested at RAF Digby and sentenced to 21 years' imprisonment. The KGB creates the 16th Directorate.

1969 The first RHYOLITE satellite is launched from Cape Canaveral. The U.S. satellite tracking station at Nurrungar becomes operational.

1970 André Labarthe dies. The USS *Tautog* collides with a Soviet submarine during a HOLYSTONE mission. The Groupement des Contrôles Radio-Électriques is absorbed by the Service de Documentation Extérieure et de Contre-Espionnage.

1971 The Danish Defense Radio Service is amalgamated with the Danish Defense Intelligence Service. The NSA facility on Easter Island is closed. Columnist Jack Anderson compromises GAMMA GUPPY.

1972 Four GCHQ intercept operators working under Cable and Wireless cover in Sinop are killed by Turkish terrorists. GCHQ's Morwenstow satellite ground station becomes operational.

1973 A SUN VALLEY mission is attacked by Libyan aircraft over the Mediterranean. All NSA personnel are withdrawn from South Vietnam. Both COBRA MIST sites are dismantled. Unit 8200 fails to provide advance warning of an Arab attack on Israel on Yom Kippur.

1974 Cyclone Tracy destroys GCHQ's intercept site at Shoal Bay, Darwin. The Agranat Commission criticizes Israeli signals intelligence. Gunter and Christel Guillaume are arrested in Bonn. Robert Lipka breaks off contact with the KGB.

1975 Large quantities of U.S. crypto equipment are seized by the North Vietnamese as Saigon is captured. United Nations convention on satellite launches requires public disclosure of orbits. The ARGUS satellite replaces the RHYOLITE system. Gough Whitlam is dismissed as Australia's prime minister.

1976 The Canadian Communications Security Establishment (CSE) is created. Christopher Boyce sells details of the RHYOLITE satellite to the KGB. The first KEYHOLE satellite is placed in orbit.

1977 The Government Communications Security Bureau is established in New Zealand. The NSA's intercept station at Kagnew is closed. The AQUACADE satellite is launched. Pierre Cot dies. GCHQ intercept sites at Flowerdown and Winscombe are closed.

1978 Death of Ramón Mercader in Cuba. Congress passes the Foreign Intelligence Surveillance Act. Geoffrey Prime resigns from GCHQ.

1979 Analytic work ceases on VENONA at the NSA. Robert Hanssen begins to spy for the GRU. The NSA loses its intercept sites at Astara, Behshahr, Kabkan, Khorramabad, and Shirabad in Iran. Trade union action handicaps GCHQ as the Soviet forces invade Afghanistan.

1980 The KGB's Victor Sheymov is exfiltrated from the Soviet Union. Geoffrey Prime is debriefed by the KGB in Vienna.

1981 Geoffrey Prime is debriefed by the KGB in East Germany. Leonid Eitingon dies in Moscow. The Ministry of Defence announces HMS *Endurance* is to be scrapped and not replaced.

1982 *The Puzzle Palace* and *The Hut Six Story* are published. The Dutch Wiskundig Centrum is renamed the Technisch Informatie Verwerkings Centrum. The Government Communications Security Bureau station at Tangimoana, New Zealand, becomes operational. GCHQ provides inadequate advance warning of an Argentine invasion of the Falklands. Geoffrey Prime is arrested. James Hall begins to spy for the KGB in Berlin.

1983 Samuel Krafsur dies. Paul Davies is arrested and charged with passing secrets from Episkopi to his Hungarian girlfriend. A Su-15 destroys Korean Airlines Flight 007, killing 269 passengers and crew. NSA intercepts prove Iranian complicity in the bombing of the U.S. embassy in Beirut.

1984 Ivor Montagu dies. The FBI's Robert Hanssen resumes his espionage. Trade unions are banned at GCHQ. Paul Davies is acquitted. Seven airmen attached to 9 Signals Regiment in Cyprus are charged with selling secrets. The FSA acquires a naval intercept platform, HMS *Orion*.

1985 NSA retiree Ronald Pelton is arrested in Annapolis. GCHQ closes its training facility at Bletchley Park. John Walker is arrested in Baltimore. Jeffrey Carney defects to East Germany and betrays CANOPY WING. NSA intercepts enable four Palestinian terrorists of the *Achille Lauro* to be captured on a flight from Cairo.

1986 President Ronald Reagan cites three Libyan decrypts as proof of Tripoli's complicity in the La Belle nightclub bombing in Berlin. Robert Lamphere publishes his memoirs and reveals FBI burglaries of Soviet diplomatic premises. The Interception of Communications Act is passed. *Spycatcher* is published.

1987 The Dutch signals intelligence organization is absorbed into the new Militaire Inlichtingendienst. The *New Statesman* magazine discloses details of the ZIRCON satellite.

1988 Duncan Lee dies. James Hall III and Huseyin Yildirim are arrested. Klaus Fuchs dies.

1989 Donald Hiss dies. The Zentralstelle für Sicherheit in der Informationstechnik is created in Germany. Isadore Stone dies.

1990 Martha Dodd dies. Alexander Vassiliev resigns from the KGB. Cedric Belfrage dies in Mexico.

1991 John Abt dies. DESERT STORM liberates Kuwait. Richard Norton-Taylor publishes *GCHQ: A Conflict of Loyalty*, Jeffrey Carney is arrested in East Berlin.

1992 Joel Barr returns to New York from Russia. Lona Cohen dies in Moscow. Colonel Alexander Simakov defects to the British. The USS *Richard Russell* removes the last IVY BELLS pods in the Barents Sea.

1993 Victor Sheymov's *Tower of Secrets* is published. Laughlin Currie dies in Bogota. The Defence Signals Directorate (DSD) station at Geraldton, Western Australia, becomes operational.

1994 The first MAGNUM satellite is placed in orbit. The Intelligence Services Bill is passed by John Major's government. GCHQ moves the Chung Hom Kok site to Geraldton.

1995 The VENONA intercepts are declassified and released. Morris Cohen dies in Moscow. John Cairncross dies in England. 13 Signals Regiment is disbanded. Gunter Guillaume dies in Germany.

1996 Chicksands Priory becomes the Defence and Intelligence Security Center. The Dutch Afdeling Verbindingsinlichtingen is reorganized. The Canadian CSE Commissioner is appointed. Alger Hiss dies.

1997 Robert Lipka is sentenced to 18 years' imprisonment. The NSG base at Edzell is closed.

1998 Joel Barr dies. The fugitive murderer Kenneth Noye is traced to Spain by GCHQ through his cell-phone use.

1999 Ted Hall makes a deathbed confession in Cambridge. GCHQ closes Culmhead. John Cairncross's ghosted memoirs, *The Enigma Spy*, are published.

GCHQ is accused of monitoring microwave links to Ulster at Capenhurst, Cheshire.

2000 The USS *Memphis* monitors the explosion aboard the Russian submarine *Kursk*. The Regulation of Investigatory Powers Act is passed in Great Britain. The NSA suffers a major computer systems failure.

2001 Robert Hanssen is arrested and confesses to having compromised MONOPOLY. NSA intercept equipment is compromised when a U.S. Navy Orion makes an emergency landing on Hainan Island. Geoffrey Prime is released on parole from Rochester prison. Al-Qaida terrorists seize airliners to attack New York and Washington, DC. David Greenglass cooperates with Sam Roberts to write a biography, *The Brother*. The NSA is given presidential authority to extend the Terrorist Surveillance Program (TSP). Ana Montes is arrested at the Defense Intelligence Agency.

2002 GCHQ moves into new headquarters at Cheltenham. The Russian intercept facility at Cam Ranh Bay, Vietnam, closes.

2003 The NSA's operation IMMINENT HORIZON identifies two Iraqi diplomats at the United Nations planning a terrorist incident. Katharine Gun is arrested in Cheltenham. Italian traffic analysts identify the CIA personnel who participated in the "special rendition" of Abu Omar from Milan. The Netherlands establishes a National Signals Intelligence Organization.

2004 Terrorists in Madrid are traced by traffic analysis of their cell-phone use. Charges against Katharine Gun are dropped.

2005 The cell phone of a putative suicide bomber in London, Osman Hussein, is traced to Rome by traffic analysts. Melita Norwood dies. The *New York Times* declines a request not to publish details of the TSP.

2006 The NSA's official history, *American Cryptology during the Cold War, 1945–1989* by Thomas R. Johnson is declassified. Harry Magdoff dies. The European Union issues the Recorded Data Directive. The U.S. Naval Security Group (NSG) is absorbed into the Naval Network Warfare Command.

2007 Two cell phones are recovered from car bombs in London and their owners traced. Aleksandr Feklisov dies in Russia. Trade unions are restored at GCHQ. The NSA declassifies a report on the Gulf of Tonkin incident. President George W. Bush authorizes the NSA's Computer Network Attack project.

2008 Russell McNutt dies. Immunity granted to U.S. commercial carriers cooperating with the TSP. Terrorists attacking Mumbai use a satphone to communicate with Pakistan.

2009 *Spies: The Rise and Fall of the KGB in America* is published.

2010 Traffic analysts in Dubai link suspects to the assassination of the Hamas military leader Mahmoud al-Mabhouh. GCHQ opens the Cyber Security Operations Centre. The BRUSA Agreement is declassified. Richard Aldrich publishes a history of GCHQ. Gareth Williams is found dead in London.

2011 Social-networking media is employed to coordinate demonstrations in Tunis, Cairo, Bahrain, and Sana'a, thereby defeating the local regimes' security apparatus. The cell phone of Osama bin Laden's courier is tracked to Abbottabad.

2012 Opponents of the regime in Syria use Internet social-networking media to evade local communications restrictions.

Introduction

Signals Intelligence (SIGINT), encompassing the various disciplines of wireless interception (Y), cryptanalysis, communications intelligence (COMINT), electronic intelligence (ELINT), direction-finding (D/F), and traffic analysis (T/A) is almost as old as wireless telegraphy itself, a technique developed by Guglielmo Marconi in 1897 with a military application in mind. Before he was able to persuade the British Admiralty of the advantages of being able to transmit messages over the horizon, he had demonstrated his apparatus to army officers on Salisbury Plain with rather less impressive results.

Since then, the potential value of being able to eavesdrop on one's adversaries has become obvious to all, and the ease with which the source can be terminated, by suddenly switching to a different method of communication, changing a cipher, moving frequency, or altering call signs, has made the technique very precarious. The slightest indiscretion, such as the assertion in September 2001 by Senator Orrin Hatch, of the Senate Judicial Committee, that the U.S. National Security Agency was monitoring the satellite phone of a terrorist linked to Osama bin Laden, can result in a valuable channel drying up overnight, so most signals-intelligence sources are cloaked in a high degree of secrecy, even within organizations well used to handling classified material of the greatest sensitivity.

The importance of signals intelligence as a reliable and valuable source can hardly be exaggerated. It was as a result of efficient cryptanalysis by the Admiralty during World War I that the German High Seas Fleet was challenged by a superior force on every occasion it ventured out of port and thus was kept bottled up for much of the conflict. It was the codebreakers of Room 40 that provided the White House with the key to the Zimmermann Telegram, the German foreign ministry message that helped bring the United States into the war in 1917.

During World War II, what was achieved at Bletchley Park hastened an Allied victory and played a crucial role in the defeat of the Afrika Korps in the Western Desert, broke the back of the U-boat fleet during the battle of the Atlantic, and ensured the success of the D-Day landings. Signals intelligence also lays behind many other successes; although, the loss of Crete demonstrated that good intelligence alone was no substitute for solid, dogged soldiering. Perhaps less well known than Bletchley Park was the similar, parallel work undertaken at subordinate satellite centers across the globe, at Sarafand in Palestine, Heliopolis in Egypt, Anand Parbat in India, Kandy in Ceylon, and Kai Cheng in Singapore. Elsewhere, the Americans invested in the Signal Intelligence Service at Arlington Hall while the Australians and Canadians established their own SIGINT organizations.

SIGINT's wholly disproportionate contribution to the ultimate Allied victory in 1945 established the discipline as a foundation for other intelligence-collection operations and guaranteed that the various organizations devoted to this esoteric field would thrive in peacetime, even if the public at large and most policymakers would never know the true scale of the clandestine battle being fought in the ether.

One of the lessons learned during World War II was the difficulty of developing a SIGINT capability from scratch. Pearl Harbor taught the U.S. intelligence establishment that mere collection was pointless without the essential attendant disciplines of processing and assessment, but the wheel had to be reinvented during the Korean War and again in Vietnam. In all three conflicts, technological breakthroughs could offer dramatic advantages, but it was not until DESERT STORM in 1991 that a truly integrated battle plan could be devised and implemented. By 2003, and the Coalition's attack and occupation of Iraq, SIGINT had become the basis upon which all combat operations were undertaken. Now widely recognized as an absolutely vital dimension to modern warfare, a little more is also known about the impact on SIGINT during the peace, mainly due to enlightened policies of declassification.

In the postwar era, signals intelligence proved to be a vital component in the counterintelligence war fought between the West and Soviet Bloc intelligence agencies, and VENONA gave a real advantage to the Federal Bureau of Investigation (FBI), the British Security Service (MI5), the Royal Canadian Mounted Police Security Service, and the Australian Security Intelligence Organisation as a means of exposing Soviet spies. More valuable and productive than any single defector, the VENONA project continued for 36 years and was the critical factor during the Cold War as the Allies sought to identify or neutralize a generation of ideologically motivated moles influenced by Communism. Painstaking cryptanalysis revealed hostile penetration of the British intelligence establishment, the Manhattan Project, and virtually every branch and level of President Franklin D. Roosevelt's administration. As well as providing reliable leads, VENONA also had the merit of offering corroboration of evidence supplied by witnesses, such as Elizabeth Bentley and Whittaker Chambers, whose veracity otherwise might have been open to question. Similarly, the source acted to verify the loyalties and activities of some double agents, such as the Hollywood movie producer Boris Morros, run by the FBI against the Soviets. Even within the FBI's counterintelligence division, the true nature of the VENONA material was never a matter of discussion, except among a small group of specially indoctrinated officers who referred to it within a classified environment only as "Source Five."

Success with VENONA encouraged the investment of more resources for what was termed "the Russian problem," and another significant milestone was BOURBON, a source that enabled Western cryptanalysts to monitor Soviet Bloc troop movements and help keep the uneasy peace of the early Cold War until the material dried up in 1948. Underlying the postwar inter-Allied SIGINT pacts, based on the BRUSA Agreements, was a common adversary, Communists in the Kremlin, supported by their adherents across the world, who demonstrably had conspired to undermine Western capitalism since the 1917 Bolshevik Revolution.

If the prospect of radio operators hunched over their intercept apparatus might be thought a rather less-glamorous existence than the occupation of a mole hunter, imagine the drama of the Berlin tunnel, dug into the Soviet zone to tap a vital landline carrying hundreds of channels of military communications. While the source was operational, it offered an unprecedented window into the Eastern Bloc's preparations for a future war and enabled the West's intelligence community to advise policymakers daily about the level of threat posed by Moscow. Since the dark days of the Berlin airlift, London and Washington, DC, had been very aware of the quadripartite city's vulnerability to a blockade, and the tunnel fulfilled a vital role in reassuring the political and military establishments that, for the moment, open conflict was not imminent. Once the tunnel was terminated, by the unexpected appearance of angry Stasi personnel, reliance switched to the Teufelsberg, the intercept facility built in the British zone, literally on top of the ruins of the devastated city, a site once earmarked for the Third Reich's military academy, with an electronic reach 50 miles over the Polish and Czech frontiers.

The National Security Agency's (NSA) Field Station Berlin was but one component in a Western mosaic of signals-intelligence coverage that had monitored Warsaw Pact wireless and telephone circuits throughout the Cold War, and the 1968 invasion of Czechoslovakia was the event that had proved the importance of electronic interception. On that occasion, the NSA had relied on its huge Bavarian facility at Bad Aibling, which had even recorded the radio traffic generated by taxis in the streets of Prague as they reported to each other the latest location of Soviet tanks. Possibly the most surprising contribution to the 1968 episode had come from the Dutch, though not directly from the Wiskundig Centrum signals organization. Its rival naval SIGINT branch had boasted a source trapped in traffic on the Czech/Soviet border who had given a running commentary over the telephone of the Russian armor queuing to cross the frontier. He was a Dutch tulip seller caught in the congestion who happened to live next door to the director of the Netherlands Naval Intelligence. Having telephoned his wife to let her know of the delay, she had informed her neighbor, who had returned the call and scooped his colleagues.

Because of the vagaries of electromagnetic propagation, some NSA sites were located in some unusual and unexpected places, among them a network of Norwegian SIGINT facilities in the far north of the country, which, since the NORUSA Communications Intelligence Agreement made in December 1954, conveniently collected the VHF transmissions of Red Army units garrisoned in the Baltic countries. Analysis of routine call signs and wireless traffic allowed the Norwegians, acting as NSA surrogates, to monitor troop movements and to offer advance warning of any sudden mobilization and deployment to the south and west.

Despite the hundreds of SIGINT facilities spread across the globe seeking to benefit from the unpredictable physics of ionospheric bounce, the vast sums devoted to monitoring the ether, and the proportion of intelligence budgets earmarked to what are simply termed "technical sources," very little has ever been disclosed about the huge organizations that often dwarf, in the size of their personnel and resources, their counterparts dealing with human intelligence. The NSA, with all its components, is easily larger than the better-known Central Intelligence Agency (CIA) and Federal Bureau of Investigation (FBI) combined, just as Great Britain's once shadowy Government Communications Headquarters (GCHQ) was always bigger by budget and staff than the relatively puny MI5 and the Secret Intelligence Service (SIS). Elsewhere, Austria, Australia, Canada, France, Germany, Italy, and Russia have all inherited long-established SIGINT organizations.

In terms of public perception, it is notable that while several thousand books have been devoted to operations undertaken by, for example, the FBI, CIA, SIS, and MI5, just two books have ever been published about GCHQ and only three about the NSA. As for the KGB, the story is much the same. Numerous histories have been written about the organization; although, only one, Victor Sheymov's *Tower of Secrets*, has been penned from inside the little-known 16th Directorate, which was created in 1968.

Given SIGINT's impact throughout the 20th century, it might be thought remarkable that so little has ever been made public about the discipline. After all, SIGINT had a history-changing effect on World War I, it served to shorten World War II, and no assessment of the Cold War could reasonably be made without giving consideration to VENONA, the source that helped expose Klaus Fuchs, Ethel and Julius Rosenberg, Donald Maclean, Alger Hiss, Ted

Hall, Duncan Lee (to name but a handful of the more notorious examples), and a veritable legion of other Soviet spies. With the benefit of declassified documents, it is now possible to move from the realm of speculation and into the field of informed analysis when weighing the role of espionage in the postwar era. Although, according to GCHQ's former director Sir David Omand, no internal history is ever likely to be commissioned and released in the way outside historians have been authorized to write about MI5 and SIS upon their centenary, the NSA has declassified almost all the VENONA decrypts, plus the three volumes of Thomas R. Johnson's *American Cryptology during the Cold War, 1945–1989*.

These official contributions to the literature have served to redress the balance during a postwar period when plenty of Communist Party supporters instinctively denied their organization's adherence to Moscow, if not their own secret membership. The innocence of the Rosenbergs was protested, and even a U.S. national-security adviser, Tony Lake, asserted that Alger Hiss might not have been guilty of espionage. Incredibly, in spite of the overwhelming evidence now in the public domain, there are still those who insist that there are other possible interpretations, even in the case of Harry Dexter White, once notoriously described as "a species of espionage." While it is disagreeable, of course, for Tony Hiss ever to acknowledge his father was a traitor or for the Meeropol children to accept that both their parents spied for the NKVD, the decrypts do not lie. Leftists may wriggle to excuse Isadore F. Stone and pretend there are other explanations for the damning VENONA references to Lauchlin Currie, but the truth is that the messages represent a veritable counterintelligence holy grail, a rock to be used as an anchor in a sea of intelligence uncertainty.

In the post–Cold War era, SIGINT continued to be arguably the most reliable source of dependable information and played a critical role as conflict enveloped the former Yugoslavia, Coalition troops assembled to liberate Kuwait, and the Israeli Defense Force confronted Hezbollah in Lebanon. Increasingly sophisticated technology, matched by a dramatic expansion in the world's use of cell phones and the Internet, opened a new dimension to the doctrine of modern warfare. Following the collapse of the Soviet Bloc, some policymakers believed that the so-called peace dividend would eliminate the need for large-scale intelligence operations. However, the atrocities committed by al-Qaida on 9 September 2001 reinforced the need for accurate, timely, actionable intelligence. Those events once again brought a new spotlight onto the international SIGINT infrastructure, which, hamstrung by obsolete legislation that never anticipated the Internet, cell phones, or emails, led the NSA to circumvent the limitations of the 1978 Foreign Intelligence Surveillance Act by persuading President George W. Bush to extend a controversial Terrorist

Surveillance Program, which, when its existence became public, was denounced as "warrantless wiretapping."

In the new, post–Cold War landscape, SIGINT became the best source of information about a transnational threat that could not be easily associated with conventional state actors. Whereas economic and criminal sanctions could be applied to countries such as Syria and Libya, which rather publicly had adopted policies to support terrorism, those measures were inadequate when failed states, among them Afghanistan, Yemen, and Somalia, provided safe havens for fugitives. However, the ease with which international terrorists moved from one jurisdiction to another made them dependent on modern technology, not just for maintaining contact with each other but also to use their cell phones as instruments of terrorism, employing the vibrate or alarm modes to remotely detonate bombs. This apparent reliance on cellular systems offered the West's SIGINT community an unexpected opportunity to monitor the movements of individual suspects, intercept their calls, develop comprehensive wiring diagrams of their contacts, and locate them in real time. Osama bin Laden, traced to Abbottabad in 2010 through coverage of his courier, Abu Ahmed al-Kuwaiti, was a high-profile example of the technique of cell-phone tracking. Thus, the organizations that had their origins in the two major world conflicts of the 20th century found themselves, once again, in the forefront of intelligence collection in the global war on terror. Similarly, the development of the Internet, with widening access in target states, provided Western analysts with the opportunity to monitor civil unrest in, for instance, the Maghreb, when text messaging and use of Facebook and other social-networking media was adopted by protestors during the Arab Spring of 2011 as a method of evading the hitherto ubiquitous control over communications exercised by the local security apparatus of totalitarian regimes.

Equally, as some states, notably North Korea and Iran, seek to isolate themselves from normal external influences and nontechnical intelligence sources become more difficult to develop, SIGINT becomes increasingly attractive. While it may be next to impossible to infiltrate agents into compartmented programs in Pyongyang and Tehran or recruit moles inside their ruling elites, the microwave systems upon which they rely to transmit telephone conversations over sometimes quite short distances are as vulnerable to interception as their satellite channels. So, even in these circumstances, no dictator can completely shut off his population from the outside world. Indeed, it can be argued that with the growth of cell-phone use and the Internet, SIGINT offers greater opportunities than ever before, even in states that are deemed to have failed. Somalia, as a coherent national entity, has ceased to exist, having fragmented into uncoordinated territories fought over by rival warlords and militias, yet its uncontrolled pirates and terrorist groups pose a regional threat. While the conventional infrastructure has collapsed, electronic communications abound, reflecting the outlaw gangs' need to conduct their business. Such communications, whether they be ransom negotiations, kidnap plans, or ship-to-shore links, are grist to the SIGINT mill, making it possible for the international community to take effective countermeasures.

The European Parliament's temporary committee report in 2000 on ECHELON failed to sort fact from myth, so ill-informed media commentators rushed to condemn a global conspiracy perpetrated by GCHQ and the NSA to eavesdrop on the innocent. Now, with the benefit of recently declassified documents, there is an opportunity to give an account of SIGINT's remarkable history.

A

ABNER. In April 1952, the U.S. **Armed Forces Security Agency** acquired a computer, codenamed ABNER, capable of processing very large volumes of cipher material at high speed. A direct descendent of **Colossus**, the world's first programmable analog computer, ABNER was the first in a series of data-processing machines designed to support cryptanalytical operations. *See also* NATIONAL SECURITY AGENCY (NSA).

ABYSSINIA. The Italian invasion of Abyssinia generated a large quantity of wireless traffic, from an estimated 200 transmitters, which were monitored at **Fort Bridgewoods**. Despite having a staff of nine, operating just four intercept sets, the **War Office Y Group** copied 16,000 messages during December 1936, of which 2,500 were encrypted. *See also* ITALIAN SIGNALS INTELLIGENCE.

ADMIRALSTAB. The Imperial German Navy established a small cryptographic branch at the Admiralstab's transmitter at Neumünster during **World War I**, which succeeded in solving the Royal Navy's two-letter code used for low-grade communications with coastal shipping. Commander H. Kraschutzki was particularly adept at analyzing the pattern of British signals.

Upon the outbreak of war in August 1914, the Admiralstab began an expansion of the Naval Airship Division and, in January 1915, commenced a series of 30 air raids against England's east coast over the year, killing 382 civilians. During the same period, three **zeppelins** were shot down by ground fire, crashing into the sea, with prisoners recovered from the *L15* and *L7*. Study of captured documents and interrogation revealed that the Admiralstab employed the *Handelsverkehrsbuch* (HVB) to communicate with its airships, the codebook seized from the *Hobart* in Australia upon the outbreak of war, and navigated using bearings from constant signals from List, Tondern, Nordholz, and Bruges.

On 14 September 1916, the *L32* was shot down over Essex by a Royal Flying Corps fighter, and the airship's slightly charred codebook, the *Allgemeinefunkspruchbuch* (AFB), was recovered from the wreckage and was found to be the replacement for the compromised HVB. Similarly, on 17 June

1917, the *L48* was brought down by another fighter at Theberton after an air raid on Harwich, and more cryptographic material found, including the AFB, three lists of recently introduced code words, and a copy of the current naval signal-book. However, the loss of the airship had been witnessed by the *L42*, resulting in Admiral Reinhard von Scheer's decision five days later to change the Admiralstab's codes. Four months later, in France, the *L49* was forced to land near Bourbon-les-Bains, and although the commander had jettisoned his codebook, it was found and delivered to the Admiralty's cryptographers.

Other cipher material was retrieved by divers led by Shipwright E. C. Miller, who visited the *UC-61*, which accidentally ran aground off Cap Griz Nez on 22 July 1917. Then, in August, divers went down 90 feet to the *UC-44*, which had been sunk the previous month by a mine in Waterford Harbor, and found a codebook.

Retrieval of cipher material from airships and submarines proved of considerable value to **Room 40** and helped fill some of the gaps left in the partial reconstructions accomplished by the Admiralty cryptanalysts.

ADMIRALTY Y SERVICE. The British Admiralty's Y Service, headed by the deputy director, Signals Directorate, designated NID 9, consisted of a worldwide network of signals-intercept stations founded on an Empire network built during **World War I**, which monitored target wireless signals and employed **direction-finding** equipment to triangulate the locations of particular broadcasts. Altogether, 17 high-frequency direction-finding (HFDF) stations were built in Great Britain during World War II, centered on **Irton Moor**, outside Scarborough, with a further 23 located around the Atlantic, from Tristan da Cuhna, St. Helena, and Simonstown in the south, and Accra and Freetown in the east, and Bermuda, Halifax, and Esquimalt in the west. In the Mediterranean, the Admiralty controlled HFDF sites at Windmill Hill, Gibraltar, Dingli on Malta, Alexandria, Benghazi, and Crete. In addition, the Royal Navy converted HMS *Hilary* and *Bulolo* to provide intercept facilities for the landings at Anzio and Salerno.

Originally planned in May 1938, the Admiralty's Signals Division could not supply the Y Service with sufficient telegraphists, causing the temporary closure of the station at Gibraltar, so a 180-strong Civilian Shore Wireless Service was proposed on the lines of the Royal Air Force's Civilian Wireless Service, which had been staffed by retirees and pensioners. *See also* FLOWERDOWN.

AGI. The acronym for Auxiliary General Intelligence, AGI was the U.S. Navy's designation during the Cold War for Warsaw Pact vessels, usually originally designed as ocean-going fishing trawlers, conducting signals-intelligence-collection operations. By 1979, more than 200 AGIs were operational across the globe. Distinctive because of their aerial antenna arrays, AGIs routinely collected **telemetry** from missile ranges, monitored the movement of **North Atlantic Treaty Organization (NATO)** warships, patrolled strategic choke points, and maintained a watch on submarine bases. *See also* NAVAL COLLECTION PLATFORMS.

AGTR. The acronym for Auxiliary General Technical Research, AGTR is the U.S. Navy designation for a class of signals-intelligence-collection platforms.

During the Cold War, two **naval collection platforms** came under attack, the **USS *Liberty***, in June 1967 by Israeli planes and gunboats, and the **USS *Pueblo***, in January 1968 by North Koreans. *See also* AGI.

AI-1(e). At the outbreak of **World War II**, Air Commodore Lyster Blandy, a Royal Air Force (RAF) officer who retired in 1928 as controller of Air Ministry communications, established an Air Intelligence (AI) intercept facility at RAF Hawkinge in Kent to monitor Luftwaffe voice channels. Designated AI-1(e), his Home Defence Units eventually covered the coastline from Scarborough, Ingoldmells, Gorleston-on-Sea, Capel-le-Ferne, and Beachy Head to Strete and channeled traffic to a disused toy factory at West Kingsdown, Kent.

AIRBORNE COLLECTION PLATFORMS. Prior to the introduction of the **RHYOLITE** satellite system in June 1969, Allied signals-collection operations were conducted from ground **interception** stations, **naval collection platforms**, and aircraft. Airborne SIGINT-collection missions consist of routine flights conducted in friendly or international airspace by distinctive reconnaissance aircraft specially adapted for the purpose, often flying a "race track" pattern on a regular basis, such as the **RIVET JOINT** operations conducted over the Pacific during the Cold War to monitor Soviet communications. The air forces of Chile, China, France, Germany, Great Britain, Russia, Sweden, Taiwan, and the United States include squadrons dedicated to signals intelligence. A second category of airborne collection involves brief but deliberate incursions, known as "ferret missions," into hostile airspace to provoke ground defenses into activating their radar systems and generating signals that can be recorded and analyzed. A third collection category conducted during the Cold War was undertaken by high-altitude reconnaissance aircraft, such as the Canberra, U-2, and SR-71 flying routes across target countries on clandestine missions.

The first airborne SIGINT-collection mission was probably flown by an airship, the LZ 130 *Graf Zeppelin*, which flew up the east coast of England at an altitude of 1,000 feet in March and August 1939 in an attempt to monitor British radar transmissions and identify the location of the ground stations. It took a close interest in the radar sites at Bawdsey and Canewdon on instructions from General Wolfgang Martini, the Luftwaffe's signals chief. Its final mission, on 2 August 1939, lasted 48 hours and traveled 2,000 miles from its base at Frankfurt am Main.

During the Cold War, the Royal Air Force, the U.S. Navy, the U.S. Air Force, and air forces of Italy, Taiwan, and Turkey suffered considerable losses of equipment and personnel while conducting high-risk airborne SIGINT-collection operations. Between 1950 and 1970, some 252 American aircrew died while flying reconnaissance missions close to, or into, Soviet Bloc airspace. The hazards involved were well illustrated in April 2001 when a U.S. Navy EP3V Orion, one of 12 of the Fleet Reconnaissance Squadron (VQ-1) at Kadena on Okinawa, made an emergency landing at Lingshui on Hainan after it had been in a collision with a Chinese Navy F-8 Finback II jet fighter. The plane's crew of 24, including 3 women, was taken into custody before all the crypto equipment could be destroyed, but the crew was released 11 days later. Much of the aircraft's sensitive hardware was presumed compromised. *See also* CENTRAL SIGNALS ESTABLISHMENT; FÖRSVARETS RADIOANSTALT (FRA); SUN VALLEY; ZEPPELIN.

AIR FORCE ELECTRONIC SECURITY COMMAND (AFESC). Created in 1947, the U.S. Air Force's Electronic Security Command was based at San Antonio, Texas, and at the height of the Cold War employed 10,000 military personnel and 877 civilians deployed across the world with a tactical headquarters at Langley Air Force Base in Virginia, with subordinate detachments at Hurlburt Field, Florida; Bergstrom Air Force Base, Texas; Nellis Air Force Base, Nevada; Shaw Air Force Base, South Carolina; Homestead Air Force Base, Florida; and Tinker Air Force Base, Oklahoma.

In the Pacific area, the AFESC was based at Hickam Air Force Base in Hawaii, with ground sites at Osan in Korea, **Misawa** and Kadena in Japan, Clark Air Base in the Philippines, and Wheeler Air Force Base in Hawaii. In Alaska, the AFESC is represented at Elmendorf and Eielson Air Force Bases. In Europe, the facilities are at Lindsey Air Base, Hahn, Tempelhof, Sembach, **Bad Aibling**, and Augsburg in Germany; Hellenikon in Greece; San Vito in Italy; Heraklion in Crete; and RAF Alconbury, RAF Mildenhall, and **Chicksands Priory** in England. *See also* AIR FORCE INTELLIGENCE, SURVEILLANCE AND RECONNAISSANCE AGENCY (AFISRA).

AIR FORCE INTELLIGENCE, SURVEILLANCE AND RECONNAISSANCE AGENCY (AFISRA). Formerly known as the **Air Force**

Electronic Security Command and prior to 1977 the **Air Force Security Service**, AFISRA is responsible for the collection of SIGINT for the **National Security Agency (NSA)**. Based at Kelly Air Force Base, Texas, with the 694th Intelligence Wing at Fort George G. Meade, Maryland, AFISRA is a key component of the NSA's Central Security Service.

AIR FORCE SECURITY SERVICE (AFSS). Established in October 1948 to monitor Soviet air operations and air defenses, the AFSS grew quickly and within a year was running 35 intercept sites in the United States and overseas, supplying raw material to the **Armed Forces Security Agency (AFSA)** until 1952, when the AFSA was absorbed into the **National Security Agency**. In 1979, the AFSS was designated the **Air Force Electronic Security Command**, a title that was later changed to the Air Force Intelligence Command, **Air Intelligence Agency**, and finally the **Air Force Intelligence, Surveillance and Reconnaissance Agency**. *See also* CHICKSANDS PRIORY.

AIR INTELLIGENCE AGENCY (AIA). Established by executive order in December 1981, the U.S. AIA's 544th Intelligence Group maintained a worldwide network of satellite ground stations located in the United States at Fort George C. Meade, Maryland; Yakima, Washington; Buckley Field, Colorado; Medina Annex, Texas; Sugar Grove, West Virginia; and Fort Gordon, Georgia. Overseas, the AIA managed sites at Sabana Seca in Puerto Rico, **Misawa** in Japan, **Menwith Hill** in England, **Pine Gap** in Australia, Leitrim in Canada, **Bad Aibling** in Germany, Guam, and Kunia in Hawaii. In 1994, the AIA was absorbed into the U.S. **Air Force Electronic Security Command**.

AIR INTELLIGENCE DIRECTORATE. In 1934, the Air Intelligence Directorate at the Air Ministry in London, headed by Air Commodore Sir Charles Blount, established a signals-intelligence branch, designated AI-4, which eventually supervised the Royal Air Force (RAF) Y Service and managed the intercept sites at Montrose in Scotland, Cheadle in Staffordshire, **Chicksands Priory** in Bedfordshire, West Kingsdown in Kent, and Waddington in Lincolnshire. In 1935, Air Commodore Lyster Blandy's nominee to head an Air Intelligence analytical branch, designated **AI-1(e)**, was Joshua E. S. Cooper from the **Government Code and Cypher School (GC&CS)**. In 1936, Cooper would return to GC&CS to establish the Air Section.

At the outbreak of **World War II** in September 1939, RAF Y was still in its infancy, and such **interception** that took place was conducted by the War Office station at **Fort Bridgewoods** where **BROWN** traffic was routinely copied; although, its source was not initially recognized as Luftwaffe.

However, by mid-1941, BROWN's origins were confirmed, and the RAF took over the task of monitoring the traffic.

ANKER. The **Bletchley Park** codename for an unidentified German traffic transmitted from a clandestine wireless in Ankara during **World War II** and reporting on Soviet and British matters in the Middle East. *See also* ULTRA.

AQUACADE. Launched in 1977 by the **National Reconnaissance Office**, the AQUACADE signals-intelligence satellite was the second generation of the **RHYOLITE** system. *See also* ARGUS; ZIRCON.

AQUARIAN. The **National Security Agency** codename for a project designed during the Cold War to identify Soviet diplomatic premises in the United States engaged in the **interception** of local telephone calls carried on microwave channels and to pinpoint the specific numbers being monitored.

ARAGO. The **Federal Bureau of Investigation** codename for a Czech cipher clerk, **Frantisek Tisler**.

ARDENNES OFFENSIVE. A large build-up of German forces in preparation for a major counterattack into Belgium on 16 December 1944 in the Ardennes, to recapture Antwerp, went undetected by Allied signals intelligence because the various components, amounting to a million men, had been assembled under the strictest wireless security. Although the unexpected imposition of **radio silence**, which occurred on 6 November, can itself be interpreted by traffic analysts as highly significant, no such clues were detected prior to the attack, which caught the Allies almost completely unawares. Helpfully, on 1 December, **Bletchley Park** cryptographers read a message in the OCELOT key from the Luftwaffe air liaison officer attached to II SS Panzer Korps announcing that radio silence had been imposed from 0730 on 30 November.

Intercepted signals over several months had indicated a surprise counterattack, and the first clues had been found in Japanese diplomatic traffic. On 21 August, **Hiroshi Oshima** referred to the "recent very thorough mobilization," which had produced 100 to 124 divisions and new fighter squadrons in anticipation of "the coming offensive," which would not be launched for at least two months. On 23 August, his naval attaché reported to Tokyo that he had been told that an offensive had been planned for November or December. On 4 September, Hiroshi mentioned a conversation with Adolf Hitler in which the Führer had described how a million new troops were being formed and that, once they were ready and the Luftwaffe had been replenished, a great offensive in the West would commence "probably from November onwards."

Separately from the Japanese diplomatic channel, there were other indications that German units were being withdrawn from the front and consolidating east of the Rhine. By the end of September, analysis of **Enigma** decrypts noted the withdrawal of the 1st, 2nd, 9th, and 12th SS Panzer Divisions and the Heavy SS Panzer Abteilungen 501, 502, and 503 for rest and refitting in Westphalia before becoming part of a new Panzer Sixth Army under General Sepp Dietrich. Early in October, the Panzer Lehr Division and I SS Panzer Korps also joined the new Panzerarmee, and at the end of the month, there was SIGINT evidence that the 3rd, 5th, and 6th Parachute Divisions were refitting in Holland. One signal, read on 18 October, showed that Hitler himself had ordered the armored refits. Other significant moves included the transfer of the Fifth Panzerarmee from Nancy to Army Group B and the creation of an entirely new Army Group H under General Kurt Student, formerly in command of the First Parachute Army. Finally, on 12 and 13 November, intercepts revealed that the entire Fifth Panzerarmee was being moved by rail to west of the Rhine.

Luftwaffe decrypts were equally informative, and on 27 October, one referred to 11 airfields north of Aachen that were to be stocked with extra fuel and ammunition for a large number of day fighters engaged in ground-support operations, in anticipation of a "*Schlagartiger Einsatz*" or surprise attack. In addition, there were references to an order from Hermann Göring for the need to convert large numbers of planes into a ground-attack role and to have made preparations by 12 November for the arrival of fighter reinforcements. According to the Air Ministry's intelligence branch (AI), which estimated that 800 to 850 fighters had been gathered, German single-engine aircraft production had reached record levels, and all the planes lost in France had been replaced, leaving it with a current strength of 2,000 fighters, a figure likely to increase to 2,500 over the next six months.

Combined with the somewhat deceptive codename WATCH ON THE RHINE, these changes in the enemy's order of battle were misinterpreted as attempts to reinforce the Reich's defenses and not seen as preparations for a massive counterattack. Finally, a Joint Intelligence Committee (JIC) assessment dated 11 November, *Recent Intelligence on German Intelligence in the West*, noted that the Sixth Panzerarmee included five recently refitted panzer divisions and that all three of divisions of the First Parachute Army had moved up to the front and were exercising radio silence. It also suggested that fuel shortages, poor pilot training, and substandard airfields would prevent any "limited spoiling attack" lasting more than two or three days. Nevertheless, the original draft of this document included the sentence,

> We conclude that the enemy are planning to spring a surprise which would almost certainly include attacks on Allied airfield.

However, this crucial sentence was omitted from the final version, which was circulated by the JIC to Supreme Headquarters Allied Expeditionary Force (SHAEF) and the prime minister, thus creating the impression that Hitler's strategy was to stymie the expected Allied offensive against the Reich by launching a spoiling attack that would serve to delay the Allies main thrust until the spring, by when he hoped to have acquired an advantage in the deployment of new weapons such as the Luftwaffe's jet fighters. In the absence of any useful signals intelligence and with bad weather that prevented any aerial photoreconnaissance flights for a month until 13 December, the Allies were blind. Indeed, German security, combined with some deception measures, was so efficient that, on 3 December, 21st Army Group complained, having lost all trace of the Headquarters Fifteenth Army, now renamed "Gruppe von Manteuffel," that it was "disquieting in the fifth year of the war to find an army HQ disappearing from the map." One of the ploys adopted by the Germans was to rename the Fifth Panzerarmee as the "Jager Kommando zbV" and the Sixth Panzerarmee as the "Rest and Refitting Staff 16."

Although military **interception** offered little, there was plenty of Luftwaffe traffic to monitor, and by 26 November, some 500 fighters had arrived at airfields earmarked for "*Jägeraufmarsch*" or fighter launch points. Although it went unremarked at the time, there was also a strange escalation in U-boat weather reports from the North Atlantic. Whereas the wolf packs usually transmitted an average of two or three weather signals a day, they were urged at the end of November to make these messages a priority, and by 15 December, they were sending 15 a day. Three days after the attack was launched, the submarines were thanked and informed that they had "contributed decisively to determining the beginning of our great offensive in the West."

There was also activity on the railway communication circuits, with constant requests for air cover at dawn to protect priority trains carrying troops and matériel toward Aachen. However, when no attack was launched, AI issued an appreciation on 6 December observing that whatever intentions the Luftwaffe

> may have for further reinforcement of the western front, the original plan for "lightening blow" and sudden attack in the west may with some certainty be said to have lapsed.

This verdict was later described by Bletchley as "a disaster," and it eloquently illustrated the danger of over-reliance on SIGINT at the expense of more conventional intelligence sources, such as prisoner-of-war interrogation, agent reports, aerial reconnaissance, and frontline patrols. In reviewing the Allied failure to anticipate the German attack and the dependency on

ULTRA, Sir Archie Nye, the vice chief of the Imperial General Staff, observed on 2 February 1945:

> There is a tendency to place too much value on this source and therefore to neglect (or not to pursue with determination) all the other sources available to us—and this has been, and I think is, the greatest weakness in our I-organization.

This view was disputed by Bletchley, which insisted that the **Geheimschreiber** traffic, euphemistically referred to as "high-grade SIGINT," had given a clear warning of the coming counteroffensive, even if the actual timing had only been available at short notice. Nye responded with:

> A good case can, of course, be made to show that the evidence was available from ULTRA sources only, if we had had sufficient clear-sightedness to see it. A very easy charge to make and impossible to disprove. On balance I should say that while it was *possible* from this evidence to draw the correct deduction from it, the odds were very much against doing so.

The debate about culpability continued in private for many years; although, relatively few of the Allied military commanders directly involved in what became known as the Battle of the Bulge had been indoctrinated into ULTRA. When the relevant decrypts were declassified, revealing that nearly 6,000 summaries had been distributed by Bletchley to the relevant commands between 1 October 1944 and 31 January 1945, at a rate of 40 to 50 a day, the controversy continued. *See also* WORLD WAR II.

ARGUS. A classified satellite-telecommunications-collection system, the ARGUS was compromised in 1976 by **Christopher Boyce**, a junior employee of a defense contractor, TRW Inc. The codename ARGUS grew from the letters "AR" standing for "Advanced RHYOLITE." *See also* AIRBORNE COLLECTION PLATFORMS; BYEMAN; LEE, ANDREW D.; PYRAMIDER.

ARLINGTON HALL. Formerly a financially troubled school for girls, Arlington Hall became the headquarters of the U.S. Army's **Signal Intelligence Service** during **World War II**, an organization that would later become the **Signal Security Agency**. *See also* ARMED FORCES SECURITY AGENCY (AFSA); BRIDE; JADE; NATIONAL SECURITY AGENCY (NSA); ONE-TIME PAD (OTP).

ARMED FORCES SECURITY AGENCY (AFSA). Created in May 1949, the AFSA supervised all United States signals-intelligence operations until it was absorbed into the newly established **National Security Agency** in October 1952. When established, the AFSA employed 4,932 civilians and 1,948 military personnel at 1,821 intercept sites across the globe, but within a year, the number of civilian staff had grown to 6,613, making a total of 8,500 employees, with a budget of $60.9 million. *See also* BRIDE; JADE; ONE-TIME PAD (OTP); VENONA.

ARMOR. Also codenamed STAMP in the **NKVD**'s **VENONA** traffic, ARMOR was a spy at the Republic Aviation Corporation factory who supplied information to Andrei I. Shevchenko, the Soviet representative at the Bell Aircraft Corporation. His true identity was withheld by the **National Security Agency** when the relevant VENONA messages were declassified in 1995. *See also* BRIDE; JADE; ONE-TIME PAD (OTP).

ARMY SECTION. In 1930, the **Government Code and Cypher School** introduced the Army Section headed by John Tiltman, a King's Own Scottish Borderers officer, assisted by Dick Pritchard, a German specialist who was responsible for liaison with the French, Frederick Jacob, later head of Combined Bureau Middle East (CBME), and Tony Dangerfield a Japanese expert.

ARMY SECURITY AGENCY (ASA). In September 1945, the U.S. Army's **Signal Security Agency**, which, until 1944, had been part of the Signal Corps, was designated the Army Security Agency. Commanded by General W. Preston Corderman, the ASA concentrated on Soviet communications traffic between September 1945 and the establishment of the **Armed Forces Security Agency** in May 1949. An essential military component of the **National Security Agency**, the ASA was merged in 1977 with the U.S. Army Intelligence and Security Command. The ASA managed intercept sites located at Amchitka in the Aleutian Islands; Asmara in Eritrea; Bellmore on Long Island; Fairbanks, Alaska; Guam; Indian Creek, Florida; New Delhi, India; Two Rock Ranch at Petaluma, California; Tarzana, at Reseda, California; and Vint Hill Farms, Warrenton, Virginia. *See also* HALL, JAMES W., III.

ARNHEM. The Allied airborne operation to capture intact the Dutch bridges at Arnhem, Eindhoven, Zon, Veghel, Grave, and Nijmegen on 17 September 1944, codenamed MARKET GARDEN, proved a costly military disaster. Of the 11,920 men of the 1st British Airborne Division and the Polish Independent Parachute Brigade Group who participated in the battle fought over nine days, 1,485 were killed, 6,525 were taken prisoner, of whom nearly half were wounded, and only 3,910 were evacuated. The U.S. 82nd Airborne and 101st Airborne Divisions suffered 1,432 and 2,118 casualties, respectively.

The Royal Air Force lost 44 Stirling bombers, 24 Dakota transports, and 152 aircrew. In contrast, according to an enemy intercept dated 27 September 1944, General Wilhelm Bittrich's II SS Panzer Korps had 1,300 dead and 2,000 wounded.

The debacle had several causes, but one was a lack of clear intelligence to indicate that the battle-hardened 9th Hohenstaufen and 10th Frundsberg SS Panzer Divisions had been deployed to Arnhem to rest and refit with new tanks. The 9th Panzers consisted of a company of Panthers, an armored infantry regiment, an artillery battalion, two assault gun batteries, and a reconnaissance battalion. The 10th Panzers included an armored infantry regiment, two artillery battalions, and a reconnaissance battalion.

Nor was there any definite advance knowledge that Field Marshal Walther Model, commander of Army Group B, had established his temporary headquarters nearby at Oosterbeek. By the time a suspicious Parachute Brigade intelligence liaison officer, Major Brian Urquhart, having been tipped off by the Dutch resistance, had requested aerial reconnaissance, with missions flown by a total of eight aircraft on 12 and 16 September, which spotted new Mark III and Mark IV panzers within 10 miles of the main road into Arnhem, it was deemed too late to change the plan. In consequence, instead of taking the enemy by surprise and seizing two lightly guarded objectives, the paratroops and glider-borne soldiers, devoid of heavy weapons, unexpectedly encountered several thousand elite troops with combat experience in Russia and Normandy, equipped with armor and artillery, who mounted the fiercest of well-coordinated defenses supervised by exceptional strategists, Model and Bittrich.

The intelligence available to the planners consisted of some prisoner of war (PoW) interrogation reports, photoreconnaissance flights, messages from the somewhat compromised local Dutch resistance organization, and MI14 enemy order-of-battle assessments based on signals. Two **ULTRA** summaries were of critical importance; although, their significance was not fully appreciated at the time. A decrypt from **Bletchley Park** on 5 September reported that elements of the 2nd and 116th Panzer Divisions and the 9th and 10th Panzer Divisions were refitting in the Venlo-Arnhem-s'Hertogenbosch area. The next day, another decrypt, attributed to a PoW as cover, revealed that II Panzer Korps had been subordinated to Army Group B and was to move to Eindhoven to supervise the refit of 2nd and 116th Panzers, 9th SS Panzers, and Heavy Assault Gun Abteilung 217. However, a Supreme Headquarters Allied Expeditionary Force (SHAEF) ULTRA summary of 12 September suggested that since the 9th and 10th SS Panzer Divisions had not been motioned in any recent decrypts, they might be in Holland but more likely had been disbanded.

> There can be either rest and refit or defense of the West Wall and Holland. There cannot be both.

On 13 September, another SHAEF summary reported that elements of the 9th and 10th SS Panzer Divisions had been seen "in small pockets" but asserted that they were in no condition to operate as a division. Accordingly, the British Airborne Division staff, who had not been cleared to receive ULTRA, were advised that "one broken panzer division" had been sent to Arnhem for refit and that the presence of II SS Panzer Korps was suspected. Then, on 15 September, SHAEF circulated a decrypt from earlier the same day of a message in the Luftwaffe's OCELOT **Enigma** key that had announced the arrival the previous day of the Luftwaffe's liaison officer at II SS Panzer Korps headquarters in Oosterbeek. Finally, on 16 September, SHAEF reported that a

> 9 SS Panzer Division, and with it presumably 10, has been reported as withdrawing altogether to the Arnhem area of Holland to be equipped with new tanks from Cleves.

Despite this advance intelligence, the briefing given to the British Airborne Division on 13 September mentioned that the German line

> is held by the remnants of some good divisions, including Parachute divisions, and by new arrivals from Holland. They are fighting well but have very few reserves. The total armored strength is probably not more than 50–100 tanks, mostly Mark IV. There is every sign of the enemy strengthening the defenses of the river and canal lines through Arnhem and Nijmegen, especially with Flak, but the troops manning them are not numerous and many are of low categories.

The British Airborne Division planners were never told that a decrypt circulated on 13 September, but intercepted on 10 September, revealed that Luftflotte 3 had announced an intention to fly reconnaissance missions to detect evidence that the Allies were about to mount an attack toward Aachen or on the First Parachute Army or on Arnhem. Another decrypt on 13 September indicated that the Luftwaffe had been requested to see if there were any gliders assembling in East Anglia, in the area around the Wash. Then, on 14 September, Bletchley Park issued an Abwehr decrypt, dated 11 September, which drew attention to air drops made to supply the Dutch resistance in the Apeldoorn-Zwolle area and warned of possible landings. The next day, Bletchley issued another decrypt, dated 9 September, warning of a possible thrust to Arnhem by the British Second Army's XXX Corps, suggesting that "there might be an operation west of Nijmegen and Wesel to surround German forces in western Holland."

After MARKET GARDEN had begun, intercepted enemy wireless traffic indicated that the Germans had been taken completely by surprise, with Field Marshal Gerd von Rundstedt noting in a message dated 1 October that there had not been any significant increase in Allied air raids on Arnhem prior to 17 September and that such attacks as had been made on local flak positions had been interpreted as attempts to destroy the bridges. *See also* WORLD WAR II.

ARSENIC. The Anglo-American codename adopted in February 1943 to denote signals-intelligence traffic originating from the Soviet Union. *See also* ULTRA.

ASCENSION. The British colony of Ascension in the Atlantic Ocean was the site of a **Government Communications Headquarters** intercept station located at Two Boats, on the opposite side of the island from the American airfield at Wideawake, which operated under Cable and Wireless cover.

ASCHÉ. The French codename for **Hans-Thilo Schmidt**, a German technician who, in 1931, began to sell copies of the **Enigma** machine's manuals to **Gustave Bertrand**. *See also* FRENCH SIGNALS INTELLIGENCE.

AUSTRALIAN SECURITY INTELLIGENCE ORGANISATION (ASIO). Created in 1949 as a result of pressure from London to staunch the hemorrhage of classified material from Canberra to the local Soviet *rezidentura*, ASIO concentrated on leads provided by **MI5**'s security liaison officer, Courtenay Young, who claimed the information had come from "the source," implying that the British had recruited a spy in Moscow.

During the first months of ASIO's existence, inquiries prompted by the **VENONA** texts identified more than a dozen spies and led ASIO to begin offensive operations, a program that included the establishment of permanent observation posts, telephone **interception**, blanket physical-surveillance cover on suspects, and the installation of listening devices into premises occupied by confirmed intelligence adversaries. Under the guidance of Young, who was given an office in ASIO's secret headquarters building, known as Agincourt, in the center of Sydney, a team was created to follow up his leads.

Some of the clues proved impossible to pursue, and ASIO simply filed the inconclusive codenames, which included SPIRITED, SANTO, GRANDSON, and a woman referred to as GLORY. Other texts provided an insight into who was of interest to the Soviets; although, their decision to conceal an identity with a codename was considered sinister by ASIO, so when Professor Sergei J. Paramonov, of the Australian Council for Scientific and Industrial

Research, was designated SCORPION, on 30 May 1947, a very negative interpretation was drawn. One VENONA text in particular, dated 5 May 1945, was to be used thereafter as a vetting instrument. The telegram from Semen Makarov listed eight members of the Federal Council of the Association of Scientific Workers of Australia who were Communist Party members, an invaluable resource when screening applicants for scientific posts with access to classified papers, not to mention compiling a list of potential candidates for detention in the event of a conflict with the Soviet Union.

Young's team, directed by Robert F. B. Wake, a veteran Commonwealth Investigation Branch officer who had served in Naval Intelligence during the war, pursued every one of the local names and traced more than a few that were not directly relevant to Australian security.

With Young's encouragement, ASIO adopted an aggressive counterintelligence approach and, having reconstructed the entire order of battle for the NKVD and GRU *rezidenturas*, embarked on some ambitious operations.

One of their principal targets was Fedor A. Nosov, the TASS news agency correspondent in Sydney since March 1943, codenamed TECHNICIAN, who, according to a VENONA dated 3 July 1945, had been put in direct contact with Detective Sergeant Alf Hughes, codenamed BEN, and had photographed documents in his flat. ASIO assigned Ray Whitrod the task of renting the flat above Nosov and tapping his telephone, and the surveillance was to prove highly productive. Nosov merited the treatment because of his close connection with a local communist, Mark Younger.

Despite his Anglicized name, Younger was of Polish origin and was the Polish honorary consul in Sydney and a frequent visitor to Nosov's apartment, as was Dr. Herbert Evatt's secretary, Allan Dalziel. All were placed under intensive surveillance to develop ASIO's knowledge of the VENONA cryptonyms, and the overall effort was given a tremendous boost when a new arrival, Vladimir Petrov, came under their scrutiny.

When Petrov and his wife, Evdokia, landed in Sydney on 5 February 1950 from the liner *Orcades*, they were greeted on the quay by the embassy's second secretary and *rezident*, Valentin Sadovnikov, and the TASS correspondent Ivan Pakhomov, who was to succeed him briefly as *rezident*, based in Sydney, when he was recalled in April the following year. The *rezidentura*'s strength had been depleted by Makarov's departure in June 1949, and Fedor Nosov was scheduled to leave in August 1950. As well as the two Russians standing on the dock, ASIO men were also watching the scene, having been tipped off by a passenger that neither Petrov nor his wife were the humble third secretary and code clerk they purported to be. Indeed, ASIO's informant was correct, for both were professional intelligence officers, having already served together abroad, in Stockholm from March 1943 until October 1947. Nine years later, when the **HASP** material became available, two 1945 VENONA intercepts from the Stockholm embassy, dated 16 July and 21 September, showed that Petrov, then codenamed SEAMAN, had been the personal cipher clerk to two *rezidents*, first Mrs. Yartseva, then Vasili F. Razin. However, their experience in Sweden had not prepared the Petrovs for the atmosphere of intrigue in Canberra.

Although some of the VENONA texts had given ASIO an indication of the turmoil that had gripped the Soviet embassy, only the members of the *rezidentura* knew the full story. ASIO had relied upon some fascinating signs of dissent and personal criticism but had never appreciated the depth of the internal strife gripping the rather isolated Soviet colony in Canberra.

Valentin Sadovnikov managed to get a subordinate, Iya Griaznova, pregnant, and had been reported by another member of the staff for staying overnight drinking at the home of an Australian acquaintance in Sydney, in contravention of the *rezidentura*'s strict rules. The consequence was an adverse report to Moscow by the ambassador, Nikolai Lifanov, which in turn resulted in Sadovnikov being recalled home, together with his unamused wife, Elizaveta.

Neither of the Petrovs could cope with the gossip and backbiting at the embassy, which only escalated when Petrov was appointed acting *rezident* upon Pakhomov's departure at the end of 1951. Petrov fell out with Lifanov, who, according to an embassy code clerk named Prudnikov, had criticized him in a secret telegram, and his relationship with the new ambassador, Generalov, deteriorated even further when Evdokia rowed with his wife. Finally, the Petrovs too were recalled to Moscow, and the last straw proved to be a surprise inspection of Petrov's office safe, undertaken by the ambassador on 31 March 1954, accompanied by the First Secretary Vislykh and his aide, Christoborodov, who found papers that should have been kept more securely. Thus, when Petrov's replacement, Evgeni Kovalenok, arrived by ship at Sydney on 3 April, Petrov accepted ASIO's offer to defect and took a briefcase of documents with him.

As well as liking life in Australia and having invested in a farm, Petrov was anxious about the reception he could expect in Moscow, which, in recent weeks, had seen Lavrenti Beria arrested and executed and a purge of all levels of the Center. Neither Petrov had any children, so there was little to return home to, or so ASIO suggested, and he had no doubt that both ambassadors had submitted adverse reports on his performance. In fact, Petrov had long been cultivated by an ASIO agent, Dr. Michael Bialoguski, whom he had met in July 1951, and his successful defection was the culmination of a long, painstaking ASIO operation codenamed CABIN. The unexpected bonus was Petrov's wife's decision a fortnight later to break free from her Soviet escorts at Darwin Airport, while on her way home, and to join her husband.

The Petrov defections were a sensation, and a Royal Commission was appointed to take his evidence and examine the documents he had purloined from his safe in the embassy. While Petrov's papers were interesting and even embarrassing, they were by themselves inconclusive and, in comparison to the VENONA intercepts, almost irrelevant. However, the Commission was briefed in private session about some of the VENONA material in order to establish a link between some of the principal suspects, such as James Hill, Ian Milner, and Walter Clayton.

In his evidence to the Royal Commission, Clayton denied ever having met either Hill or Milner and admitted only that he had been shown some low-level papers by Frances Bernie. He also denied on oath that he had ever met any Soviets.

During their debriefing by ASIO, the Petrovs gave valuable information about their previous posting in Stockholm and revealed that the assumption that Boris N. Yartsev, codenamed KIN in the VENONA traffic, had been the NKVD *rezident* was incorrect. In fact, he had been subordinate to his wife, who was the real *rezident*, a fact that did not emerge from scrutiny of the VENONA texts.

When Dr. Evatt, formerly the minister for external affairs, appeared before the commission as leader of the Opposition, he denounced Petrov's papers as forgeries and accused him of conspiring to undermine the Labour Party. He, of course, had never been indoctrinated into the VENONA program and, therefore, never realized that the accusations against Milner, Hill, Clayton, and a journalist, Ric Throssell, had predated Petrov's arrival in the country.

AUSTRALIAN SIGNALS INTELLIGENCE. During **World War II**, the Australian Army and Royal Australian Navy (RAN) developed their own signals-intelligence capability, with the RAN collecting some traffic at HMAS *Harman*, a signals station near Canberra, and the army operating a small intercept site at Park Orchards, near Ringwood in Melbourne, which provided material for a processing unit at the Victoria Barracks, Melbourne.

After the fall of Singapore in February 1942, the Allied supreme commander of the Pacific theatre, General Douglas MacArthur, in April created a **Central Bureau** at Melbourne, staffed jointly by Americans evacuated from Cavite in the Philippines, British personnel from the **Far East Combined Bureau** in Singapore, and Australians, including operators drawn from the Royal Australian Air Force center at Townsville.

In December 1944, the Australian commander in chief, General Thomas Blamey, conducted an investigation after the U.S. **Signal Intelligence Service** reported that an intercepted Japanese signal from Harbin to Tokyo on 11 November 1944 had included an Australian Armed Forces *Weekly Intelligence Summary* dated 4 November 1944, an assessment of General

MacArthur's plans to liberate the Philippines, and an editorial guidance notice that had been distributed to newspapers. According to the decrypt, the original source of these documents had been a Soviet diplomat in Australia, so General Blamey instructed a specially created organization, 1 Discrimination Unit, to monitor Soviet diplomatic wireless traffic in an effort to trace the source of the leak. The Discrimination Unit operated from January until October 1945 but failed to identify any incriminating signals. Similarly, a tap on the telephone of the TASS news agency correspondent also gave no clue as to where the Japanese information had come from.

AUSTRIAN SIGNALS INTELLIGENCE. *See* FERNMELDEAUFKLARUNGS.

***AUTOMEDON*.** On 11 November 1940, the British merchantman *Automedon*, bound from Liverpool for Singapore, was taken by surprise by an armed German supply ship, the *Atlantis*, northwest of Sumatra. The raider fired on the *Automedon*, killing 6 of the crew and wounding 12, before it was boarded and a search of the chartroom revealed copies of the new British Merchant Navy's deciphering tables (numbered 7, 8, and 9), some 120 mailbags, and a large quantity of highly secret documents destined for the British commander in chief in the Far East, Air Chief Marshal Sir Robert Brooke-Popham. The documents included contingency plans drawn up by the War Cabinet and cryptographic data that suggested the British codebreakers at **Combined Intelligence Far East** had solved an impressive number of Japanese ciphers. Having seized the confidential papers, the *Automedon* was scuttled, and, later, the surviving crew was put aboard the *Storstad* for a voyage to Bordeaux.

The captured papers were immediately transferred to the *Ole Jacob*, a Norwegian ship captured by the *Atlantis*, which landed on 4 December at Kobe, where the documents were handed to the German naval attaché, Admiral Paul Wenneker, who immediately recognized their importance, especially of a top secret, Chiefs of Staff appreciation for the War Cabinet of the situation in the Far East, and sent them by courier to the Kriegsmarine for study by analysts. In the meantime, he transmitted a message to Berlin, summarizing the windfall:

> For Naval War Staff:
>
> Of note amongst materials reaching here on 6 December from Ship "16" is a captured report of the War Cabinet about the situation in the Far East in the event of Japanese intervention against Britain, dated 15 August:
>
> 1. In present situation we (i.e. Britain) are unable to send fleet to Far East.
> 2. Japan needs Singapore in order to meet its ambitions.

3. Until the situation in Europe clarified itself, open breach between us and America and Japan improbable.
4. We must avoid "open clash" with Japan in order to gain time and promote military co-operation.
5. In the absence of Fleet, we are unable to prevent damage to our own interests. We should retire to a base from which it will be possible to restore our position later on.
6. Four possible Japanese attacks: (a) Direct attack (b) Advance into Indo-China or Siam. (c) Attack on Dutch East Indies. (d) Attack on Philippines.
7. Japanese first step either Indonesia or Siam; then Dutch East Indies before Singapore.
8. In current situation we would put up with Japanese attack on Siam or Indo-China without going to war.
9. In event of Japanese attack on Dutch, and they offered no resistance, no war between us and Japan. But if Dutch resist, then they would have our full military support.
10. Hong Kong without any significance and cannot hold out for long without presence of a substantial fleet. However, will be held as long as possible.
11. Strategy in Event of War:
    (a) Impossible to prevent Japanese gaining access to Indian Ocean.
    (b) We cannot maintain naval lines of communication with Northern Malaya.
    (c) Hope to maintain bulk of commerce by sea with Suez and eastern Australia.
    (d) Apart from cruiser raids, Japanese attack on Australia improbable without first taking Singapore.
    (e) Japanese occupation of Suva and Fiji likely in order to use as a base.
    (f) Need to defend all of Malaya and not just Singapore Island.
    (g) Holland probably willing to prepare joint plan for defense of Dutch Indies. In view of limited assistance we can give, their help unlikely if British territory attacked. Consequently, not desirable to begin Staff Talks at this time.
    (h) As long as no fleet, forced to turn to air force. But this can only be provided in a limited way. Therefore, strong land forces necessary in Malaya. Concentration there a top priority.
    (i) Borneo indefensible. Very little air power available for protection of shipping in Indian Ocean.
    (j) Until situation in Europe improves, Far East gravely threatened, especially if subjected to determined Japanese sorties, with heavy naval units.
    (l) Our construction program never intended to cover war with Germany, Italy and Japan, simultaneously. The only hope for providing a fleet for the Far East based on early, successful attack on Italy and the Mediterranean.

12. Objectives we must seek to achieve.
    (a) Commonwealth must send one Division to Malaya.
    (b) By end of 1940, two fighter squadrons and two squadrons of long-range (?) land aircraft to be sent to Far East.
    (c) Naval construction program to be accelerated.
    (d) Withdrawal of Garrison from North China and Hong Kong.
    (e) New Zealand must send one brigade to Fiji.
    (f) General Staff consultations with Holland once situation in Malaya improves.
13. Detailed information on the strengths of land, sea and airforces in the Far East.

On 12 December, the following signal from Captain Yokai, the Japanese naval attaché in Berlin, was intercepted by **Bletchley Park**:

From: Naval Attaché in Berlin.
To: Chief, Third Section Naval General Office Tokyo.
12/121630/1940 73900.
I have received from the German Navy the minutes of a meeting of the British Cabinet held on 15 August this year dealing with operations against Japan. The document will be sent by the next courier; meanwhile here are the main points:

1. Although Japan cherished the ambition of capturing Singapore, the existing situation would not allow Britain to send her fleet to the Far East, and she must defend it by sending Army and Air Force reinforcements.
2. Japan would probably invade French Indo-China or Siam as a first stop, and the Netherlands East Indies and Singapore would follow. However, Britain was not in a position to resort to war in the event of an attack on French Indo-China or Siam.
3. Hong Kong would be abandoned, but would continue resistance as long as possible.
4. If the operations against the Italians in the Mediterranean should proceed rapidly and successfully, it would be possible to send a fleet to the Far East.
5. Operations must be conducted jointly in the Netherlands East Indies.
6. Since it was probable that the Japanese would occupy Suva in the Fiji (?) Islands as a base, one Brigade must be sent there from New Zealand.

As a result of the *Automedon*'s capture, the Merchant Navy Code was changed in January 1941, but the value of the other material recovered from the ship proved of vital strategic importance to the Japanese and of great assistance in planning the successful attack on Singapore, which fell on 15 February 1942. *See also* WORLD WAR II.

B

BAD AIBLING. During the Cold War, the largest **National Security Agency** facility in Europe was located on a former Luftwaffe airfield at Bad Aibling, south of Munich in Bavaria, but after almost exactly 50 years operations, it was closed in 2002. *See also* AIR INTELLIGENCE AGENCY (AIA).

BALKANABTEILUNG. Based in Hirschberg, near Breslau, the Balkanabteilung was the Reich Foreign Ministry's principal cryptographic organization responsible for monitoring Soviet wireless traffic and breaking the enemy ciphers. Balkanabteilung cryptanalysts achieved considerable success, but in January 1945, they and their archives were evacuated to the Schloss Burgscheidungen, 10 miles northwest of Freyburg in Saxony, as the danger of being overrun by the Red Army grew.

However, as Allied troops moved closer to Leipzig, the Balkanabteilung transferred some of its records and files to a railway wagon for evacuation, but it was to remain in the local station until it was discovered in April 1945 by the U.S. 100th Infantry Brigade. Immediately, **TICOM** 6 was alerted, consisting of Colonel Paul Neff of the U.S. Army **Signal Security Agency**, Geoffrey Evans from **Bletchley Park**, and William Bundy, later the assistant secretary of state, and they quickly moved in a convoy of trucks from Paris, through Verdun, to Magdeburg to take control of the castle. The TICOM senior staff operated under cover names, with Evans adopting the pseudonym "Edwards" and Bundy calling himself "Brown."

Of greatest anxiety to TICOM was the need to debrief the two dozen civilian members of the captured German unit before Russian troops arrived to occupy the area. Arriving on 12 April, Neff had no difficulty pacifying the Schloss's aristocratic owner, the Graf von der Schulenburg, and, with the assistance of the TICOM group, supported by a German expatriate staff sergeant, and guarded by Belgian troops, supervised the removal to England by air from the nearby airfield at Kolleda of what was believed to be the Nazi Foreign Ministry's entire Balkan cryptographic archive. On 3 May, the male members of the Balkan unit were flown to England, and Neff then eliminated all trace of what the castle had been used for during the past 12 months, blowing up all the remaining tabulating machines and typewriters before they fell into Russian hands on 23 June. Much of the village was evacuated, with the civilian cryptographers transferred to an interrogation center at Frankfurt, while the documents underwent assessment. Before leaving, the Graf, who had been categorized as a non-Nazi and appointed mayor of the village, allowed TICOM 6 to celebrate the end of the war by holding a party in the Schloss's impressive paneled ballroom on the first floor and arranged for the furniture and the grand piano to be uncovered for the occasion.

Initially, it was believed that the German unit had been concentrating on U.S. diplomatic traffic exchanged between Bern and Washington, DC, and it was only much later, while the files were being studied at Bletchley Park, that a copy of a Soviet codebook was discovered. Another item was the Finnish cryptographic collection, proof that the Finns had shared the Petsamo Kod-26 diplomatic codebook with the Nazis as well as the Japanese, and this material, used in conjunction with some other documents found in Schleswig by Lieutenant Oliver Kirby, also of **Arlington Hall**, allowed more progress to be made by the **National Security Agency** from 1954 onward. Kirby, who had worked on **Enigma** at Bletchley Park, was later to head the Russian Section between 1946 and 1952.

BARNYARD COMPLEX. The generic **Bletchley Park** codename for German traffic on **Enigma** channels carrying messages relating to France. The station characteristics were designated PULLET, GOOSE, BANTAM, and other farm animals. *See also* WORLD WAR II.

BEAUMANOR HALL. The **War Office Y Group (WOYG)** intercept station located in the 300-acre park of Beaumanor Hall in Leicestershire was established in early 1941 and became one of the largest sites in Great Britain, running 160 positions simultaneously after the evacuation of **Fort Bridgewoods.** To avoid disruption from enemy air raids, the five principal set rooms and the control room were built on the perimeter, some 200 yards away from each other, disguised as farm buildings, garages, and a cricket pavilion and protected with 14-inch concrete blast roofs. Each set room accommodated 36 intercept positions and the necessary amplifiers, manual receivers, aerial switchboards, and teleprinters. Beaumanor was commanded first by John Tishart and, from 1943 onward, by Colonel Marshall J. Ellingworth, formerly a lieutenant-commander R.N., and by the end of hostilities, his headquarters ran a total of 196 intercept positions.

As the WOYG headquarters, Beaumanor monitored and copied enemy wireless traffic, conducted **direction-finding (D/F)** and **traffic analysis (T/A)** operations, and managed the other main WOYG site at **Chicksands Priory** in Bedfordshire. Local villagers, fascinated by the estate's top security and the frequent visits from staff cars carrying high-ranking officers, were encouraged to believe popular rumors that it was either experimenting with radar or masterminding plans for the Second Front.

To respond to a staff shortage from February 1942, more than 1,000 civilians, reserve officers, and Auxiliary Territorial Service (ATS) women were given 20 weeks of wireless training before being sent on a special course lasting two weeks to help them acquire the secret skills of **interception**, D/F, and T/A. Traffic analysis, initially undertaken at Fort Bridgewoods, was mechanized by the use of Powis Sarnes tabulating machinery, and the ATS women became so proficient that, in November 1942, they formed their own set room, staffed with 36 operators. At the height of activity at Beaumanor, 75 percent of the operational personnel were ATS. Only officers were indoctrinated into the decryption program, which made accurate copying necessary; although, some operators guessed their true purpose. The local T/A effort was enhanced in May 1941 by the appointment of a specialist from **Bletchley Park**, M. Spooner, and a permanent liaison post was created in September 1942 when J. Reid was appointed control liaison officer. More reinforcements arrived in 1944 in the welcome form of two American officers and three sergeants, who also acted in a liaison capacity until the end of the war. In 1943, Beaumanor accommodated an advanced-training facility, No. 6 Intelligence School, headed by Colonel Lithgow, which later moved to several requisitioned houses in Netherhall Gardens, Hampstead, to prepare personnel for D-Day and deployment to the Far East. As an intercept site, Beaumanor was closed down in 1994.

BERLIN TUNNEL. Codenamed STOPWATCH by the **Central Intelligence Agency (CIA)** and GOLD by the British **Secret Intelligence Service (SIS)**, a plan was drawn up in 1953, following the success of SILVER in Vienna, to dig a tunnel almost half a mile long extending under the border at Rudow into the Soviet zone and intercept a cable duct beside the Schönefelder Chaussee, which carried traffic between the Soviet military headquarters in Zossen-Wünsdorf and Karlshorst to Moscow.

The first planning meeting took place over three days in London in December 1953, attended by George Young as SIS's director of requirements; Tom Gimson, who headed the Y Section that had masterminded the Vienna tap; the project manager, Stewart Mackenzie, a future controller of operations; and John Taylor, a technical expert from the **General Post Office** Engineering Research Station at Dollis Hill. Having agreed the general principles, another meeting was held early in 1954 at the Y Section's office at 2 Carlton Gardens; only, on this occasion, the minutes of the discussion had been taken by a young Dutch-born SIS officer, George Blake, who the previous April had been released from captivity in Korea and had joined the section in September 1953.

In an agreed division of labor, processing the product was to be shared in Washington, DC, where a windowless building off the Washington Mall was prepared for a teletype link and a compartmented unit in Regent's Park with room for a staff of 378 linguists for the Russian and German voice traffic, working in shifts, but Blake passed a copy of the minutes to his Soviet contact on 18 January 1954.

By August 1954, the surface building, ostensibly a radar station, warehouse, and barracks, had been completed by a German contractor, and work began from the basement level to sink a shaft 30 feet and then dig a tunnel 6 feet in diameter. The U.S. Corps of Engineers completed the horizontal work on 10 March, and within three weeks had broken into the junction box housing the cables. At this point, British telephone engineers were brought in to install 25 tons of preamplification equipment, voltage stabilizers, and banks of tape recorders, together with the relay cables linking a specially fabricated tap chamber to the sophisticated hardware under the warehouse building capable of handling up to 500 separate channels. The connections were made, and on 11 May 1955, the first of many thousands of reels of tape were processed, revealing the "take" to be a veritable intelligence bonanza. The raw material ranged from indiscreet conversations among officers, complaints about the frequent mechanical breakdowns that plagued the T-52 tank, to data that helped **Government Communications Headquarters** solve some of the codes found in Soviet tactical wireless traffic. There were even sufficient references to a KGB operation to tap an American communications cable in Potsdam for the CIA to take the appropriate countermeasures.

The hemorrhage of Soviet traffic continued for 11 months and 11 days, until one o'clock in the morning of Sunday, 22 April 1955, when a team of Russian and German engineers, apparently repairing flood damage to cable conduits across the city, uncovered the tap chamber and sounded the alarm. As the astonished men ventured into the main tunnel, their conversations were picked up on ceiling microphones installed as a security precaution, and it was clear from the stunned amazement expressed by those that gathered that they at least had not been briefed on what to expect. They manifested complete surprise mixed with grudging admiration for the sophistication of the undertaking and, two days later, held a press conference to express their indignation at the intrusion.

Altogether, the operation recorded around 368,000 conversations in Russian and 17,000 in German. *See also* BOURBON.

BERTRAND, GUSTAVE. A career **French signals intelligence** officer, Gustave Bertrand headed Section D, the decryptment branch of the prewar Service de Renseignements. He had been the recipient prewar of information about the **Enigma** machine from **Hans-Thilo Schmidt**, a corrupt German cipher clerk employed by the Reichswehr's **Chiffrierstelle**. Between October 1931 and 1939, they held 19 meetings (four in Belgium, one in Copenhagen, twelve in Switzerland, one in Evian, and one in Czechoslovakia), and Schmidt, codenamed ASCHÉ, had handed over 303 documents. He also revealed that the **Forschungsamt**, where he was posted in 1934, had broken the Royal Navy's Naval Intelligence Code 1, the French Navy's BGD 30 code, and the French mobilization code, the Code d'Alerte. Schmidt would commit suicide in prison in September 1943, following his arrest by the Gestapo.

Bertrand's material was shared with cryptographers from the **Polish Cipher Bureau** at two secret conferences held jointly with the British in 1939, and an agreement was reached on the sharing of information concerning their common target, the German **Enigma**. In June 1940, after the French collapse, Bertrand's small staff moved into the unoccupied zone, continuing to work on enemy communications at Les Fouzes, near Uzes, in Provence, and in November 1942, fled to Algiers. He reached London in October 1943, via Lisbon, and thereafter cooperated closely with his British counterparts. In March 1946, he created the **Groupement des Contrôles Radio-électriques** and headed the organization until his retirement in 1949 to the Riviera, where he was elected mayor of La Napoule.

In 1968, he published *Enigma*, an account of his wartime escapades and the first disclosure of Allied success in breaking German Enigma ciphers.

BIN LADEN, OSAMA. Following the al-Qaida suicide attack on the American embassies in Nairobi and Dar es Salaam in October 1998, the **National Security Agency (NSA)** began monitoring a telephone number in Sana'a, Yemen, 200578. The landline had been contaminated by four calls from a cell phone belonging to Mohamed al-Owhali, the putative suicide truck-bomber, just before he drove into the U.S. embassy compound in Nairobi.

The landline's subscriber was identified as Ahmed al-Hada, a veteran jihadist who had fought alongside al-Qaida's charismatic leader, Osama bin Laden, against the Soviets in Afghanistan. Also living in his house was Ahmed's son Samir, who helped man the landline telephone, which became known as "the Yemeni hub" or "the Sana'a switchboard," and took messages for al-Qaida callers, including Osama bin Laden. Ahmed's other son, Abu Jafar al-Yemeni, would be electrocuted in an accident while fishing on Lake Duranta in Afghanistan in 1999. Ahmed's three daughters had all married al-Qaida terrorists, and his sons-in-law were Mustafa al-Masri, who was killed in an attack on foreign oil workers in Saudi Arabia; Abed al-Maki, killed in 1999 in Indonesia; and Ahmed al-Darbi, who would eventually be arrested while visiting his mistress in Azerbaijan.

During the nearly four years of NSA coverage, the Sana'a number received or made more than 1,000 calls, of which 221 were exchanged with a Compact-M Inmarsat satphone, 682505331, registered to Kandahar Communications in Kabul, and used by Osama bin Laden and his close associates between November 1996 and September 1998. Subsequent investigations revealed that the satphone had been purchased in the United States with $7,500 supplied by Khalid al-Fawwaz, a Saudi who had immigrated to England with his wife and three children in 1994, when he was age 32. Suspected of having headed the Al-Qaida cell in Kenya, he was placed under surveillance by **MI5** in an operation and was arrested at his home in Neasden, North London, with four other suspects in September 1998.

The actual purchase of the satphone had been made by Zayid Khalil, a Palestinian American student at the University of Missouri at Columbia, who had delivered it to bin Laden in Afghanistan in May 1998 before returning to Orlando, Florida. He was arrested in Amman in May 1999 but was released and settled in Manchester, Missouri, where he later died.

The hub was also called in October 2000 by those responsible for the attack on the USS *Cole* in Aden, prompting a decision taken in Washington, DC, not to intervene, presumably to avoid terminating what would become an absolutely invaluable source of information about al-Qaida, which continued to use the compromised landline even after testimony was given in open court in Manhattan on 6 March 2001 by Special Agent Stephen Gaudin of the Federal Bureau of Investigation (FBI) at the trial of four of the U.S. embassy bombers: Wadih el-Hage, Mohamed Sadeek Odeh, Mohamed al-Owhali, and Khalfan Khamis Mohamed. Under cross-examination, Gaudin repeatedly disclosed details of the al-Qaida telephone numbers and billing records.

The NSA's monitoring of al-Hada's landline at the end of December 1999 revealed references to a high-level meeting to be held in Kuala Lumpur. The Malay Special Branch placed the location under observation, and surveillance photographs were taken of the attendees, among them Khalid al-Mihdhar and Nawaf al-Hazmi, who on 9/11 would seize control of American Airlines Flight 77 from Virginia bound for Los Angeles; Khalid Sheikh Mohammed, the 9/11 chief planner; and Khallad bin Attash, who would become notorious because of his involvement in the attack on the USS *Cole*, in October 2000.

Following the Kuala Lumpur conference, Khalid al-Mihdhar and Nawaf al-Hazmi flew together to Los Angeles in January 2000 and shared a house in the San Diego suburb of Lemon Grove owned by an elderly FBI informant, Abdussattar Shaikh, whose registered landline, 313- 4008, was also the source of calls to the Sana'a switchboard monitored by the NSA. Due to the nature and sensitivity of the source, his information was not shared with the rest of the U.S. intelligence community.

The extent of this lapse became clear when bin Laden's bodyguard, Nasser al-Bahri, a Saudi of Yemeni parentage, known as Abu Jandal, was interrogated in Sana'a by Ali Soufan of the FBI and Robert McFadden of the Naval Criminal Investigative Service (NCIS) a week after the 9/11 attack. Actually, Al-Bahri had been in Yemeni custody since February when he had been arrested by the Political Security Organization (PCO) at the airport about to board a flight to Kabul. When questioned by Soufan in Arabic during their second interview at the PCO's headquarters, conducted on 18 September, al-Bahri gave a detailed account of his own experiences in al-Qaida and, horrified by the 9/11 atrocity, identified 7 of the hijackers from the FBI's collection of 60 al-Qaida suspects. He also disclosed that the al-Qaida leadership exchanged messages using a Yaesu radio transmitter, having encrypted their texts with a small Casio computer. In terms of evidence, this was the very first time anyone had positively linked the 9/11 hijacking to al-Qaida.

The then director of central intelligence, George Tenet, later recalled "CIA had multiple opportunities to notice the significant information in our holdings and watchlist al-Hazmi and al-Mihdhar. Unfortunately, until August, we missed them all. What if we had noticed our mistake after al-Mihdhar and al-Hazmi entered the United States, but months rather than weeks before the plot unfolded? Most likely the two men would have been deported." His FBI counterpart, Director Louis Freeh, conceded, "At the FBI we lacked technological sophistication and we let what could have been vital intelligence get trapped in the bureaucracy."

When bin Laden finally abandoned his cell phone and al-Qaida ceased calling the Yemeni hub, there were no further electronic clues to the fugitive's hiding place. Accordingly, the NSA pursued its quarry through a third party, Abu Ahmed al-Kuwaiti, who had been identified during the interrogation of al-Qaida prisoners as a close associate and courier for the terrorist group's leader. Al-Kuwaiti's cell phone was monitored through the deployment of an RQ-170 Sentinel drone from Kandahar, which loitered at 15,000 feet to maintain an electronic watch on its target.

When al-Kuwaiti was eventually traced in August 2010 to a compound in Abbottabad, Pakistan, surrounded by an 18-foot wall, some analysts came to believe that this was where bin Laden had been living in seclusion for the past five years. The unusual, complete absence of any landline or cellular activity from the site was considered highly suspicious, and bin Laden was shot dead during a night raid conducted by U.S. Special Forces on the night of 1/2 May 2011. A search of the main house revealed that bin Laden had come to rely on computer flash drives, entrusted to couriers, to direct his subordinates. Dozens were recovered, as were two telephone numbers, presumably emergency contacts, found sewn into his clothing, together with 500 Euros. Following a leak to the media, the telephone numbers could not be fully exploited.

BLACK BAG JOBS. During the Cold War, the Federal Bureau of Investigation (FBI) conducted numerous "black bag" jobs, the illicit entries mounted against target Eastern Bloc diplomatic premises. Very little has been published about these activities; although, Robert Lamphere himself has referred to the theft of Amtorg messages in his memoirs, and a **VENONA** text from New York, dated 25 July 1944, complains about the FBI entering Soviet property masquerading as certification engineers inspecting lift shafts:

> Yesterday two fellows came to see SHAH [Konstantin Chabanov] at the PLANT [Consulate General] and said that in the capacity of representatives of the City Housing Department they were to inspect the PLANT building and the neighboring house. In reply to SHAH's question about the purpose of the inspection the fellows said that they were not obliged to call. SHAH managed to show them out but they hinted roughly that they might come back with people who would ensure them access. GRANDPA [Evgeni Kiselev] [3 groups unrecovered] [8 groups unrecoverable] plausible excuse the necessity of checking the lift mechanism is very bad. GRANDPA came to an agreement that he would send out architect to the Housing Department and he would inform them about the lifts. With this the incident was closed, but it is difficult to doubt that the fellows represent the HUT [FBI]. BORIS [Aleksandr Saprygin] has no special instructions on the circumstances in which he is to destroy the ciphers without reference to anyone and when it is permissible to use weapons. Please give some guidance.

Within the FBI's Intelligence Division, it was common knowledge that such operations were undertaken throughout the 1960s, often over a weekend while the relevant building was more vulnerable, and a special laboratory was set up to photograph the pages of codebooks without disturbing the pages, and sophisticated techniques were adopted to reproduce Soviet seals, which had been broken. Apart from Lamphere's mention of illicit access to the Soviet trade data, which has been acknowledged by the KGB but denied officially by the U.S. government, there is a well-documented case of **Frantisek Tisler**, a Czech cipher clerk who defected in 1958, having given FBI "second storey men" access to his embassy's code vault. *See also* PINCH; STAFF D.

BLACK CHAMBER. Created at the end of **World War I** by a gifted cryptographer, **Herbert O. Yardley**, and sponsored jointly by the U.S. State Department and the War Department's **MI8**, the Black Chamber was a small, clandestine cryptographic organization that operated under commercial cover, the Code Compilation Company, in Manhattan. The organization was disbanded in October 1929 by Secretary of State Henry Stimson, and its records were acquired by the War Department's **Signal Intelligence Service**.

BLACK CODE. The code employed by U.S. military attachés at diplomatic missions in foreign countries, the Black Code was compromised when, in August 1941, the Italian Servizio Informazioni Militare (SIM) gained access to the safe of Colonel Norman Fiske, the military attaché in Rome. The SIM had shared its prize with the **Chiffrierabteilung** in Berlin, thus ensuring that the Axis was able to read all the traffic on the channel.

Of particular interest to the Germans was reporting transmitted in the Black Code from the American military attaché in Cairo, Colonel Bonner F. Fellers, between October 1940 and August 1942, on current and future British military operations, having attended General Claude Auchinleck's daily staff conferences.

> June 2, 1642 HRS. Hachiem is still held by the Free French who claim to have destroyed two workshops of the 21st and 15th Armd Divs. Axis lines of communications are believed by the British to be very unstable and General Ritchie is planning to push pursuits. Picture may be changed however by using units of the 1st Armd. Brigade as replacements.
> Dispositions of 1 June
> Axis Forces: Well covered by artillery and anti-tank guns in the general areas 36-40 and 36-39. Future moves are not apparent. 300 German tanks have been lost according to the British.
> British Forces: It is believed that the Southern brigade of the 50th Div. has been completely destroyed. 1st and 7th Armd Divs. In square 37-42 is the 4th Armd Brigade with one regiment of the 1st Armd Brigade. Balance of the 2nd and 22nd Brigades totaling approx. one brigade are in square 37/40. In square 38/40 is the 200th Guards Motor Brigade Exact position of the 7th Motor Brigade is not known but they moved south and west of Bir Hachiem and its car regiment of KGGs was in square U6-5 yesterday.

> June 4, 0749 HRS. On the night of June 2–3 Germans evacuated Eleut Ettamar, which position was then occupied by British Infantry Battalion, supported by remnant of 4 Armd. Brigade from the east. The position has again been attacked from the north by the German 7th Armd. Tanks with unknown results. The main German position in minefields is unchanged.
> At Bir Hachiem Free French withstood Italian attack of June 2. RAF assisted with adequate air support. The 7th Motor Brigade is in a position west and slightly north of Free French. Right flank is being covered by 29th Indian Brigade of the 5th Indian Division. Axis lines of communication are being raided from south by 4th Armd and from the north by 50th Div. and 3 South African Brigade. The 11th Indian Brigade from the 4th Indian Div. is now at Tobruk and the 10 Indian Div. is moving up in the rear of South African Div.

Fellers's accurate messages, describing his personal observations, recording his own comments, and summarizing his visits to the front, proved of immense value to General Erwin Rommel, who waited each evening for the arrival of the next intercept before issuing his own instructions to subordinates. Whenever Fellers left Cairo for an extended tour of the battlefield, Rommel was temporarily disadvantaged and starved of information but invariably rewarded soon afterward by the updated intelligence contained in his dispatches.

The leak was discovered almost simultaneously, at the end of June 1942, when **ULTRA** intercepts of Luftwaffe traffic indicated that what the Germans had referred to as "the good source" had revealed that the British had learned the precise location of the Luftwaffe's headquarters. This had led to an extensive mole hunt to find a traitor, but then in July 1942, tanks of the Australian 9th Division surrounded and captured Wireless Reconnaissance Unit 621, a forward Afrika Korps intercept station just 600 meters behind the front line, on the Tel-el-Eisa plateau, taking 69 prisoners and recovering a wealth of cryptographic material and files of past intercept traffic, including evidence that the Black Code had been compromised, leading to its replacement. Also retrieved were copies of the British call-sign book for the entire Middle East theater dated 1 May covering the period 21 to 31 May and a British map reference code issued on 9 September 1941. Supposedly protected by Italian Sabratha Division, 621's loss would have a lasting impact on the quality of SIGINT available to Rommel, even if SCORPION, the German Y Service **Enigma** key for the Mediterranean, would resist a solution until April 1943.

Among the German prisoners were 621's commander, Harald Seebohm, and his second-in-command, Leutnant Herz. Seebohm later died of wounds in the hospital in Alexandria, but Herz cooperated with his interrogators, enabling a team of **MI8** analysts in Heliopolis, led by Major Tozer, to circulate *The German Wireless Intercept Organisation*, a detailed review of the enemy's Y capability. According to Herz, his unit conducted some tactical analysis but relayed enemy signals to cryptanalytic centers in Athens and in Berlin's Mattaikirchstrasse, where specialists attacked the Allied codes. The final product was then passed to Rommel by his briefing officer, Leutnant Wischmann. *See also* ITALIAN CIPHER BUREAU.

BLACK FRIDAY. In February 1948, **William Weisband**, an **Armed Forces Security Agency (AFSA)** linguist, reported to his **NKVD** contact that cryptographers at **Arlington Hall** had solved some Soviet cipher systems and were reading the traffic. Moscow responded to the news by improving their code procedures, and the loss of access became known as "Black Friday";

although, the changes did not take place on one day but over several months. Nevertheless, the AFSA suspected a leak, and Weisband advised his NKVD contact, Nikolai Statskevich, not to make it too obvious that Moscow had been alerted.

The consequences of Black Friday were considerable, with the AFSA failing to monitor Soviet military logistics traffic that might have betrayed Josef Stalin's decision to support North Korea's plan to invade the rest of the peninsula. *See also* BOURBON.

BLACK JUMBO (BJ). The **Government Code and Cypher School**'s generic codename for diplomatic intercepts. According to **Kim Philby** and Anthony Blunt, who both had access to BJs, by 1942, **Bletchley Park**'s **Diplomatic Section** had solved sufficient ciphers to read traffic from the Bulgarian foreign ministry addressed to the ambassador in Berlin; from the Turkish foreign ministry to the ambassador in London, and from the ambassador in Kuibyshev to Ankara; from the Spanish ambassador in London to Madrid; from the Portuguese ambassador in London to Lisbon; and from the Romanian foreign ministry to the ambassador in Tokyo. In addition, some Japanese traffic to and from London, and on the Madrid to Tokyo circuit, was read, as was practically all the Chinese traffic exchanged between London and Chungking by both the ambassador and the naval attaché.

BLACK WIDOW MOUNTAIN. Located in a remote corner of Tay Ninh province near the Cambodian border, Black Widow Mountain offered the U.S. **Army Security Agency (ASA)** an attractive vantage point from which to run VHF **interception** operations during the **Vietnam War**. An ASA station was established in 1967 but had to be evacuated after just four days following shelling from the Vietcong, but the site was reoccupied in May 1968. This second attempt lasted two weeks and was withdrawn by helicopter when one ASA operator was killed and another wounded, with the assigned Special Forces protection unit taking many casualties. Despite this experience, the **National Security Agency** developed a program, codenamed EXPLORER, to build clandestine intercept stations atop strategically advantageous mountains.

By June 1970, improvements in technology had allowed the development of remote, unmanned intercept facilities, and one site equipped with four VHF receivers was built on a hill near Phu Bai. Codenamed EXPLORER I, it was destroyed a year later to keep it from falling into enemy hands. Much the same happened to EXPLORER II, but SARACEN became operational in late 1972 and, located on an almost inaccessible hill overlooking the Ho Chi Minh trail, provided valuable SIGINT without placing ASA operators at risk.

BLETCHLEY PARK. Also known as Station X, Bletchley Park was the war station of the British Secret Intelligence Service, a large mansion close to the railway junction at Bletchley in Bedfordshire, almost midway between Oxford and Cambridge. Initially mobilized during the Munich crisis of 1938, the extensive estate accommodated a large staff of linguists and cryptanalysts employed by the **Government Code and Cypher School**. At the height of its activities, some 12,500 personnel worked at the Park or in the many temporary wooden huts erected in the grounds, concentrating on encrypted enemy communications. Early breakthroughs achieved at Barnet by the **Radio Security Service** against Abwehr hand ciphers encouraged resources to be devoted to traffic encrypted on the **Enigma** machine, and the world's first programmable analog computer, the **Colossus**, was developed to race through the even more complex permutations of the 10-rotor **Geheimschreiber** machine.

Although, at the end of **World War II**, the renamed **Government Communications Headquarters** moved to Lime Grove, Eastcote, in West London, and then in 1953 to Benhall in Cheltenham, Gloucestershire, the organization retained a small training facility at Bletchley Park until 1985. *See also* BONIFACE; ULTRA; VENONA.

BONIFACE. The original **World War II** codename for signals-intelligence summaries drawn from enemy decrypts, BONIFACE was replaced in March 1942 by **ULTRA**.

BORESIGHT. The **National Security Agency** codename for a network of high-frequency **direction-finding** stations managed by the **Naval Security Group** covering the Atlantic and the Pacific designed to monitor transmissions by submerged Soviet submarines. The manual project later expanded with automation and developed into **CLASSIC BULLSEYE** and then **SAMBO**.

BOURBON. The codename for a lengthy signals-**interception** operation conducted during three years of the Cold War to monitor Soviet Bloc troop movements. The source was considered so reliable and successful that BOURBON provided what amounted to a trip wire to provide early warning of exercises and widespread mobilization until **Black Friday** in 1948 when the source was terminated by the unexpected introduction of new security procedures.

BOURBON's value was that it provided almost instant access to Soviet signal traffic concerning military logistics and provided a comprehensive picture of military supplies and their movement. The loss of BOURBON over a period of months in 1948 meant the **Armed Forces Security Agency** was unable to spot the transfer of vast quantities of matériel to North Korea in anticipation of the offensive against the south launched in June 1950. *See also* BRUSA.

BOYCE, CHRISTOPHER. In January 1977, Christopher Boyce, an employee of the California-based defense contractor TRW Inc. was arrested by the Federal Bureau of Investigation after his friend **Andrew Lee** had been detained in Mexico City for passing **Central Intelligence Agency** secrets to the KGB. For several months, Lee, a drug dealer, had been obtaining classified documents, including details of the **RHYOLITE** satellite system, from Boyce, who worked in a sensitive communications center, and selling them to his Soviet contacts for $70,000. Boyce was sentenced to 40 years' imprisonment but in 1980 escaped from prison and spent 18 months as a fugitive. When he was recaptured, a further 28 years were added to his sentence. The case was described by Robert Lindsey in *The Falcon and the Snowman.*

Although Boyce subsequently claimed to have been motivated by dismay that the Australian government had not been fully indoctrinated into the **ARGUS** satellite system, his principal compulsion seems to have been financial, to fuel a drug habit. As well as compromising the CIA's satellites, he also disclosed details of the KW-7 encryption machine and the KG-13 voice scrambler. Although the Soviets had some access to both devices because of **John Walker**, Boyce's contribution, as a means of verifying its information, must have been invaluable. *See also* PINE GAP.

BRIDE. The original codename for the **Armed Forces Security Agency**'s cryptographic attack on Soviet ciphers, the project underwent several other names before becoming **VENONA** in 1964.

BRIDE's origins can be traced back to Meredith Gardner, who spent much of 1946 concentrating on the **NKVD**'s 1944 traffic, and, by 1947, had begun to supply Lamphere with the first decrypts. At **Eastcote**, a grim **Government Communications Headquarters (GCHQ)** compound in West London, another young **Arlington Hall** cryptographer, Cecil Phillips, spent six months from June 1946 working in parallel on the same problem with Philip Howse. Phillips, who had studied chemistry at college and in June 1943 had been posted to Arlington Hall because he "knew what cryptography meant and passed the IQ test," was to become entranced with the possibilities. Initially he had concentrated on Japanese weather messages, but this work had become superfluous when an enemy codebook had been recovered intact by U.S. Marines on a Pacific island, and by October, he had prepared a briefing for Brigadier John Tiltman and Major John Cheadle on some of the

Canberra traffic, but he was unable to get anyone's attention, chiefly because the material was believed to be consular, with no hint of any NKVD involvement. It was only with the help of three truckloads of Soviet wartime bills of lading, found in a Department of Commerce warehouse by John Larkin of the **Federal Bureau of Investigation**, that persuaded the management that a comparative analysis proved that the traffic was not commercial and deserved more Anglo-American resources, including the use of IBM punch-card sorters to recover texts. The objective was to use the stereotypic cargo lists to strip away the unwanted cryptographic data and rebuild the 60-group key pages, which could simultaneously be used against the NKVD traffic. Where the same key had been used more than once, a phenomenon known as "depth," there was a real opportunity to start the last stage of a multipart decryption attack.

In the years that followed, British and American personnel were "integrated" into each other's BRIDE team, starting with the secondment for five years of an outstanding American cryptanalyst, Joan Malone, to Eastcote. A brilliant "depth reader," she was known as "Sneezy" because of her chronic hay fever, and she collaborated with a small cell, led by Wilfred Bodsworth, who had been one of the first to study the **Enigma**'s machine ciphers in 1937. Having studied Spanish at Cambridge, Bodsworth taught himself exemplary Russian while studying **VENONA** and was greatly respected as a linguist and cryptographer by his subordinates, who referred to him as "Snow White" because of his white hair. Always cheerful and charming, Bodsworth inspired his staff, who were all strongly motivated by what was then perceived as a Communist threat. Malone's initial task had been to help GCHQ set up a small unit within the Russian Trade Section to work on the pairs of matched messages.

While at Eastcote, Malone, who was to spend years liaising with the **Central Intelligence Agency** on VENONA, met and married Captain Harold Callaghan, then staff officer to Colonel Clough, the chief U.S. liaison officer to GCHQ. By the end of 1950, the BRIDE cell at Eastcote consisted of Bodsworth; Joan Malone; a retired colonel, "Happy" Plumber, who acted as a typist; and two others: "Audrey West," an Italian linguist known as "Bashful," who spoke good Russian and had worked during the war at GCHQ's Middle East base at Heliopolis, and a young Englishman, Geoffrey Sudbury.

Sudbury was to concentrate on VENONA until 1966 and continued as a consultant until, in 1969, he was appointed head of the Joint Technical Language Service, a government unit supplying linguists to Whitehall. Although he had studied modern history at Oxford, he had been sent on the Russian course at Cambridge by the Intelligence Corps, where he had been talent-spotted by Dr. (later Professor Dame) Elizabeth Hall. Originally taught

French by his bilingual mother, Doc had been posted to Paris to perfect his Russian, based at the British embassy but living among the émigré White Russian community and meeting, among many others, Prince Felix Yusupev, the man credited with killing Rasputin. At that time, GCHQ was hesitant to rely on expatriate language skills and anxious to encourage the talented children of born Britons to take up useful foreign languages, and once Sudbury had perfected his command of Russian, he was posted to Eastcote, where he was to develop a gift for cryptanalysis.

Although many others were deployed against the trade and diplomatic component, the principal target remained the NKVD traffic, to which only a very few had any access. Those focused on the very tedious trade material, for example, were sometimes told about how they had been able to help the main attack, but the work was highly compartmented, with the staff briefed on the trade problem being known as "Q" or possessing a "Q" clearance. Even Hugh Alexander, the former British chess champion, who had made a vital contribution to breaking the Enigma ciphers, and then became head of GCHQ's Cryptanalytic Division, was never fully aware, nor wanted to know, of what progress was being made on BRIDE. Still, there were a few lapses. On one memorable occasion, GCHQ's Tom Moss, who spent months concentrating on the trade traffic, was told unofficially that the arrest of Klaus Fuchs had been a direct consequence of a BRIDE breakthrough, but such indiscretions were rare.

The numbers working on BRIDE, first at Eastcote and then from 1953 onward, at the very end of GCHQ's Benhall site at Cheltenham, was a mere handful compared to the numbers deployed in the United States; although, even there, the overall figure never exceeded 125, including 10 to 15 who concentrated on the NKVD traffic, the overwhelming majority being women. The transatlantic exchange between GCHQ and Arlington soon became institutionalized, particularly when, after the return of Cecil Phillips in early January 1947, real progress had been achieved and Meredith Gardner had been given his own office to give him privacy and security. Although Philip Howse had been assigned to Arlington in a general liaison capacity, the Canberra–Moscow channel revealed the need for a British input into BRIDE, and he was integrated into the **JADE** team to look after British interests, which were also focused on the leakage attributed to HOMER at the British embassy.

Just when it looked as though further work on the NKVD's 1942–43 traffic would be worthless, Dr. Samuel P. Chew made a crucial cryptographic breakthrough that opened up more texts to attack. Chew's success occurred at the end of 1953 and allowed the cryptanalysts to apply one of the Petsamo codebooks, or what was left of it. This had been recovered at the end of the war by the **Target Intelligence Committee**, known usually by its acronym

TICOM, a highly secret Anglo-American organization of Target Reporting Officers based at **Bletchley Park**, which had been set up to exploit captured enemy cryptographic personnel and material in the hope they could be used against the Japanese.

BRITISH SIGNALS INTELLIGENCE. From the moment there was an intelligence advantage to be derived from intercepting foreign radio signals, the Admiralty was in the forefront of the effort to eavesdrop on the communications of potential adversaries. As early as 1904, an intercept operator on the cruiser HMS *Diana* at Suez criticized the poor grammar and syntax of his Russian counterpart.

During **World War I**, British cryptographers at **Room 40** and **MI1(b)** provided policymakers and military planners with information that would have a profound impact on strategy, by influencing President Woodrow Wilson about the Kaiser's true intentions, defeating the U-boat blockade, and exploiting technology that monitored the movement of **zeppelins**. Allied wartime successes ensured Whitehall would authorize the creation in November 1919 of the **Government Code and Cypher School (GC&CS)**, a civilian organization drawn from the War Office and the Admiralty. Despite a setback in 1920, and again in 1927, when politicians made public statements that disclosed and compromised British SIGINT capabilities, GC&CS broke into the Comintern's **MASK** traffic and revealed the true scale of Soviet subversion across the Empire.

During **World War II**, GC&CS, renamed **Government Communications Headquarters (GCHQ)**, expanded worldwide to link up with SIGINT satellites at Heliopolis, Sarafand, Delhi, and the Far East to give the Allied military commanders an unprecedented advantage over the Axis by reading the enemy's machine-cipher systems, hand ciphers, and **one-time pad**–encrypted communications. In consequence, SIGINT made a significant contribution to winning the battle of the Atlantic, defeating the Afrika Korps, and ensuring the success of D-Day. In the immediate postwar era, GCHQ enhanced its reputation as a source of reliable intelligence while supporting British forces in Palestine and Malaya.

During the **Cold War**, GCHQ capitalized on its access to intercept sites across the Empire to enter SIGINT partnerships first with the United States and then with Canada, Australia, and New Zealand. This unequal relationship with the **National Security Agency** gave GCHQ participation in a series of extremely expensive projects that would otherwise have been beyond its resources and ensure the organization would retain its position at the forefront of the British and Allied intelligence community while retaining a certain anonymity.

The declassification of **ULTRA** in 1974 inevitably drew attention to GCHQ, and greater public awareness of the organization led to media interest in security lapses, including several in **Cyprus** and one at Cheltenham. Further unwelcome publicity followed when trade unions took industrial action in December 1979, just as the Soviets invaded Afghanistan. Finally, when giving evidence to the Scott Enquiry, Prime Minister John Major disclosed that annually some 40,000 separate items of intelligence are delivered to the Foreign Office by the Secret Intelligence Service and GCHQ, with the latter agency supplying about two-thirds, thus confirming GCHQ as the largest purveyor of finished intelligence to Whitehall's consumers. *See also* BLETCHLEY PARK; BRIDE; BRUSA; COMBINED BUREAU MIDDLE EAST (CBME); COMBINED INTELLIGENCE FAR EAST (CIFE); COMPOSITE SIGNALS ORGANISATION (CSO); CURZON NOTE; DIPLOMATIC SECTION; ENIGMA; EWING, SIR ALFRED; GEHEIMSCHREIBER; ISCOT; ISPAL; NORTH ATLANTIC TREATY ORGANIZATION (NATO); OFFICIAL HISTORIES; PRIME, GEOFFREY; SUBMARINES; UKUSA; VENONA; WIRELESS EXPERIMENTAL STATION (WES); ZIMMERMANN TELEGRAM; ZIRCON.

BRITTEN, DOUGLAS. A Royal Air Force (RAF) chief technician and, coincidentally, the holder of an amateur wireless license, Britten sold huge quantities of classified communications data to the KGB until his arrest at RAF Digby in September 1968. According to his confession, Britten had been recruited casually in 1962, while browsing in South Kensington's Science Museum, by a Soviet diplomat based in London who addressed him by his amateur license call sign. Britten, who had joined the RAF in 1949, was posted to **Cyprus** later in 1962, and continued to maintain contact with the KGB, claiming that he had been blackmailed into supplying further material.

After he was transferred in 1966 to RAF Digby in Lincolnshire, Britten had been handled by Aleksandr Bondarenko, but he was identified as a spy by **MI5** when he attempted to deliver documents to the embassy, having missed a rendezvous. Britten was sentenced to 21 years' imprisonment in November 1968.

BROWN. The **Government Code and Cypher School** codename for Luftwaffe traffic intercepted from 1939 onward at Fort Bridgewoods, BROWN turned out to be an **Enigma** key employed by an organization conducting experimental research work in the field of navigational beams. Unlike other German networks, the radio security was abysmal, with operators often chatting in plain language, discussing their locations and cipher settings, and referring to each other by abbreviated names. This particular lapse enabled traffic analysis to identify individual stations, whatever their call signs. On one memorable occasion, a message included the Enigma rotor setting for that entire day. When Luftwaffe bombers began air raids on Great Britain, reliant on navigation beams, the BROWN decrypts attained critical importance.

Although it was not initially recognized as Luftwaffe communications, the imposition of **radio silence** during air exercises conducted between 28 July and 4 August 1939 served to confirm suspicions that had developed when some familiar air procedure signals had been used accidentally and noted.

Study of BROWN also gave access to the Luftwaffe's daily "short reports," which were transmitted from all airfields and provided data on readiness and locations, thereby offering invaluable order-of-battle intelligence. *See also* COVENTRY RAID.

BRUNEVAL RAID. Codenamed BITING, British paratroops raided a German coastal radar on the cliffs at Bruneval, 12 miles north of Le Havre, under the full moon of 27/28 February 1942 in what was probably the first **Radar Intelligence (RADINT)** operation. A force of 119 paratroops from C Company, 2nd Parachute Battalion, of the 1st Airborne Division, accompanied by a Royal Air Force radar expert, Flight Sergeant C. W. H. Cox, were dropped near the isolated site, which had been photographed extensively from the air and reconnoitered by local agents. The objective, which was accomplished, was the capture of a Luftwaffe radar operator and the removal of a Würzburg parabolic antenna, some 10 feet in diameter, a receiver, amplifier, and modulator. The prisoner and the equipment were then taken off the beach by six landing craft and shipped by motor gunboats to England where the technician underwent interrogation and the enemy apparatus was examined at the Telecommunications Research Establishment at Worth Matravers in Dorset.

The Würzburg proved to be a relatively sophisticated, short-range device developed in conjunction with the much larger Freya system that detected the approach of aircraft at much greater distance. The Würzburg complimented Freya by supplying the data required to vector fighters to intercept the intruders, and although its performance was unaffected by conventional jamming techniques, it was found to be vulnerable to WINDOW, strips of aluminum foil that reflected the Würzburg's signals. Deployment of WINDOW at altitude in large quantities had the effect of emulating the returns expected by a huge bomber force, obliterating an operator's screen.

Following BITING, other German radar installations were fortified and surrounded with very distinctive rings of barbed wire, making them much more obvious to aerial reconnaissance missions. *See also* ELECTRONIC COUNTERMEASURE (ECM).

BRUSA. The acronym used to refer to the seven-page Britain–United States signals-intelligence treaty concluded in Washington DC in March 1946 and officially known as the British–United States of America Communications Intelligence Agreement. The document built on the foundation of an earlier pact made in March 1943 between Sir Edward Travis, then deputy director (Signals) of the **Government Code and Cypher School (GC&CS)**, and the U.S. Army's deputy chief of staff, General George Strong, which had standardized SIGINT procedures, allocated **interception**, applied analytical resources, and regulated distribution. This arrangement was then reinforced by a second, signed in October 1944, relating to SIGINT operations in the Pacific. Negotiated by **OP-20-G** for the U.S. Navy and Alan Hillgarth and Travis for GC&CS, the agreement linked American cryptographers on Adak and Guam to the British network based at **Colombo** and Melbourne.

The BRUSA parties were the U.S. Army-Navy Communication Intelligence Board (ANCIB), the U.S. Army-Navy Communication Intelligence Coordination Committee (ANCICC), and the **London Signal Intelligence Board** (London SIGINT Board or LSIB), represented by Edward Travis, Eric Jones, and Harry Hinsley from **Government Communications Headquarters**. The final document, reproduced in appendix C, was signed in Washington, DC, on LSIB's behalf by the **special UK liaison officer**, Patrick Marr-Johnson.

In March 1953, BRUSA was replaced by the **UKUSA** Agreement, but the text of the original document was not declassified until June 2010.

BULLFINCH. The **Bletchley Park** codename for an **Enigma** channel carrying administrative messages relating to the Afrika Korps, BULLFINCH was accompanied by CHAFFINCH, for operational traffic, and PHOENIX, which was used exclusively by the Panzerarmee Afrika. These signals provided the very first indication of the deployment of German forces to Tripolitania, and they terminated at 0800 on 12 May 1943 after a series of poignant farewell signals, marking the surrender, and referred to within the network as KLARTEXT.

Initially designated "AF5" and then LINNET in March and April 1941, CHAFFINCH included circuits connecting Rome and Berlin to Libya and Sicily. Whereas the latter were high-power static stations, the other sites tended to be low powered and mobile. Later in August 1941, a further network, linking Rome to Benghazi and Tripoli were detected and monitored.

BUNDESNACHRICHTENDIENST (BND). The Federal Republic of Germany's signals intelligence branch, Asterhig II of the BND, was created in 1952 and operated along the West German borders with Poland from Husum in the north and the German Democratic Republic from **Bad Aibling** in the south, employing more than 2,000 staff at its headquarters at Lauf an der Pegnitz, near Nuremberg, and at 20 individual intercept sites, including the Schloss Kransberg in Hessen and a **direction-finding** facility at Augsburg inherited from the Canadian **Communications Security Establishment**. Asterhig II had begun with 14 operators collecting around 2,000 radio signals a month, but, under the leadership of Leo Hepp, the organization developed, with a commercial cover concealed as Sudlabor GmbH, a clandestine base at Tutzing in northern Bavaria to give almost blanket coverage to East Germany. Formerly the Wehrmacht chief of staff for communications security, Hepp headed the **Gehlen Org**'s Asterhig II for 10 years, from 1946, and then returned to the same role in the BND in 1967 after service in the Bundeswehr.

By 1980, the BND had combined with the Bundeswehr's own signals intelligence group, which ran intercept sites from Flensburg to Langenargen, and with the German Navy, equipped with three SIGINT vessels, to coordinate their surveillance with two Danish sites on the Baltic.

The BND's strength in signals intelligence was inherited from the Gehlen Org, which had provided an embryonic collection and analysis capability until the BND had been formally established in April 1956. Thereafter, the BND accommodated the Zentralstelle für das Chiffrierwesen (ZfCh), headed by Erich Hüttenhain, formerly head of the Wehrmacht's cryptographic research office, and later Division IV, known as TA for Technische Aufklärung. The ZfCh remained responsible for the joint functions of signals intelligence and cipher security until 1989 when it was absorbed into a new Zentralstelle für Sicherheit in der Informationstechnik.

German investment in space collection platforms with France led to the launch in December 2006 of the first in a series of five SAT Lupe synthetic-aperture radar and SIGINT satellites, which became fully operational in July 2008.

The BND's directors have been: Reinhard Gehlen (1956–68), Gerhard Wessel (1968–78), Klaus Kinkel (1979–82), Eberhard Blum (1982–85), Heribert Hellenbroich (1985), Hans-Georg Wieck (1985–90), Konrad Porzner (1990–96), Gerhard Güllich (1996–96), Hansjörg Geiger (1996–98), August Hanning (1998–2005), Ernst Uhrlau (2005–11), and Gerhard Schindler (2011–).

BYEMAN. The United States intelligence community generic codeword for intelligence derived from satellites, both signals intelligence and imagery. *See also* AIRBORNE COLLECTION PLATFORMS.

C

CAIRNCROSS, JOHN. A member of the Communist Party of Great Britain recruited by the **NKVD** while a postgraduate at Cambridge, John Cairncross gained access to signals intelligence when he transferred from the Cabinet Office to **Bletchley Park** as a German linguist in 1942. He supplied his Soviet handlers in London with original **Enigma** intercept "flimsies," which would have greatly assisted Moscow's cryptographic attack on German communications, and his contribution was considered so valuable that he was awarded the Order of the Red Banner for his role in providing the enemy's order of battle prior to the battle of Kursk. He was also rewarded with a car to assist his travel into London to attend meetings with his controllers.

Suffering from strain due to overwork, Cairncross was moved to Section V of the Secret Intelligence Service in 1943 as an analyst to concentrate on German wireless traffic in the Balkans, which, by November 1943, was being intercepted at an average of 400 a day. This too he supplied to the NKVD. Cairncross lost his access to signals intelligence when he was posted to the Ministry of Supply in 1945, a position he retained until his resignation in 1951, when he came under suspicion following the defection of Guy Burgess and Donald Maclean.

In his 1999 autobiography, *The Enigma Spy*, completed before his death in October 1995, Cairncross sought to equivocate over the scale of his espionage, but the KGB archives revealed details of his file and his activities while codenamed LISZT and MOLIERE. *See also* MANHATTAN PROJECT.

CALL-SIGN ANALYSIS (C/A). An essential component of signals-intelligence **traffic analysis** is call-sign analysis, a process intended to identify the source of individual broadcasts and link them to specific military units. The countermeasure employed to complicate an adversary's analytical work is the frequent change of call signs, or their encryption, but that behavior itself can draw the attraction of hostile intercept operators.

In 1930, the Soviet Black Sea Fleet's call-sign system was solved by British cryptographers at **Sarafand**, leading to breaks into the actual traffic itself. This success encouraged the War Office to enhance analytical resources at

Fort Bridgewoods, Chatham, then the principal intercept site for monitoring Russian traffic.

During **World War II**, the British **War Office Y Group** was able to reconstruct almost the entire call-sign system of the enemy when, in December 1941, a copy of the "D" call-sign book, which contained 40,000 trigrams organized in 200 columns and 200 rows, was captured in Libya. Study of the book confirmed that the Germans allocated each wireless unit one or more rows of trigrams, changing them daily by varying the columns, with the actual row remaining permanent. This book remained in use until it was replaced in April 1944 by a new version known as the "P Book," a reconstruction of which had begun the previous year, as soon as the first signs of its introduction emerged in the enemy's traffic.

Call-sign analysis was greatly assisted by wartime improvements in frequency measurement, employing American SCR 211B frequency meters accurate to within one kilocycle, and **Bletchley Park**'s Hector Bolitho noticed that when the Germans changed their call signs, they did not invariably switch frequency, thereby offering a clue to the true identity of the transmitter. Bolitho's team of analysts, codenamed BAS (Bolitho Analytic Section), transferred intercept log data onto punch cards and used a Powis Sarnes tabulating machine to help break the Luftwaffe call-sign system. Accordingly, frequency measurement played a significant role in assisting the reconstruction of the enemy's call-sign tables.

In August 1944, the Germans transformed their call-sign allocation system and defeated the analysts, requiring them to fall back on **direction-finding** bearings to identify particular target traffic. The new tables, which changed bimonthly, proved almost impossible to predict, but as the Allies advanced into Germany, enemy operator discipline declined, and there was more operator chat to listen to. By 24 April 1945, Berlin had gone off the air, and Prague assumed control over the entire **Wehrkreis** radio net, leaving only a small pocket of enemy transmissions to monitor and only for what proved to be a short period.

CANADIAN SIGNALS INTELLIGENCE. The principal Canadian signals intelligence organization established during **World War II** was the **Examination Unit** of the Department of External Affairs, which, in April 1946, was renamed the **Communications Branch** of the National Research Council. This was renamed the **Communications Security Establishment** in 1975. All the Canadian armed forces maintained their own small individual SIGINT capabilities, but the only overseas station was at Daniel's Head, Bermuda, a facility that continued to participate in the **North Atlantic Treaty Organization**'s **(NATO) direction-finding** operation during the **Cuban Missile**

Crisis, even after the prime minister had banned Canadian participation in the enforcement of the United States' quarantine of Cuba.

CANUSA. The acronym for the bilateral signals-intelligence agreement adopted in November 1949 between Canada and the United States covering the exchange of top-secret intercepts. *See also* BRUSA; UKUSA.

CANYON. The **National Security Agency** codename for a series of seven signals-intelligence satellites first launched in 1968. *See also* AIRBORNE COLLECTION PLATFORMS; PRIME, GEOFFREY.

CAPSTONE. In 1969, a KGB officer, Alexander Kulik, warned his Federal Bureau of Investigation (FBI) handlers that there was a Soviet mole inside the **National Security Agency (NSA)**, thereby initiating a search, codenamed CAPSTONE, that would last 16 years. The FBI made no progress with the mole hunt until 1992, when another KGB officer, Vasili Mitrokhin, defected to the British with a huge collection of notes he had copied from the organization's archives in Moscow. According to Mitrokhin's material, an NSA analyst had supplied information to the KGB from September 1965 until his retirement in 1967, and the FBI identified him as a former communications clerk, Robert Lipka. A sting operation was then mounted, and Lipka was approached by an FBI special agent, Dmitri Droujinsky, who masqueraded as a Russian intelligence officer, named Nikitin, seeking to reactivate him. He was lured from his home in Millersville, Pennsylvania, in February 1996, to a videotaped meeting, during which he made lengthy incriminating admissions.

Having enlisted in the U.S. Army in August 1963, Lipka had worked for the NSA between 1964 and 1967 but had broken contact with the KGB in 1974, and he complained freely to Droujinsky that he was still owed some money, apparently believing he was in no danger of prosecution for the crimes he had committed when aged 20 because of the statute of limitations. In total, he had been paid about $27,000 for top-secret reports from a teleprinter that he had supervised, contained in film canisters, which he had left at dead drops along the Potomac, receiving between $500 and $1,000 per delivery. Allegedly, Lipka had stuffed the documents down his shirt or wrapped them around his legs to remove them from his office at Fort Meade.

Lipka was arrested at home in February 1996, just after his two children had left for school, and in May 1997, he pleaded guilty to one count of espionage and in September was sentenced to 18 years' imprisonment.

CARNEY, JEFFREY M. In September 1985, a 19-year-old air force linguistics instructor, Jeffrey M. Carney, who had served with the 6912th

Electronic Security Group at Marienfelde between 1982 and 1984, defected to East Berlin. Later, claiming to have become disaffected through fear that his homosexuality would be discovered, Carney had crossed over to East Germany through the Friedrichstrasse border crossing in Berlin and revealed information about CANOPY WING, a plan to disable Soviet communications in the event of hostilities. According to one damage assessment, the financial implications of Carney's betrayals amounted to $14.5 billion; although, he was only paid 400 Deutschmarks on each of the dozen occasions he slipped back across the frontier to make deliveries to the Hauptverwaltung Aufklärung (HVA).

After his transfer to the Goodfellow Air Force Base at San Angelo, Texas, in 1984, Carney continued to supply the HVA with classified material, but in September the following year, in anticipation of psychological tests, he deserted to the East German embassy in Mexico City to be flown back to Berlin. In May 1997, he was granted German Democratic Republic citizenship in the name of "Jens Karney" and was decorated by Werner Grossmann, the HVA's last director. Carney remained in East Germany until April 1991, when he was arrested by the U.S. Air Force Office of Special Investigations and returned to the United States to serve 12 years' imprisonment in Fort Leavenworth for desertion and defection.

CAVIAR. Government Communications Headquarters codename, during the immediate postwar period when **BOURBON** was being read, for Soviet military high-speed teletype, sometimes referred to as "Russian **FISH**."

CELL PHONES. The widespread use of mobile cellular telephones has created a new signals-intelligence discipline that exploits the technology that provides current- and past-location data for individual cell phones. Almost universal standardization of design and roaming compatibility enables service providers to locate instruments within a few meters and offers easy access to a past history of locations and correspondent numbers for calls and text messages, as well as the time and duration of past connections, including nonbillable calls where no connection was made. Depending upon the particular legal jurisdiction, countries require cell-phone service providers to retain records of customer movements, calls, and texts for extended periods, up to seven years in the case of Switzerland. Different jurisdictions require various other personal customer details to be kept, including name, address, landline number, bank account details, and credit card data. This information can be made available to authorized recipients, principally law enforcement and intelligence authorities. Within the European Union, the Recorded Data Directive of February 2006 mandates records to be kept for a minimum of two years, and in the United States, Section 505 of the U.S.A. Patriot Act requires a similar period.

Sophisticated **traffic analysis** software has been developed that can monitor and log all cell-phone movement and use in specific locations, such as the scene of a terrorist incident, and automatically eliminate innocent subscribers with a history of past use in the area and highlight suspects whose records have been flagged because of prior contamination through contact with a potential terrorist.

Cell phones have a significant intelligence value because terrorists use them not only to communicate but also as devices to complete electrical circuits to detonate explosives, either in alarm timer mode with the vibrate option or simply as a remote receiver of an external call. While the countermeasure of battery removal disables a cell phone's automated beacon function, even that act is logged by the service provider, as is the restoration of service. In March 2004, Al-Qaida bombers in Madrid set off explosives in early morning commuter trains using cell phones to detonate improvised homemade devices concealed in abandoned rucksacks. One cell phone, programmed to a 12-hour clock option, failed to go off, was recovered by the authorities, who traced the terrorist owner.

Depending upon the make and model, the capabilities of particular cell phones have been enhanced by remotely downloaded recording software that essentially transforms them into devices capable of eavesdropping on conversations, even when they are ostensibly switched off. Access to SIM cards can also offer additional data, including deleted text messaging that was never transmitted.

In Iraq following the Coalition invasion of 2003, the entire landline infrastructure, limited though it was to just 800,000 telephones in a country with a population of 26 million, cell phones became the standard method of communication, offering a considerable advantage to the occupation forces, which concentrated on locating suspect insurgents and deploying Predator drones programmed to loiter for up to nine hours in the vicinity of a target number.

Italian traffic analysis exposed the participants in the **Central Intelligence Agency**'s **(CIA)** rendition of an Egyptian suspect in Milan and resulted in arrest warrants being issued for Jeff Castelli, the CIA station chief in Rome; his subordinate, Sabrina de Sousa; the CIA base chief in Milan, Bob Lady; and their entire team of 27, who seized Abu Omar off the street in February 2003 and spirited him away to the Aviano air base for a flight first to Ramstein, Germany, and then to Egypt for extended interrogation.

Originally owned and operated by just a handful of SIGINT agencies, this technology is now in wide circulation and helped law enforcement identify the location in Spain of a notorious fugitive murderer, Kenneth Noye, in

1998. Previously cleared of having killed a police surveillance officer, Noye had later knifed a young man, Stephen Cameron, and then fled abroad. His cell-phone use had been monitored by **Government Communications Headquarters** and led British detectives to the villa where he was living under an alias. As a result, Noye was arrested as he dined at a local restaurant, extradited to London for trial, and imprisoned in 2000.

Similarly, in July 2005, a failed suicide bomber, Osman Hussein, was traced to his brother-in-law's apartment in Rome after the Ethiopian had fled from London by train, carrying his cell phone. In June 2007, cell phones were recovered from a pair of Mercedes car bombs left in London's West End and traced through the DNA on the microphones to their owners, both jihadist doctors.

The same computer software was used in January 2010 by the Dubai authorities investigating the murder of the Hamas military leader Mahmoud al-Mabhouh. Analysis of the local cell-phone use revealed links between hotel guests who, in security cameras, had not shown any signs of knowing each other.

CENTRAL BUREAU. The principal wartime **Australian signals-intelligence** organization, the Central Bureau, was created in April 1942 and was based in a former brothel near the Ascot racecourse in Brisbane and then in Melbourne, with a diplomatic section accommodated at Cranleigh, in Domain Road, South Yarra. Commanded by General Douglas MacArthur's chief signals officer, General Spencer B. Akin, the Central Bureau's assistant directors were Abraham Sinkov, of the U.S. Signal Corps **Signal Intelligence Service** and Wing Commander Roy Booth of the Royal Australian Air Force. About half the staff of 4,000 was drawn from American personnel evacuated from the Philippines in January 1942 and **Far East Combined Bureau** analysts withdrawn from Singapore in February 1942, with Australians drafted in from the Royal Australian Navy's cryptographic unit run by two Japanese linguists, Eric Nave and Professor Arthur Trendall.

The Central Bureau's intercept facilities included stations at Beaudesert in Queensland, at Darwin, and across the Pacific, but in 1946, the American component was withdrawn, and the organization was renamed the **Defence Signals Branch**, undergoing a further change in 1949 to the **Defence Signals Division**. *See also* DEFENCE SIGNALS DIRECTORATE (DSD); FLEET RADIO UNIT MELBOURNE (FRUMEL).

CENTRAL INTELLIGENCE AGENCY (CIA). It was not until Frank Rowlett transferred to the CIA from the **Armed Forces Security Agency (AFSA)** in 1952 that the counterintelligence staff became actively involved following up leads in the **VENONA** traffic. The delay in indoctrinating the CIA, considered by some to be controversial, was primarily because the texts were regarded as belonging to the Federal Bureau of Investigation (FBI) but also because the messages showed that the Office of Strategic Services, the CIA's forerunner, had been thoroughly penetrated. By the time the AFSA had been transformed into the **National Security Agency** in November 1952, the CIA's Bill Harvey had grasped the significance of the source, and eventually it was Jim Angleton who attempted to bring in selected counterintelligence experts from the Royal Canadian Mounted Police (RCMP) Security Service, the **Australian Security Intelligence Organisation (ASIO)**, and the New Zealand Security Intelligence Service to form CAZAB and provide VENONA with global support. Eventually, CAZAB became the basis of VENONA exploitation, with individual officers from the five countries cleared to handle the material and develop it by sharing collateral, particularly that provided by defectors.

By October 1980, when work on the Stockholm VENONA had drawn to a close, with 750,000 messages examined and two million pages of key scrutinized for duplicates, a comprehensive order of battle for the NKVD had been reconstructed by the Anglo-American counterintelligence analysts, listing 24 case officers, of whom the FBI estimated only 5 could speak good English.

CENTRAL SIGNALS ESTABLISHMENT. The Royal Air Force's signals-intelligence branch, renamed the Central Signals Establishment in 1948, was based at RAF Watton in Leicestershire and, from 1951, operated **airborne collection platforms** of 192 Squadron, consisting initially of converted Avro Lincoln bombers but later equipped with four RB-29 Washingtons and, from 1953, a Special Radio Installation Flight of two converted high-altitude Canberra jet bombers. In 1958, 192 Squadron was renamed 51 Squadron.

CHAFFINCH. *See* BULLFINCH.

CHALET. The U.S. **National Reconnaissance Office** codename for a signals-intelligence satellite system launched on 10 June 1978, later designated **VORTEX**. Two further launches took place, on 1 October 1979 and 31 October 1981. *See also* AIRBORNE COLLECTION PLATFORMS; BYEMAN; MAGNUM; ZIRCON.

CHECK. The officially authorized **interception** of calls made to or from a specific telephone subscriber in the United Kingdom is known within the domestic intelligence community as a check. Prior to the passage of the

Interception of Communications Act in 1988, such operations were conducted under the Royal Prerogative by the **General Post Office** (and after privatization in 1984 by the monopoly service provider British Telecom) under the authority of warrants issued for a renewable period of six months by the appropriate Cabinet ministers (being the Secretary of State for Home Affairs, Wales, Scotland, Northern Ireland, Board of Trade, and Foreign Affairs), their having been satisfied that the relevant criteria had been met. Those statutory conditions required the minister to be satisfied that the information could not be obtained from other sources and was in support of the investigation of a criminal offense that, upon conviction for the first time, carried a prison sentence of six months or more. These restrictions were subsequently confirmed by the Regulation of Investigatory Powers Act 2000.

Various agencies are empowered to apply for warrants, which have the generic acronym HoW for Home Office Warrant, irrespective of the Whitehall department sponsoring the application. Prior to 1937, telephone interception was conducted without any warrant; although, HoWs were always required for the interception of letters carried by the Royal Mail. A HoW issued in respect of a single individual or organization would authorize the interception of all telephone lines used by that nominee. During the Cold War, these telephone warrants were often referred to within **MI5** as "**telechecks**."

No HoW was required for the employment of a listening device not attached to a telephone cable; although, the 1988 Act required a separate warrant, authorizing "the interference with property," on identical terms, for the clandestine entries required to install the necessary equipment.

Interception of telephones to target foreign diplomatic missions within the United Kingdom were authorized permanently by the Foreign Secretary, but no warrants were required by either **Government Communications Headquarters** or the Secret Intelligence Service if the target subscribers were overseas, a blanket expedient confirmed by an exemption in the 1994 Intelligence Services Act.

CHICKBEETLE. The **National Security Agency** codename for operational patrols conducted by the USS *Banner*, a signals-intelligence platform, which became operational in 1965 and, based in Yokosuka in Japan, was deployed in the Far East.

CHICKSANDS PRIORY. The family seat of the Osborne family, the entire Chicksands Priory estate was taken over by the **War Office Y Group** in June 1940 as the intercept facilities at **Fort Bridgewoods** became vulnerable to enemy air attack. In October 1941, the army operators moved to **Beaumanor**

Hall, and the Royal Air Force (RAF) replaced them, remaining in occupation with the **Admiralty Y Service** until the end of hostilities. Chicksands also provided a communications center in England for the Heavy Mobile Units, which accompanied the RAF's advance squadrons established in Europe after D-Day. In November 1950, the U.S. **Air Force Security Service** established the site as the headquarters of the 6940th Radio Squadron (Mobile). In 1996, following the American withdrawal from Chicksands, the site was selected as the newly created Defence and Intelligence Security Centre. *See also* BLETCHLEY PARK.

CHIFFRIERABTEILUNG. Known to insiders of the German High Command during **World War II** simply as "Chi," the Chiffrierabteilung, or Cipher Branch, was established in 1939 and managed intercept facilities at Treuenbrietzen and Lauf an der Pegnitz, near Nuremberg. Headed by Erich Fellgiebel, the Chi employed a staff that reached 3,000; although, its cryptographic work was largely duplicated by the much larger **Forschungsamt**. When, in August 1941, the Italian Servizio Informazioni Militare acquired a copy of the Black Code used by American military attachés, the Chi was the principal beneficiary and continued to exploit the traffic until June 1942, when the introduction of new Allied security procedures terminated the source.

The operations of the Chi were compromised at an early stage by its own interstation communications, transmitted in an **Enigma** key, which were intercepted by the **War Office Y Group** from June 1941 and designated **MUSTARD**.

CHIFFRIERSTELLE. The cipher department of the German Reichswehr, the Chiffrierstelle was a small organization principally responsible for the development of secure cipher systems, a responsibility that resulted in the adoption of the **Enigma** machine, and the study of the codes employed by potential adversaries. Headed by General Rudolf Schmidt, the organization was compromised at the outset by his younger brother, **Hans-Thilo Schmidt**, who sold manuals and other classified crypto material to the French Service de Renseignements from June 1931.

CHINESE SIGNALS INTELLIGENCE. Despite having been the subject of prolonged electronic surveillance over a long period, chiefly from **Hong Kong** and collection platforms operated by the United States from Japan, the People's Republic of China (PRC) agreed in 1973 to the establishment of intercept facilities at Qitai and Korla in the Pamir Mountains of **Xinjiang** province to monitor Soviet communications in the region.

The People's Liberation Army (PLA) General Staff's Fourth Department's participation in SIGINT was limited until 1999, when a collection site was established at Bejucal in Cuba, targeted against American domestic communications. The Fourth Department (*Dianzi Duikang Yu Leida Bu*) covers the entire spectrum of signals-intelligence operations, radar, and **electronic countermeasures**.

The PLA's most recent heavy investment in SIGINT was prompted by the electronic warfare campaign waged during **IRAQI FREEDOM** in March 2003, when Coalition forces successfully neutralized Saddam Hussein's ground defenses within a few hours of the opening offensive. Consequently, the Fourth Department, also known as the Electronic Countermeasures and Radar Department, which was created in 1990, has developed an extensive capability, including a research arm intended to improve locally produced equipment.

The first Chinese **ELINT** satellite was not launched until July 1975, when an FB-1 rocket took it into orbit from Jiuquan. These were followed by two more in December 1975 and August 1976 under the sponsorship of the Shanghai Bureau of Astronautics. The PRC's intelligence satellites are managed by the National Remote Sensing Center (NRSC) via a ground station at Lizhong. The NRSC's research branch is the Institute of Remote Sensing Application, which employs a staff of 300 in five basic research departments, three technology research departments, and the Center for Airborne Remote Sensing and the Computer Applications Center. *See also* AIRBORNE COLLECTION PLATFORMS; GOVERNMENT COMMUNICATIONS HEADQUARTERS (GCHQ); NATIONAL SECURITY AGENCY (NSA); VIETNAM WAR.

CIRCULARLY DISPOSED ANTENNA ARRAY (CDAA). Configured in a circle, the CDAA offers a compass reading from a target signal so that a goniometric bearing may be taken from the direction of the transmission. Comparison with a further bearing taken from another site results in an intersection, known as a plot, which provides an accurate geographic identification of the source, a technique known as **direction finding (D/F)**. In the pioneering days of wireless, the apparatus was known as a directional, and when more sophisticated equipment was developed to fulfill the D/F and **interception** requirement, the CDAA was known as an elephant cage.

CLARK. The **GRU** codename used by the Soviet defector **Igor Gouzenko** in his communications to Moscow when he had been a cipher clerk at the embassy in Ottawa.

CLASSIC BULLSEYE. A **National Security Agency (NSA)** project that began in 1960, codenamed BORESIGHT, the CLASSIC BULLSEYE system was a land-based high-frequency **direction-finding** network of 21 **Naval Security Group** stations spread across the globe supported by eight other sites managed by the NSA's **UKUSA** partners. Coordinated and automated, the project would later develop into a satellite-linked operation, codenamed **SAMBO**, that identified and monitored the supposedly undetectable burst transmissions of submerged Soviet submarines on patrol. *See also* PRIME, GEOFFREY.

CLASSIC WIZARD. Managed by the **National Reconnaissance Office**, CLASSIC WIZARD was an ocean-reconnaissance electronic-intelligence system run for the U.S. Navy's Space Project that monitored the movements of surface ships across the globe. CLASSIC WIZARD became operational in 1971, and a further series, completed with the launch of a final cluster of WHITE CLOUD satellites in March 1980, had been scheduled for replacement in December 1980, but the Atlas F booster had veered off its intended trajectory from Vandenberg Air Force Base and had been destroyed seven minutes after liftoff. Although the two remaining CLASSIC WIZARD systems were quite ancient, being six and seven years old respectively, they could plot the position of a surface vessel with tremendous accuracy from a distance of 2,000 miles. Each of the six WHITE CLOUD satellites, three in each cluster held in a parallel orbit, independently intercepted radio signals from shipping and made two passes each day, giving a total of four **direction-finding** fixes per day for each target, making 12 individual positions transmitted to the navy's ground stations. *See also* CLASSIC BULLSEYE; PRIME, GEOFFREY.

CLEF. The **National Security Agency (NSA)** codename for an intercept facility at Wakkanai on the northern coast of the Japanese island of Hokkaido. During the Cold War, the station monitored Russian voice channels of Soviet aircraft based at Dolinsk-Sokol, just across the water on Sakhalin Island. On 31 August 1983, the site recorded the exchanges between four Su-15 fighters as they were vectored by air defense controllers toward an airborne target, which turned out to be a Korean Airlines Boeing 747-200B from Anchorage, Alaska, destined for Seoul. Much of Flight KAL 007's progress on a route over Kamchatka, and the response of Soviet ground radar stations, had been monitored by an NSA site at Misawa, but the airliner had been off course, and interceptors had been scrambled at the Petropavlovsk-Yelizovo air base. The Soviet reaction to the accidental incursion appeared almost chaotic and would end in tragedy with the death of 269 passengers and crew.

The Korean plane had been mistaken by the Soviets for a U.S. Air Force RC-135 reconnaissance plane on a **RIVET JOINT** mission and, having fired tracer cannon rounds to attract the attention of the Korean pilots in vain, Major Gennadi N. Osipovich obeyed his order to fire two AA-3 Anab missiles to destroy the target. Twelve minutes after impact, the airliner plunged into the sea. Initially, Moscow insisted the plane had not been shot down by Soviet fighters, but the voice recordings made at CLEF proved conclusively the exact sequence of events and the very disorganized behavior of the ground controllers and left little doubt about what had happened. However, the NSA was reluctant to allow the intercepts to be disclosed publicly so as to protect CLEF and its capabilities.

KAL 007's cockpit voice and digital recorders were not surrendered to the International Civil Aviation Organization until January 1993, following the collapse of the Soviet Union.

COBRA MIST. The codename for the U.S. Air Force's FPS-95 over-the-horizon radar system designed in 1970 to use ionospheric bounce to monitor reactions to Allied ferret flights conducted along the periphery of the Soviet Union, identify missile launches from Plesetsk, and watch aircraft movements up to 2,000 miles away.

The COBRA MIST apparatus was installed at Orfordness in Suffolk and in **Cyprus**, but the scheme was scrapped in June 1973 because of cost overruns, environmental objections to the location in Suffolk, and the introduction of improved performance satellites. Technically, COBRA MIST never performed as intended, with ground clutter interfering with the received signals. Although ionospheric variations had been anticipated, the inability of the equipment to discriminate between authentic returns and background noise suggested that **electronic countermeasures** had been employed by the Soviets, but after a lengthy investigation, that explanation was rejected. *See also* GOVERNMENT COMMUNICATIONS HEADQUARTERS (GCHQ); RADAR INTELLIGENCE (RADINT).

CODE TALKERS. During the Pacific campaign of **World War II**, the U.S. Marines employed Navajo American Indians to communicate over voice channels as a countermeasure against hostile **interception** by the Japanese, who were adept at monitoring Allied traffic. The Native Americans spoke their tribal languages, an expedient that defeated the enemy eavesdroppers and provided the Marines with an element of security.

The U.S. Army had originally employed Choctaws during **World War I** after it had been discovered that the enemy was routinely monitoring field telephone conversations. During World War II, other tribes were recruited to enhance military security in the European theater, including Cherokee, Comanche, Hopi, Kiowa, and Seminole.

COLD WAR. During the period of post–**World War II** superpower confrontation, both the protagonists—the Warsaw Pact and the **North Atlantic Treaty Organization (NATO)**—relied upon SIGINT to provide their adversary's order of battle and offer an indicator of future intentions. On the basis that few military formations can be mobilized without betraying telltale communications traffic and changes of deployment are invariably accompanied by characteristic tactical radio traffic, the constant monitoring of the airwaves became an essential feature of the era. The Eastern Bloc was literally surrounded by static and mobile intercept platforms, while the **KGB** sought to discern NATO's war planning by establishing clandestine intercept facilities in strategically located diplomatic premises and in surrogate territories, such as Cuba, Egypt, Syria, Nicaragua, and Sudan, when an opportunity arose.

Even non-NATO countries, such as Austria, China, Cyprus, Ethiopia, Iran, Japan, Oman, and Sweden, provided facilities that ensured blanket coverage on all flanks of the Soviet Union. To ensure a truly global capability, the "five eyes" of the **UKUSA** partnership extended the West's SIGINT reach far beyond the Warsaw Pact.

SIGINT collection during the Cold War was not without risk, and several airborne and naval collection platforms were attacked, even when they were operating in international airspace or waters. Nevertheless, the value of the SIGINT material was sufficiently useful to justify some imaginative operations, such as the tapping of underwater cables in the Sea of Okhotsk, the Gulf of Sirte, and the Barents Sea.

In intelligence terms, the Cold War began with the defection of a **GRU** cipher clerk in September 1945 and continued until the collapse of the Berlin Wall in November 1989. During that period, the **Central Intelligence Agency**'s **Staff D** concentrated on acquiring Warsaw Pact cryptographic data, often by undertaking "**black bag**" burglaries, while their opponents achieved considerable success in recruiting Allied personnel with access to cipher systems, among them **Jeffrey Carney**, **Jack Dunlap**, **Victor Hamilton**, Robert Lipka, **William Martin**, **Bernon Mitchell**, and **Ronald Pelton** of the **National Security Agency** or its components. Other sources compromised SIGINT secrets in the U.S. Army, such as **James Hall III**, while **John Walker** ran a network within the U.S. Navy that stole crypto material. In Great Britain, **Douglas Britten**, **William Marshall**, **Brian Patchett**, and **Geoffrey Prime** compromised classified SIGINT information and passed it to their Soviet handlers.

Unquestionably, the greatest SIGINT triumph of the Cold War was the Allied success in decrypting the **VENONA** traffic that exposed large Soviet spy rings in Australia, Great Britain, Sweden, and the United States. That program, which continued until 1979, formed the basis of hundreds of counterintelligence investigations; although, very few cases resulted in criminal prosecutions. *See also* BERLIN TUNNEL; BOURBON; BOYCE, CHRISTOPHER; BRIDE; FAPSI; FÖRSVARETS RADIOANSTALT (FRA); GOUZENKO, IGOR; GOVERNMENT COMMUNICATIONS HEADQUARTERS (GCHQ); IMPULSE; IVY BELLS; *LIBERTY*, USS; NATIONAL SECURITY AGENCY (NSA); *PUEBLO*, USS; ROCKET; SOVIET/RUSSIAN SIGNALS INTELLIGENCE; ZENIT.

COLERIDGE. The **Government Communications Headquarters** codename for intercepted Red Army communications in Europe that were read during the **BOURBON** project.

COLOMBO. During **World War II**, the **Far East Combined Bureau** was evacuated first from **Hong Kong** and then from Singapore, to Pembroke College, a school in Colombo. Later in the war, HMS *Anderson*, a Royal Navy shore station on a golf course, was established and linked to intercept facilities at HMS *Allidina*, at Kilindini near Mombasa, and to Curepipe on Mauritius.

COLOSSUS. The world's first programmable analog computer, the Colossus contained 1,500 valves and was designed by Donald Michie and Jack Good. It was built by Tommy Flowers at the Post Office Engineering Research Station at Dollis Hill in 1943, becoming operational at **Bletchley Park** in January 1944. The results, which included access to some **Geheimschreiber** messages, were so impressive that a second version was ordered in March for delivery before 1 June, capable of reading Baudot-code punch tape at 5,000 characters a second. As all the circuitry was in quintuplicate, the actual reading power of the photoelectric sensors was 25,000 bits per second, a performance that would not be matched by commercially available machines for a decade. *See also* GOVERNMENT COMMUNICATIONS HEADQUARTERS (GCHQ); JELLYFISH; TUNNY.

COMBINED BUREAU MIDDLE EAST (CBME). Bletchley Park's regional branch in the Middle East, during **World War II**, CBME was based at the King Farouk Museum in Heliopolis and opened in May 1941 under the direction of Colonel Freddie Jacob, formerly the head of the **Government Code and Cypher School**'s **Army Section**. CBME was connected by radio to Whaddon Hall and received a steady stream of raw intercepts, mainly enemy **Enigma** channels, from a network of **Special Wireless Group** intercept sites that stretched from Gibraltar in the West to Sarafand in the East. Decrypt summaries, codenamed **ULTRA**, were distributed to selected military commanders across the Mediterranean theatre over a system of **Special Communications Units**, which ensured the security of the material.

CBME made a critical contribution to the Allied victory in October 1942 at El Alamein, where an Oxford don, Brigadier Edgar ("Bill") Williams, acted as General Bernard Montgomery's chief intelligence adviser and kept him supplied with ULTRA summaries that provided a comprehensive view of General Erwin Rommel's order of battle, future plans, troop strengths, reserves, ammunition stockpiles, and fuel bunkers, together with daily updates of the Afrika Korps' tank inventory, vehicles under repair, and a tally of recent losses. He even received copies of his adversary's private medical reports (describing Rommel's low blood pressure), which were transmitted to Berlin by Rommel's doctor, Professor Hörster.

In a prelude to the main offensive, Rommel attempted, at the end of August 1942, to break through British defenses al El Alamein to reach Alexandria, Cairo, and the Suez Canal, but his plans had been compromised by ULTRA. Instead of taking Montgomery by surprise and sweeping through relatively weak lines, many of Rommel's 203 tanks unexpectedly found themselves in a minefield sewn with 18,000 mines and attacked from the air at night under the light of parachute flares. Worse, Rommel had gambled on the imminent arrival of six ships from Italy loaded with fuel and ammunition, but ULTRA had identified the vulnerability, and four of the merchantmen were found and sunk, and while the battle raged, the last two tankers, the *Sanandrea* and the *Picci Fascio*, were destroyed as they approached Tobruk. After six days of intense fighting, in which the Afrika Korps lost 38 tanks and 400 trucks, Rommel began to withdraw, blaming the failure on a leak from the Italians.

When Montgomery counterattacked in October, Rommel was caught off-guard and, short of fuel and ammunition, lost most of his tanks to new Allied weaponry. Overestimating his adversary's strength and down to his last 32 tanks, Rommel led his 70,000 men on a long retreat 800 miles across the Libyan desert, constantly harassed by the 8th Army and the Royal Air Force.

This kind of comprehensive intelligence picture, a rare phenomenon, enabled Montgomery to exploit his opponent's weaknesses and mount credible deception campaigns, but the loss of Crete in May 1941 had demonstrated that, however impressive a commander's knowledge of the enemy's intentions, there was no substitute for the determined force of arms. Despite the defenders' overwhelming numerical superiority, with 27,500 British and Empire troops, supported by 14,000 Greeks, General Bernard Freyberg was forced to surrender to an airborne force of less than 16,000. Although 15,000 Allied soldiers were evacuated to Alexandria, some 1,742 were killed, 1,737 wounded, and 11,835 taken prisoner in a humiliating defeat. Yet before and during the German offensive, Freyberg had received ULTRA decrypts that had detailed the enemy's strategy and identified the precise date of the attack. He also had the benefit of the Australian 4 Special Wireless Section, an 80-strong component of 101 Special Wireless Company, that provided Freyberg with high-quality tactical intercepts throughout the fighting until the unit was evacuated from Sphakia to Alexandria. *See also* SPECIAL WIRELESS GROUP (SWG).

COMBINED INTELLIGENCE FAR EAST (CIFE). Created in 1945 as the regional headquarters of the British **Secret Intelligence Service (SIS)**, Security Service, and **Government Communications Headquarters**, CIFE was established at Chia Keng in the center of Singapore Island where it remained until 1971, when it was transferred to Kranji, in the north, overlooking the Straits of Johor. Usually headed by a senior British SIS officer, the SIGINT staff included personnel from Australia and New Zealand.

COMBINED SIGNALS ORGANISATION (CSO). The principal signals-intelligence organization in New Zealand until the establishment of the **Government Communications Security Bureau** in September 1977, the CSO was created in February 1955 upon the recommendation of a senior British officer, John Burrough, who was seconded from **Government Communications Headquarters** to advise Wellington on a SIGINT structure at a time when the country's investment consisted of a single intercept site at Waiouru and a very small staff of radio operators.

COMMUNICATIONS BRANCH, NATIONAL RESEARCH COUNCIL (CBNRC). Established in 1946, the CBNRC was Canada's principal signals-intelligence organization until the creation of the **Communications Security Establishment** in 1976. During an Allied signals-intelligence conference held in London in March 1946, the CBNRC was represented by Edward Drake, who brought the organization into the **UKUSA** Agreement. *See also* EXAMINATION UNIT.

COMMUNICATIONS INTELLIGENCE (COMINT). The term employed by Anglo-American cryptanalysts for intelligence derived from all sources of signals intelligence, including **traffic analysis**, **direction-finding**, and plain language **interception**, COMINT is a broad category and does not imply any cryptographic content. *See also* SPECIAL INTELLIGENCE.

COMMUNICATIONS SECURITY ESTABLISHMENT (CSE). Created in April 1946 as the Telecommunications Authority, the CSE is Canada's principal signals-intelligence organization and is based at the Sir Leonard Tilley Building, 719 Heron Road, in the Ottawa suburb of Confederation Heights, at the nearby Edward Drake Building, with a training center in Kingston, Ontario. During the **Cold War**, intercept facilities were located at Leitrim, south of Ottawa; at Masset on the Queen Charlotte Islands, British Columbia; and at Augsburg, Germany, a **direction-finding** facility later taken over by the **Bundesnachrichtendienst**. Other sites include an early warning station on Ellesmere Island in the Northwest Territories and a high-frequency **direction-finding** facility at Gander.

The only disclosures concerning the organization were made in 1994 by a retiree, Mike Frost, in *Spyworld: Inside the Canadian and American Intelligence Establishments*, which revealed that Canadian diplomatic premises overseas were routinely used to accommodate clandestine intercept equipment and that the **National Security Agency** discrimination program, code-named ORATORY, was designed to recognize keywords in intercepted telephone conversations. More improbably, he claimed that CSE had undertaken domestic telephone intercepts in London on behalf of its British hosts so as to provide plausible deniability for what would have been illegal if conducted by a government agency.

With a staff of around 1,000, CSE consists of five directorates (Administration, Policy and Plans, SIGINT Production, Technology, and Information Security), each headed by a director-general. The directors have been: Edward M. Drake (1946–71), N. Kevin O'Neill (1971–80), Peter R. Hunt (1980–89), A. Stewart Woolner (1989–99), D. Ian Glen (1999–2001), Keith Coulter (2001–5), and John L. Adams (2005–)

Since 1996, the CSE's operations have been conducted under the supervision of a specially appointed Commissioner. They have been: Claude Brisson (1996–2003), Antonio Lamer (2003–6), Charles Gonthier (2006–9), and Peter Cory (2009–). *See also* EXAMINATION UNIT; UKUSA.

COMPOSITE SIGNALS ORGANISATION (CSO). The collection branch of **Government Communications Headquarters (GCHQ)**, the CSO is responsible for the management of GCHQ's intercept facilities in the United Kingdom and overseas.

COMPUTER NETWORK ATTACK (CNA). At the height of the Iraqi insurgency in 2007, President George W. Bush authorized the **National Security Agency (NSA)** to deploy its offensive skills and empower the U.S. Army's Network Warfare Battalion, a component of the 704th Military

Intelligence Brigade, to conduct hacking operations against specific but undisclosed foreign targets. Hitherto, the NSA had developed a sophisticated cyber capability while building a publicly acknowledged Computer Network Defense program designed to protect vulnerable electronic infrastructure from hostile intruders, but CNA was intended to exploit the weaknesses of potential adversaries.

CONTROLLED ENEMY AGENTS. The British term for double agents coined during **World War II**, controlled enemy agents run by **MI5** and the Secret Intelligence Service remained in touch with their Abwehr and Sicherheitsdienst contacts through personal contact, secret writing, and clandestine wireless links. The manipulated radio channels provided valuable, timely information, which included questionnaires and hand ciphers. When messages written and encrypted under MI5's supervision were relayed on the Abwehr's communications channels with an **Enigma** key, **Radio Security Service** cryptanalysts were presented with an opportunity to reverse engineer the Enigma's daily key, thereby allowing recovery of the machine's settings and compromising the rest of that day's traffic. Thus, transmissions made by SNOW in September 1939 under MI5's control to a receiver aboard a spy ship in the North Sea gave access to the Abwehr's Enigma traffic exchanged between the vessel and Hamburg. By this expedient, hand ciphers, codenamed ISOS, gave rise to ISK, being machine decrypts.

COUNTER-CLANDESTINE COMMITTEE. At the end of **World War II**, much of the **interception**, cryptographic, and **traffic-analysis** assets of the **Radio Security Service** were absorbed into **Government Communications Headquarters (GCHQ)**, and the skills associated with illicit broadcast detection were retained by the creation of a Counter-Clandestine Committee to liaise between **MI5**, the Secret Intelligence Service, and GCHQ. Thereafter, the committee met occasionally to discuss a strategy for monitoring a perceived threat from covert Soviet transmissions. *See also* RAFTER.

COVENTRY RAID. On 14 November 1940, the Luftwaffe undertook a massed air raid, codenamed MOONLIGHT SONATA, on the midlands industrial center of Coventry. According to Fred Winterbotham's *The Ultra Secret* and Anthony Cave Brown's *Bodyguard of Lies*, Prime Minister Winston Churchill knew of the enemy plan from intercepted **Enigma** messages but decided not to evacuate the city in order to protect the source.

In reality, advance notice of an attack had been given two days earlier by a captured German pilot who, under interrogation, had claimed that "every bomber in the Luftwaffe" would participate in a major air operation. The codeword MOONLIGHT SONATA had been disclosed in decrypted **BROWN** messages, and the Royal Air Force (RAF) had responded with COLD WATER, a preemptive strike against enemy airfields in northern France, and ASPIRIN, the **electronic countermeasures** that were designed to "bend" the navigation beams followed by German aircraft. The COLD WATER raids proved ineffective, and the ASPIRIN jamming equipment was accidentally set to the wrong frequencies so the enemy raiders found their targets without difficulty. By the time, in the early evening, the Enigma intercepts indicated that Coventry was to be the target, RAF 80 Wing had expressed confidence that the planes would be deflected. Accordingly, the assertion that Churchill had been placed in the invidious position of having to sacrifice much of the city was erroneous.

CRIB. The employment of a shortcut, most likely access to some original plaintext during a cryptographic task, is known as a "crib."

CRYPTANALYSIS. The discipline of acquiring a plain-language version of an encrypted text without the precise knowledge or means required to do so is known as cryptanalysis. *See also* CRYPTOGRAPHY.

CRYPTOGRAPHY. The discipline of encrypting a plain-language text or converting an encrypted version into plain language is known as cryptography. *See also* CRYPTANALYSIS.

CUBAN MISSILE CRISIS. Signals intelligence played a significant role during the weeks leading up to the Cuban Missile Crisis of October 1962; although, the lead was taken by photographic imagery. Indeed, the historian of the **National Security Agency (NSA)** described the episode as "the most significant failure of SIGINT to warn national leaders since World War II."

The very first SIGINT indications of some unusual developments consisted of a marked increase in Soviet trade in July 1960, including the delivery of Czech small arms and five Mi-4 helicopters and the appearance of Cuban pilots on a piston-engined fighter training course in Czechoslovakia. In late 1961, NSA received approval to expand its coverage of Cuba, and a **naval collection platform**, the USS *Oxford*, was deployed to intercept transmissions in the Caribbean, initially seeking to establish whether Soviet pilots were flying operationally from airfields on the island.

A former mothballed wartime Liberty ship, the SS *Samuel R. Aitken*, the vessel was quickly converted by the NSA at a cost of $4.5 million. The *Oxford* was 330 feet long and designed to undertake signals-**interception** duties around Latin America, from July 1961, from Port Everglades in Florida. The ship was top heavy, with 45 antennae, and carried a crew of 285, of which two-thirds were technicians working in shifts to monitor radio traffic from the classified spaces in what had been the cargo holds. A subsurface induction aerial to tap into traffic passing along underwater telephone cables was discreetly deployed from the stern, and the resulting data was codenamed FUNNEL.

The deployment of the *Oxford* was thought, at one moment in September 1962, to perhaps be the explanation for overhead imagery of surface-to-surface missiles spotted near Banes. Ray Cline of the **Central Intelligence Agency (CIA)** speculated that the weapons had been moved to a coastal site to deter the NSA's ship from venturing too close inshore. Alarmed at the implications, the White House imposed new, tighter restrictions on the distribution of this sensitive news.

Airborne collection platforms were also flown by the CIA along the periphery of Cuban airspace and occasionally directly over the island; although, these flights resulted in three casualties. On 27 October 1962, Major Rudolf Anderson was killed when his Strategic Air Command U-2 aircraft was shot down near Banes by an SA-2 Guideline, and although the NSA concluded that the local Soviet commander had fired the weapon, there had been a dispute over control of some SAM batteries with the Cubans, ending in one incident in a firefight between the troops.

Soon afterward, another pilot, Joe Hyde, died when his U-2 failed to return from its mission. A third loss, over South America after the pilot had apparently lost consciousness, was probably not a consequence of hostile action. Nevertheless, as a result of these losses, a warning program, codenamed WHITE WOLF, was introduced to alert pilots to imminent SAM threats.

Coincidentally, the *Oxford* had reported, on 15 September, that Soviet P-12 SPOON REST emissions, associated with the acquisition radar for SA-2s, had been detected. However, the missions were highly compartmented, to the point that neither Juanita Moody, then head of the B1 division, nor her chief of staff, Harry Daniels, was indoctrinated into the first U-2 overflight until the morning of 14 October, when the NSA's William Wray disclosed details of it. The following day, CIA analyst Victor DiRenzo showed Art Lundahl, the director of the National Photographic Interpretation Center in the Steuart Building in downtown Washington, DC, pictures of six SS-4 medium-range ballistic missiles.

Moody's Cuban branch expanded dramatically as the crisis unfolded and other sources showed the arrival of Soviet MiGs and Il-28 light bombers in mid-1962. Soviet early warning and fire-control radar systems began to appear, and A6 voice-intercept desks heard heavily Russian-accented Spanish being spoken over air force channels. On 9 October, the Soviets introduced classic Eastern Bloc grid air defenses and took over the entire management on the day Major Anderson was killed, but there was absolutely no SIGINT concerning the installation of ballistic missiles. Indeed, the National Intelligence Estimate 85-3-62, issued on 19 September, concluded that the installation of offensive missiles "would be incompatible with Soviet practice to date and Soviet policy as we presently estimate it. It would indicate a far greater willingness to increase the level of risk in U.S.-Soviet relations than the USSR has displayed so far."

Once President John F. Kennedy went on national television, on 22 October, to impose a quarantine on ships heading to Cuba, the Soviet forces commanded by General Issa Pliyev introduced an entirely new radio defense network and appeared to act entirely independently of the Cubans, despite sharing a joint command center in Havana. NSA **direction-finding** stations confirmed on 23 October that Soviets ships had either stopped dead in the water or turned away from the Caribbean.

A year after the crisis, when the Soviet missiles and bombers had been withdrawn, the National Indications Center examined the whole episode, concentrating on the lack of SIGINT to reveal the huge Soviet military deployment and "faulted the entire system for failure to detect these key elements. Soviet communications security was almost perfect." The lasting impact of the crisis was the introduction of a daily *SIGINT Summary*, known as the "Green Hornet" because it was circulated under a distinctive dark green cover.

CUPAR. The signals intercept and **direction-finding** station at Cupar in Fife, established by the **General Post Office (GPO)** in 1937 as part of a GPO network, which included sites at Brora in Sutherland and **Sandridge** in Hertfordshire, was designed to monitor and trace illicit domestic wireless traffic. In the absence of any, Cupar was passed to the joint control of the **War Office Y Group**, which ran a single intercept position, and the **Admiralty Y Service**.

CURZON NOTE. During a period of civil unrest in August 1920, the British government decided to release to newspapers proof that Bolsheviks had been fomenting subversion and interfering with British domestic politics by providing the *Daily Herald* with a secret subsidy. The evidence was contained in decrypted Soviet diplomatic telegrams, and on 17 August, *The Times* reported that "the following wireless messages have been intercepted by the British government." The likelihood that such a public disclosure would jeopardize the source was acknowledged by the intelligence professionals, but Sir Basil Thomson took the view that because of the gravity of the situation, "the publication of the deciphered cables has become so imperative that we must take the risks that will be entailed." However, Cabinet Secretary Maurice Hankey disagreed, preferring not to risk losing such a "valuable and trustworthy source of secret information" and advising Prime Minister David Lloyd George:

> This particular cipher is a very ingenious one which was discovered by great cleverness and hard work. The key is changed daily and sometimes as often as three times in one message. Hence if it becomes known that we decoded the messages all the governments of the world will probably soon discover that no messages are safe.

Sure enough, by the end of December 1920, the Soviet wireless traffic was terminated, and the Kremlin adopted a system of individual couriers to deliver confidential correspondence.

In the 1921 Anglo-Soviet trade agreement, which conferred the first formal recognition of the Soviet regime, both parties undertook to refrain from fomenting subversion in each other's territories, but secret Soviet communications decrypted by Felix Fetterlein at the **Government Code and Cypher School (GC&CS)** proved that Moscow had never intended to cease spreading propaganda within the British Empire.

After much Cabinet debate in April and May 1923, Foreign Secretary Lord Curzon delivered a note to the Kremlin's Foreign Affairs Kommissar Grigori Chicherin, in which he protested about Soviet interference and support for extremists in India. However, although the director of naval intelligence, Admiral Hugh Sinclair, had warned that "if intelligence is used for publicity it will be lost to us," nevertheless Curzon's complaint referred explicitly to intercepts, and Chicherin's reply asserted that the allegation had been manufactured from "decrypted parts of telegrams tendentiously manipulated and arbitrarily extended." Just as Sinclair predicted, the Soviets quickly changed their compromised ciphers but not before more decrypts indicated that Chicherin had demanded all subversive activities should be suspended.

Thereafter, GC&CS's director, Alastair Denniston, frequently complained that politicians could not be trusted to protect cryptographic secrets. He would be proved right in May 1927, following a police raid on the Arcos headquarters in London, when Prime Minister Stanley Baldwin, in a speech to the House of Commons, announcing a break in diplomatic relations, read out four Soviet decrypted diplomatic telegrams that had been described in a Cabinet minute as "secret documents of a class which it is not usual to quote in published documents." Two days later, during a full debate on the issue, Foreign Secretary Austen Chamberlain and Home Secretary Sir William Joynson-Hicks also quoted directly from decrypted Soviet telegrams, leaving GC&CS in a state of despair.

CYPRUS. Because of its strategic location in the eastern Mediterranean and the creation, upon the island's independence in 1960, of the large Sovereign Base Areas at Episkopi, Akrotiri, and Dhekelia, Cyprus developed into a center of signals-intelligence collection sites, including the British **Composite Signals Organisation** facility at Agios Nikolaos, staffed by 9 Signals Regiment, and the **National Security Agency** intercept stations at Karavas and at Yerolakkos.

In 1983, Senior Aircraftsman Paul Davies, a 20-year-old RAF wireless operator assigned to intercept duties in Cyprus, was charged with having passed classified information to Eva Jaafar, the Hungarian-born wife of a Lebanese businessman suspected of being a Syrian spy. At his trial in 1984, Davies was acquitted when he retracted a confession allegedly made under duress and Mrs. Jaafar appeared as a defense witness to deny any involvement in espionage.

In another similar case, in June 1985, five young airmen, Senior Aircraftsmen Geoffrey Jones, Adam Lightowler, Wayne Kreihn, Christopher Payne, Gwynfor Owen, and two soldiers from 9 Signal Regiment, Martin Tuffy, and Anthony Glass, were charged under the Official Secrets Act. Charges against Signalman Hardman were dropped on medical grounds.

According to the prosecution, the men had disclosed vast quantities of information about the Agios Nikolaos base to Soviet intelligence officers who supposedly had entrapped and blackmailed them after they had attended homosexual orgies. The case was almost wholly dependent on confessions extracted by the RAF Provost Police, and after a trial lasting 115 days, the longest and most expensive espionage case in English legal history, the jury acquitted all the defendants in October 1985.

There followed three inquiries, one held by the Security Commission, chaired by Lord Griffiths, which found weaknesses in personnel management on the base, in that officers and NCOs took no interest in how the junior ranks spent their free time, and in the handling and destruction of classified documents. A second review, conducted by David Calcutt QC, pursued the issue of the men's confessions and found that, while they had been subject to interviews that produced statements that were inadmissible, nothing improper had taken place. The findings of a third investigation, by the Ministry of Defence, were not published.

D

DANISH DEFENSE INTELLIGENCE SERVICE (DIS). Denmark's principal signals intelligence organization, the DIS supplied advance warning in April 1940 of a build-up of German troops on the border, but the government ignored the advice offered by Colonel H. M. Lunding. The DIS remained in Copenhagen during the Nazi occupation, operating under the commercial cover of Nielson and Moller until August 1943, when it was evacuated to Stockholm.

In November 1946, the DIS's vice-chief, Poul Adam Mørch, signed the Danish–United States Agreement on the exchange of intelligence and identified Poland as a priority target for Danish intercept stations located at Åkirkeby on Bornhol, in Copenhagen's suburb of Amager, and at Gedser, Hjørring, and Løgumkloster. Supported by another naval officer, Find Haugsted, Commander Mørch expanded the DIS by creating a Defense Radio Service in 1957; establishing a covert headquarters at Vennerslund, a mansion in a fashionable district of Copenhagen, which previously had been owned by the Danish millionaire H. N. Andersen; and building an intercept facility in Greenland to monitor Soviet naval traffic.

A veteran intelligence officer, Haugsted employed his wife Else, who spoke 22 languages, as a linguist and his mathematician son as a cryptanalyst. He would also install Denmark's first computer, the DASK, to aid the cryptographers, and in 1961, he reported to the government that his organization, known as "the villa," was routinely reading the diplomatic communications of nearly two dozen countries, including the Vatican's, using a clandestine exchange located under Copenhagen University's East Asian Institute, equipped with 20 teleprinters, to intercept the cable traffic. This site would be closed abruptly in 1969 when radical student demonstrators compromised it in the mistaken belief that their activities were being watched.

In 1957, Mørch's organization was renamed the Defense Radio Service and was moved to Amager but, in February 1971, was reabsorbed by the DIS. Three years later, Haugsted died.

DARWIN. The British government radio relay station at Shoal Bay, in the Northern Territories, provided cover for the **Composite Signals**

Organisation, which used the site as a wireless-intercept facility from 1939. In 1975, following the devastating Cyclone Tracy the previous December, the antenna field was moved to Berrimah, close to the airport. *See also* DEFENCE SIGNALS DIRECTORATE (DSD).

DEFENCE SIGNALS BRANCH (DSB). Australia's principal signals-intelligence agency from 1949, the DSB was renamed the **Defence Signals Division (DSD)** in 1964. Although ostensibly a national organization, the DSB was actually a joint British, Australian, and New Zealand structure with integrated personnel from **Government Communications Headquarters** and New Zealand's **Combined Signals Organisation**, a distinction drawn in appendix J of the 1955 **UKUSA** Agreement, which noted the difference between the DSD and the **Communications Branch** Ottawa.

DEFENCE SIGNALS BUREAU (DSB). Australia's principal signals-intelligence agency during the immediate postwar era, the DSB was established in 1947, replacing the **Central Bureau** and the **Fleet Radio Unit Melbourne**, and would be renamed the **Defence Signals Branch** two years later. The DSB's principal intercept facilities were located at RAAF Pearce, near Perth, opened in 1946, and at Cabarlah, which became operational the following year. *See also* AUSTRALIAN SIGNALS INTELLIGENCE.

DEFENCE SIGNALS DIRECTORATE (DSD). Based until February 1991 at the Victoria Barracks in St. Kilda Road, Melbourne, a facility known as the Joint Telecommunications Unit Melbourne (JTUM), and in Russell Hill, Canberra, the DSD is Australia's principal signals-intelligence organization and replaced the **Defence Signals Division** in 1978. Prior to that, the organization had been named the **Defence Signals Branch** until 1949 and, during **World War II**, had been simply the **Central Bureau**. In 1964, it underwent a further name change, to the Defence Signals Division.

The DSD's main intercept facilities are at Kojarena, outside Geraldton in Western Australia, a site established in 1995 after the withdrawal from Chung Hom Kok in **Hong Kong**, and at Shoal Bay, near **Darwin**. The DSD also has technical responsibility for **Pine Gap**, the U.S. **National Security Agency** station near Alice Springs. The DSD is a vital component of the **UKUSA** structure, which links Great Britain's **Government Communications Headquarters** to the New Zealand **Government Communications Security Bureau** and the Canadian **Communications Security Establishment**. *See also* AUSTRALIAN SIGNALS INTELLIGENCE.

DEFENCE SIGNALS DIVISION (DSD). Australia's principal signals-intelligence organization, the DSD was established in 1964 and was renamed the **Defence Signals Directorate** in 1978. *See also* AUSTRALIAN SIGNALS INTELLIGENCE.

DENMARK HILL. The Metropolitan Police wireless station at Grove Park, Denmark Hill in South London, was headed by Commander Harold Kenworthy and between 1934 and 1937 intercepted Moscow's clandestine communications with the Communist Party of Great Britain, a signals-intelligence source circulated as **MASK**. During **World War II**, the site accommodated seven intercept positions for the **War Office Y Group**.

DeSOTO. The **National Security Agency (NSA)** codename for signals-intelligence collection operations conducted in 1962 off the coasts of China and North Korea involving U.S. **naval collection platforms** carrying special portable compartments containing intercept equipment and personnel. The DeSOTO containers were moved from ship to ship, and in July 1964, one such mission was undertaken by the USS *Maddox* near the island Hon Me in the Gulf of Tonkin, eight miles off the Demilitarized Zone between North and South Vietnam, when the intercept operators reported that they had picked up North Vietnamese VHF naval traffic indicating that permission had been granted for an attack on an enemy. The message, transmitted from Port Wallut to a squadron of *Swatow* gunboats and intercepted at two shore stations, USN-414T at Phu Bai and USN-27 at San Miguel in the Philippines, was unequivocal:

> USE HIGH SPEED TO GO TOGETHER WITH THE ENEMY FOLLOWING TO LAUNCH TORPEDOES.

Another signal identified the target as being in the *Maddox*'s exact location, so the crew was put on action stations and was ready at 1600 when three North Vietnamese torpedo boats attacked. The engagement lasted half an hour, but surface fire from the destroyer kept the smaller vessels at a distance, and when air support from the carrier USS *Ticonderoga* appeared, they withdrew, one having been sunk and another damaged.

Despite this episode, the *Maddox*'s patrol continued, reinforced with a destroyer, the *Turner Joy*, as an escort. Intercept operators at San Miguel reported that the surviving torpedo boat, the *T-142*, had claimed to have shot down two American aircraft, asserting "WE SACRIFICED TWO COMRADES BUT ARE BRAVE AND RECOGNIZE OUR OBLIGATIONS."

Two nights later, the *Maddox*'s **radar** operators warned of several small vessels approaching at high speed, and a message was sent to a carrier, the USS *Ticonderoga*, which launched fighters to support the *Maddox*. However, on the moonless, cloudy night, the pilots were unable to find the attackers, and the destroyer slipped away into international waters unscathed, having fired on radar targets for two hours. Nevertheless, the version of events sent to Washington, DC, suggested a significant naval engagement had taken place and led Congress to pass the Gulf of Tonkin resolution, which formalized the conflict with Hanoi and authorized President Lyndon B. Johnson "to take all necessary steps, including the use of armed force" against North Vietnam.

In January 2008, the NSA declassified a report on the incident, which concluded that there had not been any North Vietnamese attack on 4 August 1964 and that inexperienced radar operators had mistakenly identified "ghost radar returns" as hostile torpedo boats. *See also* VIETNAM WAR.

DIEGO GARCIA. Part of the British Indian Ocean Territory but leased to the United States in November 1965, Diego Garcia is the site of a large **Naval Security Group** satellite ground station and intercept site. The facility became politically controversial because the British government removed the indigenous islanders to Mauritius.

DIPLOMATIC PREMISES. Although protected by the Vienna Convention, diplomatic premises represent attractive sites, and targets, for the collection of signals intelligence. During the **Cold War**, British, American, and Canadian embassies in Moscow accommodated collection facilities. The British embassy, directly across the Moscow River from the Kremlin, conducted operations codenamed TRYST while the **National Security Agency** occupied the top floor of the U.S. embassy and supervised a collection program codenamed BROADSIDE. A Canadian version was codenamed STEPHANIE. Soviet countermeasures included a bombardment by microwave transmissions and attempted forced entries into classified spaces by personnel posing as firefighters during staged fire emergencies.

Communication vaults in both British and U.S. embassies have been targeted by intelligence personnel posing as rioters during demonstrations, and in 1957, the U.S. embassy code room in Taipei was entered by protestors who demolished a wall to breach the security. Similarly, in August 1967, the British embassy in Beijing was ransacked, resulting in the loss of a **Rockex** cipher machine. In 1979, following the fall of the Shah of Iran, students entered and occupied the U.S. embassy in Tehran and looted the classified spaces before all the confidential papers and equipment could be destroyed.

In Allied and neutral countries, the **Staff D** of the U.S. **Central Intelligence Agency** and the British Secret Intelligence Service's Y Section conducted penetration operations intended to collect information from target code rooms. According to George Blake, Y Section ran successful clandestine operations against the Bulgarian embassy in London, several other premises in Cairo, and in European capitals. *See also* REPRIEVE; TEMPEST.

DIPLOMATIC SECTION. The Diplomatic Section of the **Government Code and Cypher School (GC&CS)** was responsible for breaking the ciphers employed by target foreign countries. The Diplomatic Section had been created in **Room 40** and subsequently was divided into individual country subsections, the most heavily resourced having been the Soviet branch, which achieved considerable success until the Kremlin ordered an improvement in communications security following the disclosure in the notorious 1927 white paper released by Stanley Baldwin's government that Russian telegrams had been intercepted and read.

When GC&CS moved to **Bletchley Park** in August 1939, the Diplomatic Section was kept deliberately isolated from the remainder of the estate by being accommodated temporarily at a local school, Elmers, and in 1942 was moved to offices above Peggy Carter's hat shop in Grosvenor Street and headed by Commander Alastair Denniston. A separate subsection was also established at Alford House, on nearby Park Lane, which concentrated on Comintern traffic, codenamed ISCOT, and Jewish communications to and from Palestine, codenamed ISPAL.

DIPLOMATIC WIRELESS SERVICE (DWS). Based at Hanslope Park and ostensibly created to provide communications facilities for British diplomatic premises overseas, the DWS also provided **interception** facilities for **Government Communications Headquarters**, with around half the staff working abroad on intercept duties.

A series of interception operations conducted from the British embassy in Moscow, strategically positioned directly across the river from the Kremlin, was codenamed TRYST. The Soviets attempted to disrupt the embassy's activities in September 1964 when firemen attempted to enter the building's classified areas while ostensibly fighting an accidental conflagration. *See also* MARSHALL, WILLIAM M.

DIRECTION DE LA SURVEILLANCE DU TERRITOIRE (DST). Following **World War II**, the French security service established a small radio-monitoring unit, the Police de Contrôle Radio-électronique (PCR), at its headquarters in the rue des Saussaies in Paris and constructed two intercept facilities, at Boullay-les-Troux and at Limours. *See also* FRENCH SIGNALS INTELLIGENCE.

DIRECTION FINDING (D/F). The triangulation technique of identifying or "fixing" the location of a signal's origin is known as D/F and was pioneered by the British Admiralty during **World War I** when enemy transmissions were monitored from 19 geographically distant sites, known as "directionals," and bearings taken to trace the source, known as a plot, usually a vessel or a **zeppelin**.

Conventional D/F involved string to plot an azimuth to obtain "lines," "cuts," and "cocked hats" but was replaced in **World War II** with the "Rose-Barfield method," which introduced the concept of "probability plotting" that was drawn graphically on Admiralty gnomonic charts as ellipses. The equipment, manufactured by Marconi, boasted a range of up to 1,000 miles but an accuracy of only within three degrees, so Dr. R. H. Barfield of the National Physical Laboratory's Radio Research Board was commissioned to improve the performance. In addition, a **General Post Office** D/F expert, F. R. Thornton, was seconded to the **War Office Y Group**, and the result was the resiting of some poorly positioned high-frequency stations, the introduction of new procedures, and the construction of a duplicate system. Other innovations included radio-frequency amplifiers so up to 18 receivers could be connected to one aerial and various types of recording devices, such as cylinder, wire, disc, and, from June 1943, film.

Upon the outbreak of World War II, the Admiralty established the Shore Wireless Service (SWS), which was composed of civilian operators, Wrens (Women's Royal Naval Service), and naval ratings based at 12 SWS stations: **Cupar** in Fife; Bowermadden, near Wick; Oban; **Irton Moor** (outside Scarborough); Pembroke in South Wales; Cooling Marshes in the Thames estuary; Ford End in Essex; Goodhavern in Cornwall; **Flowerdown**, near Winchester; Portrush in Northern Ireland; and Lerwick in the Shetlands. Overseas. the principal sites were located at Daniel's Head, Bermuda; Jamaica; Georgetown, British Guiana; Iceland; Freetown, Sierra Leone; Dingli in Malta; Powder Island, Trincomalee; and Alexandria in Egypt. Some D/F operators were assigned frequencies used by fixed targets, such as the Kriegsmarine transmitters at Wilhelmshaven and Saint-Nazaire, while others scanned the airwaves and reported intercepted signals to the Admiralty's D/F Control Room at Irton Moor, where the bearings coordinates were plotted on a gnomic chart. In 1964, the Admiralty's Shore Wireless Service was absorbed into **Government Communications Headquarters**.

The British Army's parallel D/F organization, run by the War Office Y Group, consisted, in 1939, of stations at Montrose, Scotland, Helston in Cornwall, and Sutton Valence in Kent, which provided a baseline of 600 miles, which was extended when the British Expeditionary Force arrived in France in 1940 and a new site was built at Chartres. Completed in March 1940 by John Tiltman, from the **Government Code and Cypher School**, and the Admiralty's Commander Marshall Ellingworth, this was hurriedly dismantled three months later by Major A. E. Barton as the German invaders approached.

Later in the war, additional D/F facilities were constructed at Perton, near Wolverhampton, and, in March 1942, at Thurso. In October 1942, the Helston site was moved to Chacewater, and in 1944, a final station was added at Croft Spa, near Darlington.

D/F achieved some impressive results, such as the location of Adolf Hitler's Wolfsschanze, the German military headquarters at Goldap in East Prussia; the transfer in 1943 of Army Group Center from the Russian Front to Italy and the Balkans; and the movements of the Wehrmacht and Luftwaffe headquarters between Zossen, East Prussia, Salzburg, Weimar, and Flensburg.

During the **Cold War**, D/F technology advanced, and huge circular "elephant cage" Wullenweber AN/FLR-9 antenna arrays were familiar sights at Anglo-American intercept stations across the globe. *See also* ADMIRALTY Y SERVICE; CIRCULARLY DISPOSED ANTENNA ARRAY (CDAA); DISCRIM; TRAFFIC ANALYSIS.

DIRECTION GÉNÉRALE DE LA SÉCURITÉ EXTÉRIEURE (DGSE). When the Service de Documentation Extérieure et de Contre-Espionnage (SDECE) was renamed the Direction Générale de la Sécurité Extérieure in April 1982, the **French signals intelligence** agency, the **Groupement des Contrôles Radio-électriques (GCR)**, fell under its responsibility. During the 1980s, the DGSE expanded its activities in Latin America, Africa, and the Middle East, as reflected by the establishment of intercept sites at Djibouti, French Guiana, New Caledonia, and on Reunion in the Indian Ocean.

DISCRIM. An essential **traffic analysis** discipline, discrimination is the technique intended to isolate target traffic from the thousands of other signals exchanged over the ether. During **World War II**, discrim was more of an art than a science, but analysts provided an essential service by separating Luftwaffe from Wehrmacht signals, no easy task after new security procedures were introduced by the enemy in November 1944, which made the signals difficult to disentangle.

During the **Cold War**, automated systems, including the **National Security Agency** computer program designated **ECHELON**, were shared with **UKUSA** partners to pinpoint specific telephone subscribers while their traffic was carried by satellite or microwave channels. Without ECHELON and similar discrim techniques, **interception** would present an impossible task in an era of burgeoning domestic and international communications. *See also* ADMIRALTY Y SERVICE; BEAUMANOR HALL; CALL-SIGN ANALYSIS (C/A); WAR OFFICE Y GROUP (WOYG).

DOLPHIN. The **Government Code and Cypher School** codename for the Kriegsmarine's Heimisch Home Waters **Enigma** cipher key, later known as HYDRA. DOLPHIN, which was broken briefly by the Allies in June 1941, following the capture of cryptographic material from the *U-110* and the weather ship *München,* was the principal naval Enigma key used by U-boats and surface ships in the Atlantic during 1940 and 1941, until the introduction of **SHARK** in February 1942. *See also* OP-20-G; PINCH; ULTRA.

DRUID. Within the **UKUSA** partnership, the codename DRUID was the designation of intercepted traffic that originated from third parties, usually **North Atlantic Treaty Organization** countries, where the source could not be disclosed without the prior consent of the originator.

DUNLAP, JACK. A U.S. Army sergeant who had been posted to the **National Security Agency (NSA)** in April 1958, Jack Dunlap had acted as chauffeur to the assistant director before being promoted to a courier. After his suicide, by gassing himself in his car in July 1963, when he was under investigation as the possible source of a leak to the KGB, it was established, with the help of documents found by his widow, that he had been selling secrets to the Soviets for at least four years; although, the exact extent of his betrayal could not be determined.

Dunlap had been drafted into the U.S. Army in New Orleans in June 1952 after he had joined a merchant ship as a seaman to avoid the draft. He was then sent to Camp Roberts in California for basic infantry training before being posted to Korea. While he underwent training, he gained a reputation as a "fixer," someone who could procure almost anything for his officers and fellow soldiers.

Sergeant Dunlap spent six years in the U.S. Army before he was transferred to the NSA and had served with a combat unit in Korea. It was not until 1960 that evidence emerged of his high expenditure; although, his ownership of a 30-foot cabin cruiser, paid for in cash, a Jaguar, and two Cadillacs failed to raise any suspicions. In 1963, Dunlap had opted to resign from the army and work for the NSA as a civilian, but when he was subjected to a routine polygraph, he committed suicide. A subsequent investigation suggested that Dunlap had volunteered his services to the KGB at the Soviet embassy in Washington, DC, in about June 1960 and had received some $30,000–40,000 during the first year of his espionage.

DUTCH SIGNALS INTELLIGENCE. In January 1947, the Dutch government established a signals-intelligence organization, the Wiskundig Centrum (WKC), headed by Colonel J. A. Verkuyl, a former wartime liaison officer with **Arlington Hall**. Initially, the WKC operated under naval-intelligence cover as Marine Inlichtingendienst, designated MARID VI, and was based at the Kattenburg naval barracks in Amsterdam, with intercept facilities at Hellevoetsluis, near Rotterdam. Later, the Dutch national security service, the Binnenlandse Veiligheidsdienst (BVD), an organization created in 1949, would provide additional sites, including five direction-finding stations, with American technical support, and Holland's foreign intelligence service, the Inlichtingendienst Buitenland (IDB), provided additional intercept material. In 1950, Verkuyl was appointed head of the Code Coordination Bureau and replaced by Colonel J. Spanjaard, an East Indies Army retiree.

Apart from conventional trade and diplomatic targets, the WKC concentrated on Indonesia and achieved considerable success by employing cryptographers and other personnel who had served in the Far East in the Royal Netherlands Navy. The existence of the WKC became public in October 1954 when a **National Security Agency (NSA)** cryptographer, **Joseph Petersen**, was arrested in the United States and imprisoned for passing classified information to Giacomo Stuyt of the Dutch embassy. This embarrassing episode compromised the previously good relations between the WKC and the NSA, which had been built on having shared an embassy codebook provided by a Czech defector in The Hague.

Under Spanjaard's leadership, the WKC expanded from a staff of 92 to 164 and took on Warsaw Pact targets as well as processing data supplied by other agencies, including the various SIGINT-collection units of the armed forces, such as the army branch, the 898th Radio Battalion, at Gorkum; the air force 1st LVG station at Zeist; and the facilities operated by the Dutch Post Telegraph and Telephone Service (PTT) at Goes en Ockenburg, Kerkrade, and Waalsdorp, which monitored Eastern European diplomatic traffic. Overseas, the WKC established sites in New Guinea, Surinam, and, in 1963, at Sint Joris on Curacao. In addition, the NSA provided the WKC with printers, recorders, Collins receivers, Marconi **direction-finding** apparatus, and ZEBRA, an early computer that would be replaced by IBM's TEMPUS.

In 1962, the WKC opened a new, purpose-built site at Eemnes, to replace Zeeburg, and it was equipped with a GRANGER antenna donated by the U.S. **Naval Security Group**. In 1982, the WKC was renamed the Technische Informatie Verwerkings Centrum (TIVC), the technical-information-collection center, and managed a new satellite-antenna field at the PTT site Burum in Friesland to assist in the **interception** of telex traffic in a controversial project codenamed 231. This would be replaced by Project 532 at Zoutkamp, a facility run exclusively by the TIVC, which would later develop into DENA, a major satellite ground station.

A major reorganization in January 1996 established the Afdeling Verbindingsinlichtingen (AVI) to coordinate all Holland's signals-intelligence operations and, since the creation in 1987 of the tri-service Militaire Inlichtingendienst (MID), fulfills a wider analytical role; although, the total staff is declared at 195. In 2003, in a further restructuring, the National Signals Intelligence Organization (NSO) was formed to manage the facilities at Burum and Zoutkamp.

E

EASTCOTE. A grim compound of government buildings at Lime Grove in West London, Eastcote was the site of **Bletchley Park**'s principal external cryptographic center and accommodated **Government Communications Headquarters** until 1953, when it was relocated to Benhall in Cheltenham.

EAST GERMAN SIGNALS INTELLIGENCE. *See* MINISTERIUM FÜR STAATSSICHERHEIT (MfS).

ECHELON. A joint **U.S. National Security Agency** and **Government Communications Headquarters** computerized discrimination program intended to assist in the identification of target signal traffic, ECHELON allows **UKUSA** intercept operators to screen satellite and microwave carrier voice and data circuits so as to intercept, and record, specific channels. The automated **discrim** is undertaken by a FLINTLOCK processor that searches for the designator signal in the preamble to every electronic communication passing over commercial circuits that routes the call and identifies the initiator for billing purposes.

In 2000, the European Parliament set up a temporary committee to investigate ECHELON, and the rapporteur, Gerhard Schmid, produced a 194-page report that concluded that there was no evidence the system had done anything illegal or had ever engaged in industrial espionage.

EDZELL. A U.S. **Naval Security Group** intercept site active during the **Cold War**, Edzell was located near Aberdeen in Fife on Scotland's east coast and was closed in October 1997.

ELECTRONIC COUNTERMEASURE (ECM). Although a separate and distinct branch of modern warfare, ECM is the deliberate exploitation of identified signals and thereby part of the SIGINT spectrum. Developed as a separate technique during **World War II**, ECM was designed to defeat the accuracy of enemy radar and included WINDOW, strips of electronic foil that emulated the reflection of authentic aircraft on cathode-ray tube receivers, transmitters set to interfere with the Luftwaffe's night navigation beams, and the apparatus created to jam selected frequencies, thereby neutralizing the target systems.

During the **Cold War**, as radar-protected ground defenses became more sophisticated, ECM was developed to give an advantage to planes and helicopters vulnerable to electronically guided missiles. Ultimately, the absorbent coatings and designs associated with stealth technology were realized in the hope of making aircraft, submarines, and surface vessels electronically invisible.

In the post–Cold War era, ECM has been deployed passively to detect the transmissions of an unseen adversary and actively in counterinsurgency scenarios to jam frequencies carrying terrorists' remotely controlled detonation signals to roadside culvert bombs. *See also* IMPROVISED EXPLOSIVE DEVICE (IED).

ELECTRONIC INTELLIGENCE (ELINT). Intelligence from all non-communications channels, including **radar** and **telemetry**, is broadly defined within the Anglo-American signals-intelligence community as ELINT.

***ENDURANCE*, HMS.** A **naval collection platform**, HMS *Endurance* was also the Royal Navy's Antarctic guard ship, responsible for patrolling the British dependencies of South Georgia, South Sandwich, and Falkland Islands until the Ministry of Defence announced in 1981 that the distinctive ship, painted bright orange and equipped with two Wasp helicopters, was to be scrapped to cut costs. The decision was opposed by the *Endurance*'s colorful commander, Captain Nick Barker, but his protests in January 1982 were undermined by the views expressed by the British ambassador in Buenos Aires, Anthony Williams.

When evidence emerged that an Argentine invasion fleet had put to sea and troops were to land near Port Stanley on 1 April 1982, *Endurance* was reduced to adopting a deception role, pretending to exchange messages with HMS *Superb* as though it were in the vicinity. The futile objective was to give Argentine intercept operators the impression that the hunter-killer submarine was in the vicinity and therefore a potent threat to any vessels assembling in anticipation of an amphibious assault, but the attempt went unnoticed.

The catastrophic intelligence failure was largely the fault of the Joint Intelligence Committee's assessment staff, which had placed too much reliance on the ability of **Government Communications Headquarters (GCHQ)** to monitor the wireless traffic of the Argentine Special Forces units considered the most likely to participate in any attack on the disputed islands. However, GCHQ's ability to provide coverage was limited. Aside from the Two Boats facility on Ascension, there was only the intercept equipment aboard HMS

Endurance, which was anyway scheduled for scrapping. Fortuitously, the intercept site at the HMNZS *Irirangi* station, located just outside Waiouru, and run by GCHQ's New Zealand partner, the **Government Communications Security Bureau**, was operational and cooperative. Originally established in 1942 as a naval wireless station, Irirangi had been running a fairly relaxed intercept program targeted against South America, but after the campaign began, the facility went into overdrive to assist Cheltenham.

For the **National Security Agency (NSA)**, collection in the region was a sensitive subject, as there were no overt intercept bases in Chile or Argentina, and the State Department, terminally nervous about discovery and accusations of undiplomatic behavior, had baulked at any expansion of the embassy facility. The official line was that the nearest NSA outpost was the **Naval Security Group** site on Galeta Island, off Panama, but in fact, the closest was a clandestine station in Chile operating under the synthetic cover of a National Aeronautical and Space Administration (NASA) tracking facility at Peldehue, 13 miles north of Santiago, and understandably, the NSA had no wish to compromise this very useful source.

ENIGMA. The electromechanical cipher machine marketed by Arthur Scherbius in 1923 was used widely by Continental banks to encrypt confidential communications. In 1930, the German Reichswehr adopted the device, and it would become standard equipment throughout the German armed forces, with an estimated 30,000 units distributed. In 1932, the British Admiralty purchased a commercial version and used the design for the **Typex** machine, which became the standard British encryption system.

For added security, the German military model included three rotors, chosen from a selection of eight, which offered a total of 17,576 possible permutations, making a total of 150,000,000,000,000, and had an added refinement, a plugboard, which increased the permutations to a challenging 158,000,000,000,000,000,000.

A variant with four rotors was introduced by the Kriegsmarine's U-boat arm in February 1942. When used correctly, an Enigma-generated cipher was considered practically unbreakable, and the only intrinsic flaw that assisted cryptanalysts was the impossibility of the machine ever choosing the identical letter as the original plaintext. The advantage of the Enigma was its simplicity of use, resembling a portable typewriter with a standard keyboard without numbers, and the opportunity to regularly change the rotors, each with their 26 individual starting points, and to rewire the plugboard.

Although some limited Enigma traffic was broken during the Spanish Civil War by the **Government Code and Cypher School**, and the Polish cipher bureau succeeded in reading some prewar Wehrmacht signals, the real breakthrough occurred in 1940 when British mastery of Abwehr hand ciphers, codenamed ISOS, allowed some access to the Abwehr's Enigma key, codenamed ISK. Thereafter, many Luftwaffe, Wehrmacht, SS, and some Kriegsmarine **Enigma** keys succumbed to a concentrated and sustained Anglo-American cryptographic attack aided by racks of an ingenious electromechanical device, the bombe, which was designed to race through thousands of potential rotor-setting permutations to recover an individual key.

As soon as the first Enigma keys were recovered, more resources were devoted to copying the traffic, and by the end of hostilities, a vast organization had been created, employing some 500 intercept positions at **War Office Y Group** stations dedicated to the traffic 24 hours a day, every day of the year.

With so many variables, mere possession of an Enigma machine did not greatly assist cryptanalysts seeking to recover a key, but numerous examples did fall into the hands of the Allies, and the Polish AVA firm in Warsaw used manuals supplied by **Hans-Thilo Schmidt** to **Gustave Bertrand** to manufacture several working replicas. Among the complete examples known to have been captured were those recovered from the *U-110* in May 1941, the *U-559* in October 1941, and the *U-205* in February 1943. Individual rotors and other cryptographic data were also retrieved from other submarines, and in addition, an Enigma machine was seized by Indian troops when they captured a Horch *Funkwagon* at Sidi Rezegh during the battle of El Alamein in July 1942. *See also* BLETCHLEY PARK; GOVERNMENT CODE AND CYPHER SCHOOL (GC&CS); SHARK; ULTRA.

EWING, SIR ALFRED. Born in Dundee in March 1855, the son of a Free Church minister, Ewing graduated in engineering from Edinburgh University and then worked at the University of Tokyo until he returned to Edinburgh in 1883, moving to Cambridge six years later. In 1902, while professor of applied mechanics at King's College, Cambridge, Alfred Ewing was appointed the Admiralty's first director of education, and upon the outbreak of war in August 1914, he was invited to **Room 40** to work on enemy communications that had been intercepted by the Naval Intelligence Division.

With a small staff, and assisted by Brigadier Anderson, a sapper who had achieved considerable success with ciphers during the Boer War, Ewing approached the Eastern Telegraph Company for copies of past German messages and received current intercepts from a station built at Hunstanton. The volume of raw material from the airwaves was greatly increased when underwater telegraph cables, which linked Berlin to German possessions in Africa, China, and the Pacific, were cut off at Emden.

In December 1927, Ewing gave a lecture to the Edinburgh Philosophical Society in which he gave an account of his wartime cryptographic work and, in describing the way the **Zimmermann Telegram** had been intercepted, mentioned that one of the copies had been sent "ostensibly as a Swedish cipher" over British cables from Stockholm to Buenos Aires. Another, he said, had been attached to a lengthy telegram sent by the American embassy in Berlin directly to Washington, DC.

Both copies had been intercepted and read by Room 40, but Ewing's disclosure, that the British had been routinely intercepting Swedish and American diplomatic communications, caused consternation in Whitehall, and requests were made as recently as 1979 to omit these references from printed versions of Ewing's lecture.

Knighted in 1911, Ewing retired as vice chancellor of Edinburgh University in 1929, and died in 1935.

EXAMINATION UNIT. During **World War II**, Canada's wireless **interception** and cryptographic operations were conducted by the Examination Unit of the National Research Council, which ran intercept sites at Rockville in Ottawa; West Point Barracks, Victoria; Amherst, Nova Scotia; and Riske Creek, British Columbia. After the war, the organization was renamed the **Communications Branch** of the National Research Council. *See also* CANADIAN SIGNALS INTELLIGENCE; COMMUNICATIONS SECURITY ESTABLISHMENT (CSE); YARDLEY, HERBERT O.

F

FALAISE GAP. In August 1944, following the capture of Alencon and Argentan, the Allied invasion force almost surrounded 10 enemy divisions of the German 7th Army near Trun, and Field Marshal Günther von Kluge reported to Adolf Hitler that he needed to withdraw his troops through Falaise before they were completely cut off. Plans were prepared for the 9th SS Panzer Division to counterattack at Écouché while the 2nd SS Panzer Division seized Trun. However, all the enemy communications were intercepted by 110 Special Wireless Section, and the enemy's entire order of battle was revealed, including the location of the 12th SS Panzer Division east and southeast of Thury-Harcourt, with the 1st Leibstandarte Adolf Hitler and the 2nd SS Panzer Divisions in defensive positions north of Ranes, behind the 9th SS Panzer Division, which was deployed to attack the Americans from the south. By 2115 on the evening before, the offensive intercepts had disclosed every detail, including the start lines and timing, thus allowing the Allied forces to anticipate every move. Hitler later said, "the 15th of August was the worst day in my life" and replaced von Kluge with Field Marshal Walther Model, commander of the Fifth Panzerarmee, two days later. Von Kluge committed suicide, and the German losses included 344 tanks and armored vehicles, 2,447 trucks, 252 artillery pieces, 10,000 killed, and 50,000 taken prisoner. The 12th SS Panzer Division, which had amounted to 20,000 men and 150 tanks before the campaign, was reduced to 300 men and 10 tanks. In total, six infantry divisions were destroyed, and the route opened for the liberation of Paris on 25 August.

FALKLAND ISLANDS. The invasion by Argentine forces of the Falkland Islands on 1 April 1982 was predicted by **Government Communications Headquarters (GCHQ)** but only after the troops had embarked and were only two days away from landing Special Forces near Port Stanley.

Although Argentina had long claimed sovereignty over the islands, the Joint Intelligence Committee (JIC) analysts responsible for monitoring the threat misread the political situation in Buenos Aires, particularly in respect of the Argentine troops likely to be deployed for an amphibious attack on the Falklands. The JIC had decided that the professionals of VI, VIII, and XI

Mountain Infantry Brigades would be the junta's preferred choice because they were well equipped and experienced in operating in the extreme weather conditions to be expected in the South Atlantic. Both units were identified in positions along the Chilean frontier by the **interception** of their communications and the acquisition of their individual radio call-signs. Once the JIC had tasked GCHQ to concentrate on these units, it was assumed that any unexpected redeployment would be spotted, thereby giving London at least a fortnight's notice of any hostilities, sufficient time to dispatch a nuclear submarine into the area as a deterrent.

The first serious alert was sounded by GCHQ on 26 March 1981, which noted the withdrawal of two Argentine frigates, the new French-built A-69s, ARAs *Drummond* and *Granville*, from joint naval exercises with Uruguay. According to the intercepted message, Admiral Jorge Anaya has personally ordered the two ships to set a new course toward the south. In fact, they were to join Task Force 40, a fleet of six other ships, which was to undertake the amphibious landings, supported by Task Force 20, a further six ships intended to provide a protective screen for the invaders. The hurried embarkation of 900 troops and their 20 American FMC amtrac armored personnel carriers was started in Puerto Belgrano on Friday 26 March and completed the following evening, in time to sail at midday on Sunday morning. The troops chosen were the 2nd Marine Infantry Battalion, based conveniently in the town, who only five months earlier had been exercising in Patagonia with the U.S. Marine Corps. In addition, they were to be accompanied by a single platoon from the 25th Regiment, which had been flown up from their barracks in Colonia Sarmiento. While none of this activity was spotted, not even the participants themselves believed the cover story that had been prepared, of an antisubmarine exercise with the Uruguayan Navy and of more seasonal maneuvers in Patagonia, this time with teams of officer cadets.

At GCHQ, the second significant alert was given with the interception of a signal transmitted to the submarine *Santa Fe*, which was ordered to land a reconnaissance squad of a dozen men from the Buzos Tacticos on the beach at Mullett Creek to survey it. The only purpose for such an assignment was an imminent invasion, and this was the intercept that was forwarded by a secure landline teleprinter from Cheltenham to the Joint Intelligence Committee, arriving at teatime on Wednesday, 31 March. It was recognized for precisely what it was by a retired Royal Navy officer on the Assessment Staff who took it straight to Robin O'Neill. Visibly shaken by its implications, he passed copies immediately to the prime minister's private office at 10 Downing Street, to the intelligence coordinator and to the cabinet secretary, Sir Robert Armstrong. Soon afterward, a warning was relayed to the governor, Rex

Hunt, but the invasion flotilla was at sea and Operation ROSARIO ("rosary") had been transformed into Operation AZUL ("blue").

Thereafter, GCHQ was able to pick up a veritable crescendo of signals, including ship-to-ship communications as well as messages exchanged between Rear-Admiral Gualter Allara on his flagship, the Type-42 destroyer *Santisima Trinidad*, and his chief of naval operations, Vice Admiral Juan Lombardo at Fleet Headquarters in Puerto Belgrano. Much of the traffic concerned the need to delay the assault for 24 hours because of inclement weather, but there were other messages, too. One informed the senior army officer in command of the Malvinas land forces, General Osvaldo Garcia, that the element of surprise had been lost and the party of Royal Marines in Stanley, which now amounted to double the number originally anticipated, had been warned to expect an amphibious assault. This meant a change in plans that had to be authorized by the junta, and there were several other unexpected developments to be reported: A Puma helicopter on the polar ship *Almirante Irízar* had broken loose in the storm and was too badly damaged to ferry troops, as originally intended, to Goose Green and Darwin. The prospect of armed resistance, even by 80 men, also prompted a last-minute alteration to the master plan.

While GCHQ eavesdropped on the messages circulating between Allara on his flagship and the officer commanding the marines, Rear-Admiral Carlos Busser, who was aboard the landing ship *Cabo San Antonio*, a further channel was monitored, which was a clandestine link between Stanley and the mainland. The spy already in the town was an Argentine Air Force intelligence officer, Vicecomodoro Hector Gilobert, who knew the islands well, having previously operated under the commercial cover of the local manager of Argentina's domestic airline Líneas Aéreas del Estado, which is actually a branch of the air force. He had been replaced recently, after a tour of duty on the Falklands lasting 12 months, by Roberto Gamen, but on 31 March, Gilobert had returned unexpectedly, ostensibly to conduct an office audit. In reality, he was on a mission to report on the situation and confirm the presence of the Royal Marines in their barracks at Moody Brook. Instead, he signaled the unanticipated presence of the change-over troops and Governor Rex Hunt's order to block the runway, but he failed to spot the precautions taken to defend Government House until too late.

Accordingly, when Lieutenant Diego Garcia of the navy's amphibious Special Forces unit approached the building, he was not expecting any opposition, least of all from the Royal Marines. Although Garcia's briefing had appeared comprehensive, complete with a photograph of the Government House drawing room, taken during a reception held there, the assurances that he had received regarding complete surprise and the British troops safely in their barracks proved entirely fallacious, and in consequence, he was one of the first casualties of the conflict, requiring evacuation to the hospital later the same morning. Having been hit by two bullets, his life was saved because one of them was deflected by a Swiss Army knife in the pocket of his tunic, and he made a full recovery.

Operación AZUL had been planned in two phases, the first being the initial assault to seize Government House and the airfield, the second consisting of a C-130 air bridge to bring over the main occupation force of the 25th Regiment from Comodoro Rivadavia. As soon as the town was secured, General Garcia had been replaced by Brigadier-General Américo Daher of the IX Brigade, and the garrison settled down for the occupation, again generating a large volume of wireless traffic for GCHQ to intercept.

During the course of the conflict, which lasted until the Argentine surrender on 14 June, GCHQ supplied the British forces with some valuable intelligence, enhanced by tactical support provided by a Special Task Detachment (STD) created from the Intelligence Corps' Communications and Security Group UK, which operated from the amphibious assault ship HMS *Intrepid*. Usually based at Woodhouse in Loughborough, the STD successfully intercepted the radio channel on which the Argentine ground forces exchanged messages between their occupation command headquarters in Stanley and the garrison at Port Howard on East Falkland. This traffic included detailed logistical information and a daily schedule of helicopter flights and their routes. One intercepted signal from an outstation on East Falkland requesting rations listed the strength and map coordinates of every defensive position. Analysis of this material assisted in the compilation of a comprehensive order of battle, which later proved to be entirely accurate.

Following the conflict, the British government impaneled a committee, chaired by Lord Franks, a retired senior civil servant, to undertake a review of the events that led up to the invasion and the perceived intelligence failure. GCHQ's director, Sir Brian Tovey, gave evidence to the Franks committee in private, but the final report contained no reference to the organization, and the trenchant criticism of the intelligence services was restricted to a classified annex that was never published and was restricted to a limited circulation within Whitehall. *See also* OFFICIAL HISTORIES.

FAPSI. The Russian abbreviation of the Russian Federation's principal signals-intelligence organization, Federalnoie Agentsvo Pravitelsvennoi Sviazi i Informatsi (Federal Agency for Government Information and Communications), formerly the KGB's Eighth Chief Directorate. Western knowledge of this highly secret agency is largely based on information supplied by a defector, Victor Sheymov, who was exfiltrated from the Soviet Union across the Finnish frontier in 1980. He had joined the unit in 1971, and his defection was not disclosed until 1992.

FAPSI's two principal overseas intercept bases, at Lourdes in Cuba, and at Cam Ranh Bay in Vietnam, were closed in December 2001 and February 2002, respectively, on grounds of cost. The site at Lourdes had been opened in 1964 and had been run jointly by FAPSI's Third Directorate, jointly with the GRU's Sixth Directorate, and had employed an estimated 1,100 personnel at an annual cost of $300 million.

Having been separated from the remainder of the KGB upon reorganization in 1992, following the unsuccessful coup against Boris Yeltsin's government in August the previous year, FAPSI was absorbed into the Russia's Federal Protective Service in 2002.

FAR EAST COMBINED BUREAU (FECB). In 1934, Paymaster Captain Arthur Shaw RN was posted to Stonecutter's Island, **Hong Kong**, to establish a regional British cryptographic organization, the Far East Combined Bureau. Linked by radio to Admiralty stations on Horsea Island in Portsmouth Harbour, **Flowerdown**, and Cleethorpes, FECB monitored Japanese naval traffic and supplied intercepts to the **Government Code and Cypher School**, where John Tiltman achieved considerable success with **JN-25**. Shaw was followed by Captain John Waller and then Captain F. J. Wylie.

Almost the entire FECB staff was evacuated to Kai Cheng, Singapore, shortly before the outbreak of **World War II** and, in January 1942, was forced to make a further move to a Royal Navy shore establishment, HMS *Anderson*, at **Colombo**, in Ceylon. Soon afterward, some of the FECB staff was dispersed to HMS *Allidina*, a shore station near Mombasa, and to the **Wireless Experimental Station** at Delhi. *See also* COMBINED INTELLIGENCE FAR EAST (CIFE).

FERNMELDEAUFKLARUNGS. Established in 1955, the Austrian signals-intelligence Fernmeldeaufklarungs battalion at Königswart, just outside Pressburg on the Slovakian border, made a valuable contribution to the **National Security Agency**, acting as a surrogate and using American-supplied intercept equipment but staffed entirely by ostensibly "neutral" Austrians.

FERNMELDEREGIMENT 70. The principal signals-intelligence organization of the Federal German Republic's Bundeswehr, Fernmelderegiment 70 collected traffic during the **Cold War** and provided raw intercepts to other West German intelligence agencies, principally the **Bundesnachrichtendienst**.

FIELD STATION BERLIN. The **National Security Agency (NSA)** intercept facility at Teufelsberg in the British Sector of Berlin was strategically situated on the summit of a man-made mountain constructed over what had been intended to become the Third Reich's military academy. Bombed into total destruction in 1945, the location was used to dump rubble from 800,000 buildings and proved to be an ideal site for a joint NSA / **Government Communications Headquarters** listening post, which, through its isolation from the city, offered an almost ideal electronic environment, free from interference, and with its tall aerials reached to Czechoslovakia and 50 miles into Poland.

The U.S. **Army Security Agency** Field Station operated from 1957, when it was known as the 280th Army Security Agency Company, and had proved its worth by monitoring the military VHF traffic generated by 600,000 Soviets troops during the Prague crisis of 1968, and the number of masts and distinctive golf-ball antenna radomes had increased continuously until 1977, when the site was absorbed into the NSA as the U.S. Air Force's 6912 Electronic Security Group at Marienfelde, with more than 1,000 Americans and 100 British technicians from the Royal Air Force's 26 Signal Unit and the Royal Signals 13 Signal Regiment maintaining 24-hour cover in three shifts on East German and Soviet telephone, telex, and radio circuits until closure in 1989. *See also* HALL, JAMES W., III.

FINNISH FRONT. During the end of 1939 and much of January 1940, British **War Office Y Group** intercept operators monitored Soviet military traffic transmitted from Finland at night. Although the task was not considered a priority, the experience allowed traffic analysts to learn about Russian procedures, call signs, and frequency allocations, all of which would later prove invaluable when, at the end of hostilities in Europe, the signals became highly significant. *See also* BOURBON.

FLEET RADIO UNIT MELBOURNE (FRUMEL). The principal U.S. Navy signals-intelligence organization in Australia during **World War II**, FRUMEL operated in parallel to the **Central Bureau** and operated as a regional **interception** and cryptanalytical branch of **OP-20-G** from April 1942 until mid-1945, when it was absorbed into the **Defence Signals Bureau**.

FLORADORA. The **Government Code and Cypher School** codename for the decrypts produced from the German Foreign Ministry's wireless communications encrypted on **one-time pads**. *See also* PANDORA.

FLOWERDOWN. The principal **Admiralty Y Service** intercept station in southern England, Flowerdown was operational before and during **World**

War II. After the war, Flowerdown came under **Government Communications Headquarters** and was closed down in 1977. *See also* BLETCHLEY PARK; ULTRA.

FLOWER KEYS. The **Enigma** keys for Luftwaffe administrative traffic designated DAFFODIL, NARCISSUS, PRIMROSE, and SNOWDROP were known at **Bletchley Park** as flower keys. *See also* ULTRA.

FOREIGN INTELLIGENCE SURVEILLANCE ACT (FISA). Upon the recommendation of the U.S. Senate and House committees impaneled in 1973 under the chairmanships of Frank Church and Otis Pike, Congress passed the Foreign Intelligence Surveillance Act in 1978 to establish a court that would regulate warrants authorizing the domestic **interception** of the mail, fax, and telephone communications of individuals suspected of espionage and related criminal offenses. FISA's limited scope appeared to exclude Internet traffic, text messaging, and satellite and cell-phone use and failed to address the issues of American citizens residing abroad and suspect calls made to or from subscribers overseas. Rather than amend FISA and hold public hearings that would risk alerting terrorists to the technical capabilities of investigating agencies, President George W. Bush used his executive powers to issue a decision directive that authorized the appropriate agencies to intercept the anomalous traffic. When the existence of these so-called warrantless wiretaps was disclosed, a political controversy ensued, which was concluded by Congress approving an indemnity for cell-phone service providers that had cooperated with federal agencies.

FORSCHUNGSAMT. The signals-intelligence branch of the Reich's air ministry, the Forschungsamt was created at Behrenstrasse in Berlin in April 1933 under Hans Schimpf with a staff of six to provide the Nazi government with a SIGINT capability and, in April 1935, moved into Berlin's Schillerstrasse to provide information from two sources, being "A" telephone landlines and "B" wireless communications. Six Forschungsamt bureaus handled personnel, requests, processing, cryptography, analysis, and equipment, with a network of intercept stations across the Reich at Templin near Berlin; at Glienecke, Lubbe, and Lissen; Cologne-Deutz; and Constance.

By December 1935, 2,200 decrypts were produced, mainly of French, British, and Italian diplomatic traffic, and in that year, 34,000 internal German communications and 8,500 external signals were intercepted. The organization eventually employed some 6,000 interpreters, analysts, and operators who were transferred to Breslau when Berlin came under threat from air raids. Decrypt summaries were distributed to recipients personally authorized by Hermann Göring on distinctive brown paper that was thoroughly impossible to photocopy.

In April 1935, Schimpf committed suicide over the loss of some classified documents during the move from Behrenstrasse to Schillerstrasse, and was replaced by Prince Christoph of Hesse and then by Gottfried Schapper. Later, the headquarters would be forced to move again, first to Bavaria and finally to Schleswig-Holstein. *See also* BALKANABTEILUNG; SCHMIDT, HANS-THILO.

FORSCHUNGSSTELLE. The Deutsch Reichspost's signals-intelligence service, the Forschungsstelle succeeded in 1941 in decrypting the A-3 radiotelephone scrambler developed by Bell Laboratories in 1939, which was employed to prevent the **interception** and decryption of sensitive transatlantic conversations conducted between the White House and 10 Downing Street by inverting the wavelengths and jumping frequencies. The objective, to distort the voices beyond recognition, failed on 7 September 1941, when the Forschungsstelle, a small laboratory sponsored by the Sicherheitsdienst, succeeded in monitoring the traffic from a dedicated intercept station built at Eindhoven in Holland. The cabling in the United States was routed to the American Telegraph and Telephone exchange in Walker Street, New York, and in London the equipment was installed in the basement of the Selfridge's department store in Oxford Street.

The decryption process became fully operational in March 1942 but was compromised because the decrypts were communicated to Berlin by a **Geheimschreiber** channel that itself had been compromised.

FÖRSVARETS RADIOANSTALT (FRA). Created in June 1942 from the Crypto IV branch of the General Staff, the Swedish National Defense Radio Institute is Sweden's principal signals-intelligence organization. Prior to 1942, the Radio Control Department (RKA) had detected some illicit wireless transmissions, and in July 1941, Bertil Eriksson, codenamed ERIK, was arrested in Midsommarkransen, Stockholm. His accomplice was identified as Albert Elming, codenamed JIM, and their **NKVD** contact was Pavel Formenko. Soon afterward, another network was discovered in Gothenburg, led by Karl Frantzen and Sven Rydstedt, codenamed GUSTAV. Also arrested was Anton Strand, codenamed MAX, who acted as a courier for Stefan Astimiev, based at the Soviet embassy. An analysis of the spy ring's activities suggested that the members had reported on German railway traffic transiting through Sweden.

In May 1942, the FRA, based on the island of Lidingö, monitored signals from two Norwegians, Karl Frantzen, codenamed SVANTE, and Robert

Hansen-Nygaard, codenamed FRED. They were both arrested, together with a trade unionist, John Wivegh, codenamed BERTIL, and Hans-Erik Rydman.

As well as monitoring Soviet traffic in support of the Security Police's counterespionage operations, the FRA's mathematician, Professor Arne Beurling, succeeded in intercepting and reading some 296,000 wartime cipher traffic transmitted by landline from Germany to the German military headquarters in Oslo and encrypted on the **Geheimschreiber** machine. When work of this breakthrough leaked to the Germans in 1942, some improvements were made in the German encipherment procedures, but they proved insufficient to prevent both the FRA and the Allies from exploiting the source.

After **World War II**, having moved to the island of Lovön, the FRA continued to concentrate on Soviet traffic and, in 1947, achieved some success with messages sent to and from the embassy. This material was codenamed SKIM and, in 1961, was offered to the British, who referred to the source as **HASP**. The FRA also operated **airborne collection platforms**, the Flygvapnet, over the Baltic, and in June 1952, a Dakota DC-3 was shot down by Soviet MiGs with the loss of all eight crew. A few days later, MiG fighters shot down a Convair Catalina near the island of Hüumand while searching for survivors. As well as flying Dakotas, the Flygvapnet was equipped with Vickers Varsity aircraft, and in 1960, the Swedish Air Force purchased two Canberra T11 signals-intelligence aircraft, which were based at Malmslätt. In 1971, they were replaced by a pair of former Scandinavian Airline System Tp85 Caravelles, and the most recent intercept platforms include two Gulfstream IV jets and a naval vessel, the *Orion* (A201), which was launched in 1984.

Public concern about the FRA's activities led to a debate in which the FRA's director-general, Anders Wik, persuaded the Riksdag to pass controversial legislation, which became active in January 2009, authorizing the agency to intercept domestic communications. *See also* TUNNY; VENONA.

FORT BRIDGEWOODS. Built in 1883 as one of "Gladstone's Follies" to guard the River Medway, Fort Bridgewoods began wireless-intercept operations in March 1926 and remained active throughout **World War II** as part of the **War Office Y Group (WOYG)**. Prior to September 1939, when there was little German military radio traffic to monitor, the site concentrated on diplomatic communications; although, it was agreed that this activity would cease in the event of hostilities.

Upon the outbreak of World War II, three of the station's 132 civilian operators were detached from **interception** duties to conduct **traffic analysis (T/A)** operations in the Compilation Records and Reports section and were soon reinforced by six reserve officers. As the intercept burden of what turned out to be mainly Luftwaffe signals grew, more staff was required, and, in the absence of suitably trained operators, recruitment visits were made to local schools, and some Boy Scouts and Merchant Navy retirees were also enrolled, together with three members of a London dance band. Altogether, 120 schoolboys were trained as intercept operators, and when they were old enough to be called up for military service in 1945, 78 of them, known as "the gallant 78," opted to become radio operators.

Until the middle of 1941, the facility, which was next to a Royal Air Force (RAF) airfield, concentrated on Luftwaffe traffic because, apart from some Soviet traffic from Finland, there was little else to cover, and RAF Y was still embryonic, uncertain of what were definitely enemy air signals. Because of its proximity to the RAF airfield, which often came under attack, a large control room, 90 feet underground, was constructed, and although some incendiary bombs were dropped inside the fort's perimeter, it was only once put out of action, and that incident lasted just four hours until some cables could be repaired.

Finally, in January 1942, following more unwelcome attention from the Luftwaffe, much of the T/A task was transferred to the WOYG's headquarters at **Beaumanor** in Leicestershire. *See also* ENIGMA; ULTRA.

FRENCH SIGNALS INTELLIGENCE. Before **World War II**, French signals-intelligence operations were conducted primarily by the French Navy, which ran a series of **direction-finding stations** in the Mediterranean, with some cryptographic work undertaken by a section of the Service de Renseignements headed by **Gustave Bertrand**, studying German traffic intercepted by sites at Strasbourg, Metz, and Mulhouse, survivors of a chain that had been built during **World War I** by Major François Cartier.

After the fall of France, the Vichy government's Deuxième Bureau established in October 1940 a cryptanalytical center at Uzès, codenamed CADIX, which was headed by Paul Paillole until he fled to Rabat in January 1942, carrying much of his organization's files in tennis ball boxes. Chosen because Bertrand and his wife, Mary, owned a villa in the neighborhood, CADIX consisted of seven Spanish republican refugees and 15 Poles and was equipped with two **Enigma** machines reconstructed by the **Polish Cipher Bureau**. CADIX would remain in operation for 610 days, until it was detected by **Funkabwehr direction finding**, and accumulate 4,679 Enigma decrypts, including 3,097 from the police referring to anti-resistance activities undertaken by the German Armistice Commission in Vichy. Much of this material was transmitted to London, and between March 1941 and November 1942, it was on the air for up to three hours a day and sent a total of 2,748 messages.

At the end of World War II, the **Groupement des Contrôles Radio-électriques** was established at Domme, which became the French government's principal signals-intelligence organization. It was supported by a large military **interception** capability concentrated in Germany and represented by the 44th Transmission Regiment, based initially at Landau but later withdrawn to Mutzig, and the 54th Transmission Regiment, with sites at Bad Lauterberg, Furth im Wald, Goslar, and Achern, with a facility at Tagel in Berlin shared with the **Bundesnachrichtendienst**.

The French ran SIGINT operations during the war in Indochina, and after the defeat in July 1954, the equipment, mainly obsolete American high-frequency direction-finding apparatus, located at Rula Ray, Ta Hien, and on Con Son Island, was donated to the administration in Saigon. During the Algerian campaign, only limited SIGINT operations were undertaken, the Groupement des Contrôles Radio-électríques (GCR) only acquiring an **airborne collection platform** capability in 1964, with eight converted Nord 2501 Nordatlas transports flown by the Escadron électronique 54 "Dunkerque." In 1977, this capability was modernized with the introduction of a pair of DC-8 aircraft based at Evreux, headquarters of the Escadron électronique 51 "Aubrac," and in 1989 by a pair of C-160 "Gabriel," based at Metz-Frescaty. In addition, the French invested in a **naval collection platform**, the 2,700-ton *Berry*, commissioned originally in 1958 as the *Médoc*. A replacement, codenamed ARGOS, was planned to replace the elderly vessel.

Heavy French investment in satellite technology culminated in the 1995 launch of the Helios-1A, which was followed by more in the same series but supported in 2004 by Spain and Italy. A second-generation satellite, Clementine, was placed in orbit in 1999. *See* DIRECTION DE LA SURVEILLANCE DU TERRITOIRE (DST).

FRIEDMAN, WILLIAM. A Jewish refugee born in September 1891 in Bucharest, William F. Friedman was the civilian selected to head the Signals Corps' Code and Cipher Section after his return at the end of **World War I** from France, where he had been General John Pershing's personal cipher clerk. In 1921, appointed the War Department's chief cryptographer, Friedman helped create the **Signal Intelligence Service (SIS)**, which, in October 1929, acquired the redundant files from the recently disbanded State Department **Black Chamber**. Under Friedman's direction, the SIS recruited several mathematicians and linguists, including Abraham Sinkov, Solomon Kullback (an interpreter in Japanese), John Hurt, and Frank Rowlett, with whom he developed the SIGABA cipher machine.

In December 1940, Friedman experienced a nervous breakdown while working on a project to reconstruct the **PURPLE** machine used to encrypt top-level Japanese diplomatic communications; although, the task was completed by Leo Rosen, who built the device, using telephone-exchange switching equipment, working entirely from the intercepted traffic.

After his recovery, Friedman returned to work, visiting **Bletchley Park** in April 1943 and in March 1946 participating in the negotiations for the **BRUSA** Agreement. In 1949, he was promoted head of the newly created **Armed Forces Security Agency**'s cryptographic division and, in 1952, became the **National Security Agency**'s chief cryptologist, a post he retained until his retirement four years later. He died in November 1969, having written most of the textbooks on cryptography and having compiled his famous 1920 *Index of Coincidence*. Friedman's wife, Elizabeth, was also a distinguished cryptanalyst, having worked for the Federal Bureau of Investigation and the U.S. Coast Guard.

FUNKABWEHR. Created in 1940 and headed by Hans Kopp, the Funkabwehr was the German High Command's principal signals-intelligence organization responsible for the monitoring of illicit broadcasts. In July 1941, the Funkabwehr station at Cranz intercepted some clandestine traffic, which turned out to be transmissions from an extensive **NKVD** network, later designated the ROTE KAPELLE.

The Funkabwehr achieved considerable success with intercepting and breaking Soviet encrypted wireless signals and, after D-Day, was able to read daily U.S. Military Police reports on Allied vehicle movements in liberated areas, which were transmitted in a very primitive Slidex hand cipher.

The Funkabwehr devoted very limited cryptographic resources to attacking Allied **Typex** traffic as several of the cipher machines had been captured at Dunkirk and had been recognized as a version of the **Enigma**, which was considered unbreakable. *See also* BALKANABTEILUNG; FRENCH SIGNALS INTELLIGENCE; *FUNKSPIEL*.

FUNKBEOBACHTERDIENST (B-DIENST). The signals-intelligence branch of the Kriegsmarine, B-Dienst cryptographers solved the Royal Navy's Naval Cipher No. 3 during the Spanish Civil War and continued to read British communications until July 1943. Employing 8,000 analysts and cryptographers, the B-Dienst was based in Berlin's Tirpitzufer, until it was bombed in December 1943, when it moved to a bunker outside the city, codenamed BISMARCK, and communicated to the Kriegsmarine in an **Enigma** key codenamed FREYA. Later, following heavy Allied air raids, the headquarters was evacuated to Eberswalde, 50 miles northeast of Berlin in Brandenburg. The B-Dienst's **interception** network consisted of seven stations in the Baltic and four on the North Sea, with a clandestine site at Langenargen on Lake Constance, which monitored Italian traffic.

Initially, German success with the Royal Navy's ciphers followed the deployment of a squadron of the Mediterranean Fleet to Aden between June 1935 and March 1936, when British warships were operational but dependent on a rather primitive code introduced in 1933. Much repetition of the same words and phrases, such as "7761" for a new paragraph and "4834" for a full stop, provided the B-Dienst with the patterns that cryptographers crave. Their success was assisted by the lack of an interservice communication system, leaving the Royal Navy to depend on a cipher, long compromised, that was employed by consular offices in neutral countries to report shipping movements. The B-Dienst acquired more British valuable cryptographic material in January 1940 when three submarines, HMS *Undine*, *Seahorse*, and *Starfish*, were sunk by depth charges in the Heligoland Bight. The wrecks of all three were visited by Kriegsmarine divers, who salvaged the Admiralty's subtraction tables. Three months later, during the Norwegian campaign, more confidential papers were retrieved from HMS *Hardy*, a destroyer grounded on 10 April in the Ofotfjord in the battle for Narvik, and the Merchant Navy Code was lost as Bergen was being evacuated, together with a copy of the Foreign Office's Inter-Departmental Cipher.

The Admiralty distributed a new four-digit Naval Code No. 1 and Naval Cipher No. 2 in August 1940, designated BLUE and BROWN by the B-Dienst, which experienced difficulty in distinguishing the two. Nevertheless, until a Combined Operations code was introduced, after the raids on Saint-Nazaire in March 1942 and on Dieppe in August 1942, the B-Dienst could read much of the Admiralty's traffic, an advantage exploited during the Norwegian campaign, especially following the recovery of HMS *Sikh*'s codebooks from the destroyer after it had been sunk in Tobruk harbor in September 1942.

In February 1942, the B-Dienst's success in reading the wireless traffic exchanged between British, Canadian, and American ships in the Atlantic in the Convoy Cipher gave the U-boats an advantage that masked the coincidental introduction of the **SHARK** cipher that would defeat **Bletchley Park** for 10 months. German failures, at TORCH and Anzio, resulted from the Allied use of special codes, which were dedicated to specific operations. Prior to Anzio in January 1944, the B-Dienst intercepted 138 messages but could break none of them.

Because of the lack of long-range reconnaissance flights and the Luftwaffe's poor record in giving accurate navigation reports of sightings of enemy shipping, Admiral Karl Dönitz arranged for aircraft to harass Allied vessels until they transmitted a radio message, giving their exact position, that could be intercepted and read by the B-Dienst.

A postwar analysis of the B-Dienst's success, *The German Navy's Use of Special Intelligence*, completed for the Naval Intelligence Division by R. T. Barratt, concluded that by the middle of 1940, the Germans could read between 30 and 50 percent of the Royal Navy's encrypted wireless traffic and was able to read virtually all of it during the Norwegian campaign. It was only the replacement of the Convoy Cipher, introduced in June 1941, with Naval Cipher No. 5 in June 1943, which in turn was superseded by the introduction of unbreakable **Typex** machines in November 1943, that finally denied the B-Dienst access to Allied signals.

Although several security reviews were conducted during the course of **World War II**, especially when it was suspected cryptographic material might have fallen into the hands of the enemy after the capture of a submarine or surface vessel, the B-Dienst never found evidence that the Kriegsmarine's cipher systems had been compromised. Whenever there appeared to be good reason to suspect that a communications breach had occurred, the B-Dienst invariably found persuasive arguments to support the view that the Enigma's integrity had remained intact. For example, the successful escape of the *Scharnhorst* up the English Channel from Brest in February 1942 was regarded as eloquent proof that the British had not recovered cryptographic material off the *U-559* the previous October.

The B-Dienst was headed from November 1941 to January 1944 by Heinz Bonatz, who was then transferred to command the port of Rotterdam. His replacement was Max Kupfer, who remained in command until the end of hostilities and surrendered his men to the British Army at Flensburg.

FUNKSPIEL. Literally "radio game," *funkspiel* was the German term applied to the playback of captured enemy wireless sets during **World War II**. The **Funkabwehr** proved highly adept at manipulating and exploiting Allied radio links with clandestine networks operating from Nazi-occupied territory, most notably in Holland, with Operation NORDPOL, and in Belgium where a combined Abwehr and Sicherheitsdienst investigation of the ROTE KAPELLE provided an opportunity to develop several signal channels to Moscow. The objective of the *funkspiel* is to convey deception material, collect information through the receipt of intelligence questionnaires, and expose previously undetected networks. *See also* CONTROLLED ENEMY AGENTS.

G

GAMMY GUPPY. The **National Security Agency** codename for an **interception** program undertaken at the U.S. embassy in Moscow that monitored the voice channels used by Kremlin limousines, GAMMA GUPPY was compromised by premature disclosure in Jack Anderson's column in the *Washington Post* in September 1971.

GARDNER, MEREDITH. A brilliant cryptographer, Meredith Gardner made the first breakthroughs into the Soviet intercepts while working at **Arlington Hall** for the U.S. Army's **Signal Intelligence Service**. Until the outbreak of **World War II**, Gardner had been teaching German at Akron College in Ohio, speaking several languages fluently and being able to read some Sanskrit, until the outbreak of war when the college dropped German and he took a job at Arlington Hall, teaching himself Japanese in just three months, and thereafter concentrating on Tokyo's signal traffic.

Gardner remained a linguist and cryptographer for the rest of his life, and stayed in the **National Security Agency**. *See also* ARMED FORCES SECURITY AGENCY (AFSA); BRIDE; JADE; ONE-TIME PAD (OTP).

GARNET. Bletchley Park's codename for a German **Enigma** communications channel linking Oslo and Stavangar to Berlin, GARNET was first intercepted in 1941 and was later found to have been extended to provide a channel for military traffic from all occupied territories. Although many of the messages bore a prefix that indicated the signal was a practice transmission, the Norwegian GARNET was usually found to carry operational texts. *See also* ULTRA.

GEHEIMSCHREIBER. The 10-rotor, automatic long-distance encryption machine manufactured by Siemens, codenamed STURGEON, and Lorenz, codenamed **TUNNY**, and employed, initially at Berlin, Vienna, and Greece, to encrypt high-level messages during **World War II**, *Geheimschreiber* means "secret writer" in German. Allied cryptographers at **Bletchley Park** seeking to solve the high-speed, non-Morse Baudot teletype traffic, which they codenamed FISH (because the Germans referred to it as *sagefisch*), developed the world's first programmable analog computer, the **Colossus**, in 1943 at the **General Post Office** Engineering Research Station at Dollis Hill to race through the machine's 893,622,318,929, 520,950 permutations. In January 1944, the apparatus was installed at Bletchley, where it provided instant results, supplied with intercepts from the Foreign Office Research and Development at Knockholt Pound in Kent, where 600 operators copied the traffic. *See also* FÖRSVARETS RADIOANSTALT (FRA); GOVERNMENT COMMUNICATIONS HEADQUARTERS (GCHQ).

GEHLEN ORG. Built by Reinhard Gehlen on the remnants of the Wehrmacht's Fremde Heere Oste, the Org was West Germany's principal intelligence agency from 1946 until the formal establishment in April 1956 of the **Bundesnachrichtendienst** and was responsible for the collection of information about Soviet strengths and intentions. Based initially at Oberursel, and then from 1947 at a secure compound at Pullach, south of Munich, the Org depended on sources behind the Russian Front developed during **World War II** and, under the sponsorship of the U.S. **Central Intelligence Agency**, built a signals-intelligence capability, designated Asterhig II, headed by Leo Hepp, who had served with Gehlen in the Reichswehr in 1933. Hepp's unit, amounting to only 18 analysts, achieved considerable success when it identified and monitored the Soviet 24th Air Army during the 1949 Berlin Blockade. By June 1950, Asterhig II's staff had grown to 50 and was providing useful material to its **North Atlantic Treaty Organization** partners. *See also* COLD WAR.

GERMAN POLICE. The network employed by the German Police for internal communications became well known to British intercept operators before the outbreak of **World War II** because the transmitters were organized in groups of 15, used fixed call-signs, and usually were linked directly to a headquarters station in Berlin. As the war progressed, more traffic appeared, often from occupied countries, a useful indication of a local SS presence.

From July 1941, intercepted **Enigma** traffic from the Eastern Front revealed certain wireless channels being used by Nazi security police battalions containing references to atrocities. Further study of this very distinct category of signals at **Bletchley Park** disclosed messages bearing the prefixes HOR and HUG, which contained concentration-camp statistics. Generically, this material was designated the innocuous title "German Police," but in reality, it amounted to the first evidence of systematic genocide. Against advice, Prime Minister Winston Churchill depended on these decrypts when he broadcast a veiled warning in September 1941 in which he alluded to "the crime that cannot speak its name," thereby resulting in the immediate circulation over the network of a directive prohibiting any further transmissions of these data. Nevertheless, the German Police networks continued to exchange details of prisoner deaths, as was documented after the war by Peter Calvocoressi and Eric Jones, who wrote the relevant Volume XII of **Government Communications Headquarters**' internal classified wartime history. By 1943, some 36 sets were devoted to covering the traffic, which averaged 600 messages a day.

GERMAN SECTION. In May 1938, a German Section was established at the **Government Code and Cypher School**. *See also* BLETCHLEY PARK; ENIGMA.

GERMAN SIGNALS INTELLIGENCE. At the outbreak of **World War II**, the German Army's organization committed to signals intelligence, the **Funkabwehr**, consisted of six permanent ground intercept sites, located at Königsberg, Breslau, Treuenbrietzen, Stuttgart, Munster, and Sternberg, with an additional 18 **direction-finding** facilities and some mobile stations. There were nine radio-intelligence companies, each of three officers, 54 NCOs, and 155 men, amounting to total of 1,900 personnel. They all played an active role in the campaigns in Poland and France, with 3 Company, Reconnaissance Battalion 24 establishing a permanent facility in Le Havre to monitor traffic from the British Isles. A detachment from this unit was posted to Tripoli in February 1941 and then participated in the occupation of Crete in May, but in July 1942, an entire **interception** and direction-finding company, Wireless Reconnaissance Unit 621, was captured by the Australian 9th Division near El Alamein, forcing a reorganization and the transfer of a replacement company from Frankfurt-am-Main.

In 1944, following the Allied invasion of Italy, a further reorganization of the Wehrmacht's SIGINT capability under General Fritz Boetzel created 17 Communications Reconnaissance Battalions in eight regiments, each assigned to a particular Army Group, with 2 Regiment serving Army Group Center on the Russian Front and 5 Regiment, commanded by Colonel Baron von Oer, at Army Group West in Saint-Germain-en-Laye. Battalion 12 was based at Euskirchen, concentrating on American traffic, with Battalion 2 and 13 at Port Marly, monitoring British signals.

In contrast to the German Army, the German Navy had developed a sophisticated **Funkbeobachterdienst** organization during the Spanish Civil War that achieved considerable success in solving Royal Navy ciphers and exploiting lax radio security procedures. Similarly, the Reich Air Ministry's **Forschungsamt** had been monitoring target wireless traffic since 1933, and the Deutschen Reichspost established a cryptographic Forschungsstelle in 1941 to break scrambled transatlantic radio-telephone channels.

The very fragmented, uncoordinated German signals-intelligence effort was handicapped by a lack of resources and a prevalent misconception that the British **Typex** cipher machine, an obvious variant of the **Enigma**, enjoyed the same invulnerability. *See also* BALKANABTEILUNG; BLACK CODE; WORLD WAR II.

GENERAL POST OFFICE (GPO). Authorized by the Wireless Telegraphy Act 1904 to license and supervise Great Britain's radio traffic, the GPO, headed by the postmaster-general, established two Marconi stations, a receiver at Devizes, and a transmitter at Wychwood, near Leafield in Oxfordshire, and by July 1914, the network included 14 sites across the country, linking Guernsey in the south to the Orkneys in the north. Upon the outbreak of **World War I**, most were closed down, along with 2,500 licensed sets in private hands, but Caister, Niton, and Fishguard were kept open "to check illicit correspondence." In monitoring the airwaves, the GPO's Wireless Branch, headed by Commander Frederick Loring, discovered and dismantled a further 750 unlicensed transmitters. By September, Loring controlled 10 GPO sites, supported by 24 Admiralty stations and a further 10 run by Marconi. In addition, the GPO was responsible for supervising the country's 13 underwater cable terminals and passing the traffic to **MI8**.

Upon the outbreak of **World War II**, the GPO had a network of monitoring stations at Baldock and Burnham, linked to the GPO's Engineering Research Station at Dollis Hill in North London, but in the absence of any clandestine transmissions, the organization was absorbed into **Bletchley Park**'s **interception** capability.

GGG. Originally the call sign of the Abwehr wireless outstation in Algeciras, GGG became **Bletchley Park**'s designation of the **Enigma** key employed by two other transmitters, at Tetuan and Ceuta, which each sent a daily report to Madrid, which then relayed the messages to Berlin using the main Abwehr Enigma channel. The traffic was considered a priority because the Abwehr concentrated on monitoring aircraft and shipping movements in and out of the Straits of Gibraltar and activity on the Rock itself, and it was broken in February 1942. GGG proved invaluable because the "turnover" messages, which were repeated on the main channel, provided a crib to the Abwehr's daily setting for **ISK**.

GIANT LEAP. The codename for the flights conducted by the high-altitude SR-71 Blackbird **airborne collection platform** from Griffiss Air Force Base in New York in October 1974, during the Yom Kippur War, to the Middle East. The missions were flown from the United States because Prime Minister Edward Heath banned U.S. reconnaissance flights from British bases at Wideawake on Ascension, Akrotiri in Cyprus, and Mildenhall in Suffolk. Each GIANT LEAP operation required 19 airborne refuelings attended by 38 tankers but produced imagery and signals intelligence considered vital to Israel's survival. When poor weather closed Griffiss, the later flights were made from Johnson Air Force Base in North Carolina.

GOUZENKO, IGOR. A **GRU** cipher clerk based at the Soviet embassy in Ottawa, Igor Gouzenko defected in September 1945, having removed 109 documents from the *referentura*. His information, and testimony before a Royal Commission impaneled to investigate Soviet espionage, would have a lasting impact on how Soviet intelligence agencies protected their cipher systems and personnel. Until September 1945, in the absence of any cribs, the American interest in Soviet ciphers had been pursued in an almost academic environment, but Gouzenko's dramatic defection was proof once again that it is by no means the most senior ranks that possess the greatest knowledge. Gouzenko's application for asylum, and for his wife and their daughter, accompanied by the promise of a batch of stolen documents, brought the GRU and the **NKVD** close to global paralysis and inflicted irreversible damage to Soviet prestige.

Under cross-examination by the Royal Commission, Gouzenko was the first wartime Soviet to describe the GRU's cipher procedures, explaining how he had worked alone in a tiny room with a barred and shuttered window on the second floor, entered through a steel double door, receiving telegrams through a small hatch from his colleague Aleksashkin:

> Colonel Zabotin wrote out the telegram in Russian and I coded that telegram, making a draft copy first, and then typed it out on Canadian Pacific or Canadian National telegram forms, which went to Aleksashkin for dispatch to the telegraph company. Colonel Zabotin's original telegram or document was placed in the bag, in the sealed bag, and handed to Aleksashkin.

Even his superior, the GRU *rezident* Nikolai Zabotin, was forbidden to enter the secret cipher room until Gouzenko had completed the first of the three-part coding procedure, which turned the text, handwritten on special blue paper, into cipher groups. Once transmitted, the original would be sent to Moscow by couriers using Soviet cargo ships that sailed from North American ports two or three times a week, a hazardous sea journey that in wartime sometimes took up to five months. When he communicated with his superior, Colonel Baltenko, who was the GRU's Director of Special Communications, he adopted the cryptonym **CLARK**.

Although he subsequently claimed a political motive, asserting that he was disenchanted with the regime, the catalyst for Gouzenko's dramatic departure had been a minor administrative blunder that had resulted in his recall to headquarters. Fearful of the consequences, and anyway reluctant to swap a life of relative comfort in Canada for one of hardship and shortage in Moscow, Gouzenko had taken the precaution of marking some of the most sensitive papers in the *referentura*'s cipher room. When, on the night of 5 September, he made his move, he grabbed the preselected files, stuffed them into his clothes, and left the building undetected.

Gouzenko's contribution to the VENONA program was threefold: there were the documents themselves, which consisted of original plaintexts, and pages extracted from the GRU *rezident*'s daily diary, and also there was his personal knowledge of the GRU's cipher procedures. Unfortunately, none of the plaintexts matched any of the material in the VENONA collection, which contained only some of Ottawa's NKVD traffic, to which Gouzenko had never enjoyed access. The clerk also described the GRU's emergency cipher system, and although this was considered at the time to have potential, it was never found to have been used. Some years later, a single VENONA, dated 25 January 1945, was found to have a reference to Gouzenko, codenamed CLARK, who evidently had made a mistake in the encipherment of a previous cable:

> In our No. 18 I transmitted the figure text of another party's cipher telegram. To ensure there is no difficulty [3 groups unrecovered] 18, here are the first and last groups of the figure text: 74389 and 48817. The figure text begins with the hundredth group of telegram No. 18. [2 groups unrecovered] telegrams [3 groups unrecovered] pad 1471, although in [3 groups unrecovered] pad 68 [2 digits unidentified]. In order to keep the sequence after using pad 1471 please [1 group unrecovered] to pad 68 [2 digits unidentified].

GOVERNMENT CODE AND CYPHER SCHOOL (GC&CS). Established on 1 November 1919, replacing the Admiralty's **Room 40**, GC&CS was based initially at Watergate House, Marconi's old offices in the Adelphi, and headed from 1923 by Commander Alastair Denniston, a nominee of the Director of Naval Intelligence. His deputy was Edward Travis, and his four senior assistants were Lieutenant Fryer from **MI1(b)**; Dillwyn Knox; Oliver Strachey; and Nobby Clarke from Room 40. There were 21 junior assistants, 33 women clerks, four female translators, one radio expert, three typists, and one junior administrative assistant, with an annual budget of £21,217.

In 1922, the organization, with an increased staff of 91 split between cryptographers, linguists, and clerks, moved under Treasury pressure to find less expensive premises, to Queensgate, South Kensington, before transferring in June 1926 into the second and third floors of Broadway Buildings, Victoria, directly underneath the three floors that accommodated the headquarters of the Secret Intelligence Service (SIS). Britain's principal signals-intelligence organization between the wars, GC&CS achieved considerable success by breaking some Soviet diplomatic and military ciphers. However, the British government's decision in May 1923 to disclose to the Kremlin that some of its communications had been intercepted and read effectively closed the source to the **Diplomatic Section** for the following decade.

Having successfully monitored the diplomatic communications of a large number of countries, circulating the decrypts to authorized recipients as **BLACK JUMBO**s, GC&CS solved the Comintern's clandestine wireless traffic, distributed between January 1934 and January 1937 as **MASK**.

GC&CS created the **Naval Section** in 1924, the **Army Section** in 1930, and the Air Section in 1936, exploiting the huge increase in wireless traffic following the Italian invasion of Abyssinia in 1935 and then the outbreak of the Spanish Civil War.

For raw material, GC&CS was dependent on the Metropolitan Police facility at Grove Park, Denmark Hill, on intercept stations managed by the **General Post Office** and the three armed services, and on a steady supply of cables submitted by the commercial carriers that were required, under the terms of the 1920 Official Secrets Act, to supply copies of all their international traffic within 10 days of transmission. The principal intercept sites were the Admiralty's station at **Flowerdown** and the War Office facility at **Fort Bridgewoods**, Chatham. GC&CS also received intercepts from overseas satellites—at **Sarafand** in Palestine and a chain of military positions on India's Northwest frontier.

In 1938, Admiral Sir Hugh Sinclair bought **Bletchley Park** with £7,000 of his own funds as a war station for SIS in anticipation of enemy air raids on London as demonstrated by the Luftwaffe during the siege of Madrid, and from mid-August 1939, the estate accommodated GC&CS, now being referred to increasingly by a new title, **Government Communications Headquarters**. *See also* BONIFACE; CURZON NOTE; ENIGMA; ULTRA; WAR OFFICE Y GROUP (WOYG).

GOVERNMENT COMMUNICATIONS HEADQUARTERS (GCHQ). The principal British cryptographic organization, GCHQ replaced the **Government Code and Cypher School** after it had moved in 1939 from Broadway Buildings in Westminster to **Bletchley Park**.

During **World War II**, GCHQ underwent a dramatic increase in size in response to the greater volume of intercepted wireless traffic and the possibility of exploiting **Radio Security Service** success with the enemy's hand ciphers by solving some **Enigma** keys. Work on the Abwehr's Enigma messages, codenamed **ISK**, led to breaks in the Luftwaffe's RED signals, and thereafter significant numbers of keys were broken using specially designed electro-mechanical devices, known as bombes, which raced through the possible permutations of key settings. While some keys, such as the Kriegsmarine **SHARK**, proved hard to solve, some others succumbed relatively quickly, allowing high-grade intelligence to be distributed to a strictly limited number of authorized recipients. Initially circulated to them over secure radio links designated **Special Communications Units** under the codeword **BONIFACE**, the system was renamed **ULTRA** in early 1942, following the entry of the United States into the war.

The ULTRA source proved exceptionally valuable and reliable and made a major contribution to the successful prosecution of the war by the Allies. Known also as "Station X" and "BP," GCHQ was supplied with intercepts by a network of stations managed by the **Admiralty Y Service**, the **War Office Y Group**, and the **Air Intelligence Directorate** and created subordinate organizations in Cairo, **Sarafand**, and Delhi.

By the end of hostilities, some 15,000 linguists, clerks, cryptographers, and analysts had passed through Bletchley Park, which established satellite establishments on the estates of local country houses and a purpose-built facility at Eastcote in northwest London. Automation, with the introduction in December 1943 of the Colossus machine, which read the German **Geheimschreiber** traffic, greatly assisted the cryptanalysts and placed GCHQ at the forefront of technical research to develop advanced, fast processing systems. Subsequent computers included the Ferranti COLOROB and then, in 1945, OEDIPUS, a machine capable of storing prodigious quantities of information magnetically.

Although Great Britain could not hope to compete with the massive SIGINT resources of the U.S. **Armed Forces Security Agency** (from November 1952 the **National Security Agency**), the relationship forged with the 1943 **BRUSA** Agreement and then enhanced with **UKUSA** became a partnership with shared facilities established on conveniently located overseas territories, including Aden, Ascension, Bermuda, Ceylon, Cyprus, Diego Garcia, Grand Turk, **Hong Kong**, Iraq, Masirah, Newfoundland, St. Helena, Singapore, and the Seychelles, as well as the bases in the United Kingdom, at **Edzell**, Brawdy, and **Menwith Hill**. In addition, the link with GCHQ provided a link to the SIGINT agencies of Australia, Canada, and New Zealand.

During the **Cold War**, when the Warsaw Pact became the principal target, GCHQ acquired its own intercept capability through the **Composite Signals Organisation** and harnessed the collection assets of the armed forces, with certain dedicated units, such as 9 and 13 Signals Regiments and RAF 192 and 162 Squadrons, becoming surrogates. However, with the introduction of satellite collection, microwave carriers, and computerized data processing, the need to maintain so many ground sites diminished, and many of the intercept stations established during World War II were closed down in preference to greater investment in more sophisticated technology.

The traditional secrecy that cloaked GCHQ was stripped away in the 1970s, partly as a consequence of disclosures about wartime cryptographic achievements but also by radical journalists suspicious of unfettered Whitehall eavesdropping. Thereafter, trade union activism and proof of hostile penetration by the Soviet spy **Geoffrey Prime** exposed the organization to unprecedented external scrutiny, which included a series of budget reviews by Treasury ministers less willing to accept the Secret Vote unchallenged. Although successive administrations found GCHQ invaluable, and appreciated the need for secrecy, some projects, such as a nuclear-powered **naval collection platform** and the **ZIRCON** satellite, attracted considerable skepticism on cost grounds. The ban on trade unions at GCHQ, imposed by Prime Minister Margaret Thatcher's administration in 1985, was rescinded by Tony Blair's government in 1997.

Internally, GCHQ emerged from World War II with sections devoted to Whitehall customers, serving the three armed forces, **MI5**, the Secret Intelligence Service, the Foreign Office, the Home Office, and the Board of Trade, with technical sections applied to particular problems, such as machine automation. By 1946, reorganization split GCHQ into five groups, being Technical; Traffic Analysis and Task Control; Cryptanalysis; Intelligence; and Cipher Security. In 1970, two directorates were introduced, being SIGINT Operations and Requirements—incorporating S (statistical operations and traffic analysis), Z (foreign liaison), W (communications), U (search technology), J (Soviet operations), K (general SIGINT), H (cryptanalysis), X (computing), and T (ELINT)—and the Organization and Establishment Directorate, which included C (overseas staff), E (personnel), F (finance), Q (technical), G (management), M (mechanical engineering), and R (security).

This structure lasted almost two decades, but greater openness, the development of the Internet, cyber warfare, statutory accountability through the Intelligence Services Act, and a widened remit to support criminal investigations led GCHQ to undergo a further transformation into four divisions, being SIGINT Missions (consisting of sections dealing with mathematics and cryptanalysis; information technology; linguistics and translation; intelligence analysis; and open sources); an enterprise division (applied research, corporate knowledge, commercial supplier liaison, and biometrics); a corporate management division (with enterprise resource management, human resources, internal audit, and new SIGINT, known by the acronym SINEWS); and the communications-electronics security division, made up the XA sections of mathematics support (X1), international (X3), policy (X7A), and methodology (X7B), and the XB sections of products (X2), **TEMPEST** (X4), business planning (X5), consultancy (X6), communications certification (X8), and financial services (X9).

Following legitimization through the Intelligence Services Act in 1994, GCHQ's roles in support of counterterrorism in Northern Ireland, the military in Bosnia, and the police in exploiting opportunities offered by communications intelligence through the expansion of cell-phone and Internet use, expanded dramatically. However, greater media awareness of GCHQ's capabilities drew unwelcome attention to an intercept program based at Capenhurst, Cheshire, which monitored microwave communications to and from Ulster.

A move from Oakley to a new, purpose-built circular headquarters, known as "the doughnut" on the Benhall site in 2003 marked a transformation for the organization, once a shadowy branch of Whitehall, into a modern corporate entity complete with a public-relations staff advising on media relations and document declassification.

GCHQ's directors have been: Alastair Denniston (1921–43), Sir Edward Travis (1943–52), Sir Eric Jones (1952–60), Sir Clive Loehnis (1960–64), Sir Leonard Hooper (1965–73), Sir Arthur Bonsall (1973–78), Sir Brian Tovey (1978–83), Sir Peter Marychurch (1983–89), Sir John Adye (1989–96), Sir David Omand (1996–98), Sir Kevin Tebbit (January 1998–July 1998), Sir Francis Richards (1998–2003), Sir David Pepper (2003–8), Iain Lobban (2008–). *See also* FALKLAND ISLANDS.

GOVERNMENT COMMUNICATIONS SECURITY BUREAU (GCSB). New Zealand's principal signals-intelligence agency, GCSB was established in September 1977 with a staff of just 55 and is now based on the top four floors of the Freyberg Building in Aitken Street, directly opposite the Parliament in Wellington. It consists of four divisions, being O, Operations; T, Technology; X, Information Systems Security; and C, Corporate Services. As well as conducting routine **interception** and other work as a **UKUSA** partner, GCSB also has responsibility for conducting inspections or electronic "sweeps" of New Zealand's diplomatic premises overseas and supplied SIGINT suites and personnel for naval collection platforms, including Royal New Zealand Navy frigates patrolling in the Pacific and HMNZS *Monowai*, an oceanographic research vessel, which began operations in 1977 and was decommissioned in 1997 and replaced by HMNZS *Resolution*.

GCSB maintained intercept facilities at HMNZS *Irirangi* station, located just outside the Waiouru township on North Island, which was replaced in 1982 by a large, new satellite ground station at Waihopai, near Blenheim on the South Island, another site at Tangimoana Beach, equipped with **direction-finding** apparatus, and a single overseas facility at Kai Cheng, Singapore, which later moved to Kranji. Originally established in 1942 as a naval wireless station, Irirangi played a major role during the 1982 campaign fought by British forces to liberate the **Falkland Islands**; although, its main **Cold War** target was mainly Soviet and Warsaw Pact shipping. Since then, GCSB has concentrated on Chinese, French, Argentine, Japanese, Philippine, Russian, and Vietnamese traffic, as a regional partner in the **UKUSA** structure linking the organization to **Government Communications Headquarters**, the U.S. **National Security Agency**, the Australian **Defence Signals Directorate**, and the Canadian **Communications Security Establishment**.

During **World War I**, the New Zealand Post Office conducted some radio interception from the wireless station at Awanui, and additional work was undertaken at stations in Wellington and Suva; although, there was no local analytical capability at that time, so the traffic was processed in Australia. However, the Royal Navy's **Admiralty Y Service** extended its reach to Musick Point, Auckland, and during **World War II**, a further four naval sites were added at Awarua, Northland, Mount Crawford, Rapaura, and at Tamavua on Fiji, with a fifth, additional facility in the Radio Wellington premises. From October 1941, the navy stations were coordinated by the Combined Operational Intelligence Centre with an army signal establishment at Nairnville Park, Wellington, which concentrated on Japanese traffic.

New Zealand's first postwar intercept site was "Navy Receiver 1" at Waiouru in central North Island, close to the Royal New Zealand Navy's NR-2 wireless facility.

Prior to 1977, responsibility for the collection and analysis of New Zealand's signals intelligence was vested in the army's **Combined Signals Organisation**, which had been created in 1955. In 2003, the Government Communications Security Bureau Act gave the GCSB independence from the Ministry of Defence and by statute required reporting only to the prime minister. With a staff of around 500, GCSB's directors have been drawn from the New Zealand Air Force and the army and have been Colin Hanson (1977–88), Ray Parker (1988–99), Warren Tucker (1999–2006), and Bruce Ferguson (2006–).

GROUPEMENT DES CONTRÔLES RADIO-ÉLECTRIQUES (GCR). The principal **French signals-intelligence** organization based at Domme in the Dordogne, the GCR was established in March 1946 at Hauterive, near Vichy. Headed by **Gustave Bertrand**, the GCR built wireless-intercept stations in Algeria, Corsica, Saigon, Nouméa, Tunisia, Rabat, Dakar, and at Tegel in West Berlin. Domestic **direction finding** and monitoring was conducted at regional facilities located at Lille, Marseilles, Nancy, Rennes, and Toulouse. In 1949, Bertrand retired and was replaced by his deputy,

Colonel Black, who would later be suspected of having been recruited as a Soviet spy before the war.

In July 1970, much to the dismay of some of its staff, the GCR was absorbed into the politically controversial Service de Documentation Extérieure et de Contre-Espionnage (SDECE). Designated Service 5 and headed by André Benoit, its operations center was moved to SDECE sites at Mont-Valérien and to the fort at Noisy-le-Sec.

GROUP I TRAFFIC. Originally all Abwehr hand-encrypted wireless traffic was designated Group I until successful cryptographic attack by the **Radio Security Service** produced **ISOS**.

GROUP II TRAFFIC. The **Radio Security Service** designation of Abwehr wireless traffic linking Berlin to most Abstellen in Western Europe was Group II, a source broken in November 1940.

GROUP III TRAFFIC. The **Radio Security Service** designation of Abwehr wireless traffic from France was Group III.

GROUP V TRAFFIC. The clandestine wireless link between Hamburg and an Abwehr station on Long Island produced encrypted communications that were solved by **Radio Security Service (RSS)** cryptographers in 1940. Study of the hand ciphers, which was designated Group V, revealed the existence of an extensive spy ring based in New York, and this information was disclosed to Hugh Clegg and Clarence Hince of the Federal Bureau of Investigation (FBI) in December 1940. Subsequently, the two visitors to the RSS headquarters at Barnet revealed that the Group V traffic had been generated under the FBI's control as the Abwehr network, headed by Frederick Duquesne, had been penetrated by a double agent, William Sebold. At **MI5**'s request, the FBI agreed not to close down the organization in a way that would compromise the source. Accordingly, Duquesne and his organization were not arrested until June 1941. The Group V traffic was but one of several different Abwehr channels employing hand ciphers that were designated with Roman numerals and were known by the generic codename **ISOS**.

GROUP VII TRAFFIC. The **Radio Security Service** designation of Russian wireless traffic was Group VII. Individual channels were identified by a numerical suffix, such as VII/23.

GROUP VIII TRAFFIC. The **Radio Security Service** designation of Italian wireless traffic was Group VII.

GROUP X TRAFFIC. The **Radio Security Service** designation of Abwehr wireless traffic encrypted on a machine cipher was Group X, and the resulting decrypts were codenamed **ISK**.

GROUP XIII TRAFFIC. The **Radio Security Service** designation of Abwehr wireless traffic on a radio link between Berlin and Tetuan was Group XIII.

GROUP XIV TRAFFIC. The **Radio Security Service** designation of a German wireless traffic from Istanbul was Group XIV.

GRU. The acronym for the Soviet military intelligence service, Glavnoe Razvedyvatel'noe Upravlenie, an organization that operated independently from, but often in parallel with, the **NKVD**. The decision by Lavrenti Beria in November 1942 to combined the efforts of both to concentrate on ENORMOZ, the collection of atomic information from the **Manhattan Project**, reduced the intrinsic security of compartmentalization, and reliance on a common cipher system assisted Western counterintelligence analysts to study the **VENONA** traffic and expose wartime GRU spy rings in the United States, Great Britain, and Sweden.

GUILLAUME, GUNTER. A long-term spy who worked for the East German Hauptverwaltung Aufklärung (HVA) since 1955, Gunter Guillaume moved to Frankfurt as a refugee in 1956, joined the Social Democrat Party, and was appointed a party functionary before becoming a civil servant in 1970. Eventually, Guillaume earned the trust of the Federal Republic's chancellor, Willy Brandt, and was invited to be his personal adviser. Meanwhile, his wife, Christel, also an HVA mole, was promoted to be an aide to Brandt's defense minister, Georg Leber.

Some of the HVA's signals to Guillaume had been intercepted by the Bundesamt für Verfassungsschutz (BfV), which had monitored East German messages, dating back to 1957, in which a source codenamed GEORG had been sent birthday greetings on 1 February, his wife had received the same on 6 October, and they had been congratulated on the birth of their son on 8 April 1957. This clue had led the BfV mole hunters to conduct a lengthy trawl for anyone with the same birthday, and eventually in 1971, the field had narrowed to Guillaume, his wife, and their son Pierre, and all three were placed under surveillance. He was arrested by the BfV early in the morning on 24 April 1974, thus provoking a major political scandal that led to Brandt's resignation just 12 days later.

Guillaume was sentenced to 13 years' imprisonment in Rheinbach prison, outside Bonn, and Christel received eight. Suffering from kidney disease, he was released in October 1981 in a spy swap and returned as a hero to East Germany, where he died in April 1995.

GUN, KATHARINE. The arrest of the **Government Communications Headquarters (GCHQ)** linguist Katharine Gun in March 2003 and the British government's decision 11 months later not to prosecute her became highly controversial because her offense, which she admitted, was to leak a sensitive **National Security Agency (NSA)** email to a London newspaper.

Given the contentious political background to the case and questions over the legality of the Coalition's invasion of Iraq while this rather less-momentous drama was being played out, there were plenty of critics ready to assert that the government had avoided a potentially embarrassing public trial because it feared disclosure of the attorney general's apparently shifting legal views on whether Prime Minister Tony Blair might be vulnerable to accusations of war crimes if he joined the White House's rush to war. Every time Blair was asked to publish his legal advice, he insisted on confidentiality, citing precedents to protect what the government's top law officer had told him. Despite the pressure, that 13-page document remains undisclosed.

Gun, fluent in Mandarin and recently married to a Turkish asylum seeker, had worked at GCHQ for two years when she copied an email from the NSA's Frank Koza in which the senior Fort Meade official announced that some United Nations Security Council members would be targeted for electronic surveillance. Gun felt this action was wrong, if not illegal and, covering her tracks through an intermediary, delivered the document to the *Observer*, which published it a month later, still not entirely sure of its authenticity.

Gun would assert that her admitted conduct was driven by a wish to save lives, and in the political climate of the time, with the country largely opposed to the quagmire of the Iraq occupation, the prosecuting counsel warned the Crown Prosecution Service that a jury was likely to acquit Gun. Her case was documented by Marcia and Thomas Mitchell in *The Spy Who Tried to Stop a War* in 2008.

H

HAGELIN. The Swedish Hagelin C38m cipher machine, a hand-cranked commercially available device designed by Boris Hagelin and marketed by his Swiss company Crypto A. G., was employed by the Italian Navy upon the advice of the German Kriegsmarine in preference to the **Enigma**. However, the traffic encrypted on it was read continuously at **Bletchley Park** during **World War II** from May 1941, when the radio channel between Genoa and North Africa was broken. To reduce the Italian reliance on cable communications, the Royal Navy trawled for, and cut, the underwater link between Malaga and Genoa.

HALL, JAMES W., III. A career soldier, James W. Hall III was posted to Schneeberg, an **Army Security Agency (ASA)** intercept facility on the Czech frontier in 1977, subsequently worked at the Teufelsberg in West Berlin, and was later transferred to the 513rd Military Intelligence Brigade at Fort Monmouth, New Jersey in 1985.

A year later, he volunteered to return to Germany, to a signals-intelligence post with the 205th Military Intelligence Battalion in Frankfurt, but, by January 1988, was back in the United States with a top-secret clearance and working at a succession of headquarters military-intelligence posts at Fort Stewart in Georgia, having undertaken training courses at Fort Still in Oklahoma, Fort Huachuca in Arizona, and Fort Devens in Massachusetts.

When, in July 1987, Hall was transferred with the rank of warrant officer to the 24th Infantry Division at Fort Stewart, Georgia, he came under surveillance from both the Federal Bureau of Investigation (FBI) and the 766th Military Intelligence Detachment when he held three meetings in November and December 1988 with a Turkish car mechanic, Huseyin Yildirim, who had worked at the Teufelsberg and had acted as his courier.

On 21 December 1988, Hall boasted on videotape to Dmitri Droujinsky, a Federal Bureau of Investigation special agent posing as a Soviet intelligence officer, that he had supplied top-secret information to the East Germans for the previous six years. In his subsequent confession, he admitted that he had begun his espionage in November 1982 when serving at the Teufelsberg in West Berlin and had dropped a letter to the Soviet consulate in Grunewald, offering a meeting. Hall attended the proposed rendezvous and was driven to a safe house in Karlshorst for a detailed interview, at which he persuaded the KGB that he was not a plant or double agent. As a young 25-year-old soldier with Heidi, his German wife, and their first daughter, he was short of money and was driven to espionage by his wish to earn some extra cash by exploiting his access to hugely valuable signals intelligence. Among the items he compromised was a Berlin Field Station manual, the *Allied Communications Publication*, the *National SIGINT Requirements List*, a huge, 3,000-page loose-leaf binder containing requests for NSA support from various branches of the U.S. government, and details of TROJAN, a **National Security Agency (NSA)** program to identify the locations of Soviet weaponry in the event of hostilities.

Hall also admitted that, during his year at Fort Monmouth, he had maintained contact with the KGB through a series of dead drops in New York City and New Jersey, and on one occasion had flown to Vienna to be debriefed by a Soviet signals-intelligence expert.

When reassigned to Frankfurt, Hall had continued to sell information, for an estimated $300,000. A search of his home in Richmond Hill, Georgia, revealed a British passport in the name of Robert S. Hillyer, a duffle bag filled with classified files, and $5,000, with a further $4,150 in his pickup truck. Following a plea bargain to avoid a death sentence, Hall was sentenced by a court-martial at Fort McNair in March 1989 to 40 years' imprisonment at Leavenworth in March 1989 and fined $50,000. His plea bargain ensured the briefest of courts-martial and his evidence against a Turkish coconspirator, Huseyin Yildirim, then aged 62 and living in Belleair Beach, Florida. When questioned, Yildirim agreed to accompany his interrogators back to a wine cellar under his former apartment in Berlin where they recovered a cache of some 10,000 NSA and ASA documents. After a trial lasting just two days, Yildirim was found guilty of espionage and sentenced to life imprisonment at Lompoc, California.

The NSA's damage assessment, verified by polygraph examinations and conducted at Fort Meade, suggested that, over six years, Hall had made between 30 and 60 deliveries of classified documents, some of them in packages three inches thick, to the Soviets and to the East Germans. Only in June 1985, following a meeting in Vienna, had he broken with the KGB but kept in contact with the Hauptverwaltung Aufklärung (HVA) through Yildirim, information that was later to be confirmed by the interrogation of former East German intelligence personnel. The quantity of documents taken by Hall at the Teufelsberg was so great that he feared he would exceed the limits on the use of his office photocopier, so he installed another one in a van he parked in the base's car park. Later, in Frankfurt, he rented an apartment solely for the purpose of installing a copier. On two occasions, the KGB dispatched signals-intelligence experts to question him and to direct his document acquisition, which included details of a highly classified black operation initiated in 1984, which was to be published in 1990 in the German magazine *Quick* when the relevant papers were found in the HVA archives.

HAMILTON, VICTOR. A **National Security Agency (NSA)** linguist of Libyan descent, originally named Hindali, Victor Hamilton had been fired in June 1959 on psychiatric grounds and later traveled to Czechoslovakia. Hamilton was not seen publicly following his visit to the Soviet embassy in Prague in 1962, and his request for political asylum was announced in *Izvestya* the following year. Once in the Soviet Union, he also denounced the NSA's operations, identifying those countries he claimed had been targeted for **interception** and claiming that the communications of several Middle East countries and their U.N. missions had been routinely read in his section. In 1992, he was found in a Russian mental institution by the Ark Project, a search for Americans who had gone missing in the Soviet Union.

HARPENDEN. The Rothamsted Experimental Station at Harpenden, Hertfordshire, was one of the first signals-intercept sites of **World War II**, but in the absence of any illicit clandestine traffic to monitor for the **Radio Security Service**, control was transferred briefly to the **War Office Y Group**, which ran 30 intercept positions there until handing the entire facility over to the **Polish Cipher Bureau**. *See also* MI8.

HASP. In 1959, the Swedish **Försvarets Radioanstalt (FRA)** revealed to the British that it had retained copies of some of the Soviet embassy's wartime Stockholm–Moscow traffic, and negotiated with **Government Communications Headquarters** to release its archive, codenamed SKIM, to the U.S. **National Security Agency** via Cheltenham. Codenamed HASP, this unexpected windfall consisted of 390 partially deciphered messages, exchanged with Moscow between December 1940 and April 1446. The FRA had succeeded, as early as 1947, in reading a few messages, and between 1957 and 1959, some 53 texts were broken out. Information identifying individual Soviet spies had then been passed to the Allmänna Säkerhetstjänsten (General Security Service), which conducted investigations that effectively neutralized them without compromising the source.

HASP provided the stalled **VENONA** project with further "depth" and mentioned **NKVD** sources, including Ada, Adam, Alma, Anton, Arkadij, Artist, Arvid, Assistant, Birgit, Burimova, Colett, Croat, D, Dima, Dobrovoljets, Dora, Dorin, Drug, Edward, Elita, Erika, Fenya, Frost, Godin, Gradus,

Greek, Grisha, Gromov, Idler, Irina, Istorik, Jan, Jeanne, Khochev, Kin, Klara, Klim, Knave, Kok, Kolmar, Kon'bobezets, Kuznet, Lan, Leaf, Mak, Markus, Max, May, Metalist, Metr, Mink, Misha, Moritz, Moroz, Mountaineer, Nik, Orestes, Oscar, Patriot, Paul, POET, Polyarnik, Robert, Roman, Rose-Marie, SENATOR, Soldier, Talant, Terentij, Tim, Tsilindr, Ulrik, UNC17, UNC25, UNC36, Valerian, Valet, Viktoria, Wax-works, and Zhak.

Among the **GRU** codenames were Admiral, Akasto, Akua, Albin, Anderson, Anger, Arel, August, Austrian, Baldin, Barbro, Bard, Berns, Braido, Brando, Branjberg, Bratya, Bunden, Edgar, Edvin, Enar, Feliks, Floro, Gans, Garin, Garry, Gent, Georgij, Greger, Gurvin, GUSTAV, Ispanet, Jim, John, Karl, Kurbo, Laurin, Lebeden, Levanger, Matson, Mauritz, MAX, Merchant, Metron, Michail, Mitron, Modeller, Morits, Mostovik, Narym, Oskar, Per, Philip, Samuel, Sand, Saturn, Seaman, Serzhant, Smit, Stege, Stroitel, Svante, Sven, Svensson, Tyura, UNC33, Valden, Vetrov. In addition, there was some GRU Naval traffic that mentioned A, Aksel, André, Anker, Antoni, Baltazar, Brita, Burgen, Chief, Count, Dal, Dimau, ERIK, Fogel, Fritz, Georg, Gunar, Hans, Harald, Ibsen, Josephine, Klass, Knudsen, Kryger, Leva, Ludvig, Michailov, Name BK, Name BV, Name BX, Name 42, Name 59, Name 62, Olav, Oskar, Oskar II, Petrus, Prince, Roland, Runar, Stig, Valer, Verner, Viktoria, Vinter, Yanek, and Zhan.

Of course, the Swedish security police had long been aware of Soviet espionage and, in 1940, had rounded up a large Comintern sabotage organization headed by Ernst Wollweber and his wife. In August 1942, Lennart Katz, a source run by a contact working under diplomatic cover named Scheptkov, was arrested. On the same day, a married couple, Ture Georg Eriksson and Signe Eriksson, had also been taken into custody. Then in November 1942, Knut Nyblad, codenamed EDGAR, was arrested with three others. In June 1943, Per Meurling, codenamed JACK, and Sonja Strand, codenamed SONJA, were arrested, followed by May Jarke, codenamed KLARA.

So many Soviet spies were arrested in wartime Sweden that it was not evident that the Security Police had broken any Soviet signals, and when some, such as Wollweber, were quietly deported to Moscow in 1944, the NKVD failed to conclude their cipher system had been compromised.

HOLYSTONE. The U.S. Navy codename for a signals-intelligence collection program undertaken by submarines penetrating Soviet territorial waters, HOLYSTONE continued throughout the **Cold War** under exceptional conditions of secrecy. From 1969, the American operations, codenamed BARNACLE, were supervised by the National Underwater Reconnaissance Office. Only one really serious incident occurred during the HOLYSTONE era, when in mid-August 1949 the USS *Cochino*, on a mission from Portsmouth,

England, to the North Cape, suffered a schnorkel failure and sank with the loss of five of the crew and a civilian engineer. In July 1957, Soviet destroyers forced the USS *Gudgeon* to surface off Vladivostok, and in another episode, the USS *Tautog* collided in 1970 with the *K-108*, a Soviet Echo-class diesel-electric, but there were no casualties. Four other minor incidents were also reported, but no details were ever disclosed publicly. Both British and Norwegian submarines have also participated in similar operations. *See also* IVY BELLS; NAVAL COLLECTION PLATFORMS.

HONG KONG. The home before **World War II** of the **Far East Combined Bureau**, which was evacuated in August 1939 to Singapore, Hong Kong became an important postwar signals-intelligence center with the Royal Navy managing a SIGINT facility on Stonecutter's Island, the **Royal Air Force**'s 367 and 743 Signals Units operating intercept sites at Chung Hom Kok and at Old Belvedere, atop Victoria Peak on Hong Kong Island, with two smaller stations at Tai Mo Chan and Kong Wei in the New Territories. In addition, there was an Australian detachment from RAAF Pearce and a site run by the **National Security Agency**, which was later transferred to Okinawa.

In August 1954, a total of 15 wireless positions were operational at Old Belvedere, with a further 23 sets working at the other stations, producing 30,315 intercepts. By December, the intercepts grew to 43,782 generated by a total of 50 sets. By the following year, 53 sets, working for 5,038 hours a week, produced 49,804 intercepts. In March 1957, 64 sets, averaging 7,003 hours a week, produced 61,149 intercepts, mainly of mainland Chinese military communications.

In 1961, Chan Tek Fei, a locally employed Chinese linguist, was arrested following an ostensibly routine customs search at the border and charged with espionage on behalf of the People's Republic of China. Allegedly, he was found to be carrying classified documents, lists of **Government Communications Headquarters (GCHQ)** personnel with descriptions of their vulnerabilities, and details of an affair between his wife and a senior GCHQ officer, who was promptly transferred. In another case, two GCHQ clerks defected to Taiwan after they came under suspicion of leaking secrets.

Civilianization in January 1964 passed responsibility for signals-intelligence collection and analysis to GCHQ, which was established at Little Sai Wan. The facility acquired some notoriety in May 1980 when Jock Kane, a 32-year veteran of the organization, complained publicly about wide-scale corruption at Little Sai Wan, and in 1984, after his retirement, the British government injuncted him on national security grounds to prevent publication of his memoirs, *GCHQ: The Negative Asset*.

In 1962, 367 Signals Unit was disbanded, and Little Sai Wan closed down in 1982 and moved to a purpose-built facility at Chung Hum Kok, on the southern side of Hong Kong Island, and, in January 1995, was transferred to the Australian **Defence Signals Directorate** base at Kojarena, near Geraldton in Western Australia.

HOOVER, CALVIN. A member of the Office of Strategic Services (OSS) station in Stockholm, which was headed by Wilho Tikander, a lawyer of Finnish origin from Chicago, Calvin Hoover was an economist from Duke University who used his position in OSS's Secret Intelligence branch to collect cryptographic information on the Soviets. Having spent five years in Moscow, Hoover was very antagonistic toward Josef Stalin and masterminded several important acquisitions from Finnish refugees, including one military and three diplomatic codes, the Red Army's complete order of battle, the register of the Red Banner Fleet, and, in November 1944, the **Petsamo codebook**.

This latter purchase was to prove highly controversial, for although General Bill Donovan approved of it, the State Department did not, and when Assistant Secretary of State Edward Stettinius, discovered that his ban on collecting intelligence against the Soviets had been ignored, he protested to the White House. Donovan had reported his coup to President Franklin D. Roosevelt on 11 December 1944, describing the material as the key to both military and **NKVD** codes, but word of it reached Stettinius four days later. The result was that the OSS liaison officer in Moscow, General John Deane, informed his counterpart, General Pavel Fitin, on 9 January 1945, that OSS wanted to return some 1,500 NKVD papers and associated material, and by 15 February, they had been delivered to Andrei Gromyko, then Soviet ambassador in Washington, DC.

Precisely what this OSS codebook really was and whether it was copied before it was returned to the Soviets is unknown, but the angry exchanges between Donovan, the State Department, and the White House are well documented. However, it is equally clear that the **Arlington Hall** cryptographers working on Soviet traffic relied on the **TICOM** material rather than anything provided by OSS, and a summary of what the Finns had accomplished, dated 17 December 1946, indicates that they had worked on Soviet military, police, diplomatic, and trade ciphers. It also mentions a four-digit NKVD code, which may be a reference to the *Pobeda* ("Victory") code known to have been used between 1941 and October 1943, when a new system, designated 075-B, was introduced. The switch was disclosed in a **VENONA** text dated 31 October 1943, from Grigori Kheifets in San Francisco to Moscow, in which the *rezident* acknowledged receiving copy No. 27, the new codebook, the B in the title being an abbreviation of *Bezopasnost* (Security), and undertaking to destroy the old *Pobeda* version.

HORCHDIENST. The Luftwaffe's principal signals-intelligence organization, the Horchdienst routinely monitored Royal Air Force (RAF) wireless traffic and voice channels, providing the Reich with highly effective early warning of imminent Allied air raids. One Horchdienst technician, Hauptmann Kuhlmann, appeared so frequently in the MUSTARD **Enigma** key read at **Bletchley Park** from June 1941 that the Horchdienst later became known as the "Kulhmannmeldungen." Warned by Allied cryptographers that poor RAF security was jeopardizing operations, steps were taken to improve wireless discipline.

HUNSTANTON. Probably the world's first signals-intelligence intercept facility, the coast guard hut at Hunstanton was the site of the Admiralty's experimental wireless station, monitoring German naval and **zeppelin** traffic.

I

IBIS. The **Bletchley** Park codename for traffic transmitted by an unidentified German wireless in Istanbul reporting on Syria and Palestine during **World War II**.

IGLOO WHITE. The **National Security Agency** codename for a program conducted during the **Vietnam War** that employed remote sensors inserted into territory occupied by the Vietcong to collect seismic and electronic data, which was transmitted to overhead **airborne collection platforms**.

IMMINENT HORIZON. On 5 March 2003, just as Coalition forces began to attack Iraq, the **National Security Agency** gathered evidence in an operation codenamed IMMINENT HORIZON that implicated two Iraqi diplomats accredited to their country's mission to the United Nations in a plot to mount a terrorist atrocity in New York. Both were declared *persona non grata* and expelled.

IMPROVISED EXPLOSIVE DEVICE (IED). The British experience in Northern Ireland, where the Provisional Irish Republic Army routinely resorted to manufacturing their own explosive devices from homemade ingredients, stolen components, or ingeniously converted commercially available gadgetry, provided the foundation for a new SIGINT discipline. Having learned the disadvantage of command wires, which provided a telltale route to whatever forensic evidence had been left behind by terrorists operating by line of sight, paramilitaries opted to adapt remote-control equipment to detonate IEDs. While this expedient eliminated the risks associated with command cables, it opened the security authorities, and specifically **Government Communications Headquarters** technicians, to an opportunity to scan the relevant frequencies to warn of a potential danger and, more controversially, to initiate a premature detonation.

The skills acquired in Ulster proved highly effective in Iraq where, following the Coalition invasion in 2003, insurgents took advantage of a ready supply of abandoned ordnance and adopted roadside bombs as a very effective tactic against lightly armored target vehicles. Similarly, no foot patrol or convoy in Afghanistan would contemplate venturing into hostile territory without being equipped with the sophisticated apparatus designed to identify culvert bombs and block command transmissions. *See also* ELECTRONIC COUNTERMEASURE (ECM).

IMPULSE. The Soviet codename for the signals-intelligence cubicle usually attached to individual *rezidenturas* to monitor transmissions of the local security apparatus. Staffed by specialist technicians from the KGB First Chief Directorate's Sixteenth Department, IMPULSE operators scanned communication channels known to be used by hostile surveillance teams to warn the *rezident* of exceptional activity or the deployment of adversaries in locations where operations were imminent. *See also* ROCKET; ZENIT.

INDICATOR. Within the preamble to most encrypted wireless signals is a prearranged sequence, known as the indicator, which identifies the cipher system that has been adopted. Although not easily discernible from the other five-figure groups in a message, the indicator will appear at the beginning of any text and provide the receiving cipher clerk with the information required to return the remainder of the content into clear language. Accordingly, indicators are of immense assistance in the **traffic analysis** process, and can be vital to cryptographers. *See also* JADE.

INSECT KEYS. The **Enigma** keys designated COCKROACH, GADFLY, HORNET, LOCUST, and MOSQUITO were categorized at **Bletchley Park** as INSECT and contained low-echelon Luftwaffe traffic.

INTELSAT. The first two transatlantic International Telecommunications Satellite Organization's commercial communications satellites, in geosynchronous orbit 22,375 miles over the ocean, each carrying 4,000 voice circuits, were linked in 1964 to ground stations at Andover, Maine; Yakima in Washington state, southeast of Seattle; and Etam, West Virginia, on the United States side, and to Madley, Herefordshire, and Goonhilly, Cornwall, in England. However, by 1967, the **National Security Agency** had built facilities at Winter Harbor, Maine, and Sugar Grove, West Virginia, with **Government Communications Headquarters** installing radomes at **Menwith Hill** in Yorkshire and at Morwenstow, near Bude in Cornwall, also capable of receiving the INTELSAT traffic.

INTERCEPTION. The acquisition of target communications is known as interception and broadly falls into three categories: the monitoring of telephone landlines at the central exchange, known in the British parlance as a

check, usually requiring official authorization through a court-issued warrant to gain the cooperation of the service provider; the eavesdropping of voice and fax calls relayed on satellite or microwave channels remotely; and access to Internet communications at choke points.

Interception of radio channels consists of strategic interception, usually from large sites close to antenna fields, and tactical interception, often from mobile facilities, which concentrate on low-power local traffic. Even if the traffic on these channels is encrypted, various techniques such as **traffic analysis**, **direction finding**, and **call-sign analysis** can offer very useful information, which becomes part of the signals-intelligence matrix.

To avoid interception, various countermeasures can be taken, including changing frequencies, employment of frequency-hopping equipment, directional antennas, and use of minimal power. *See also* SKIP ZONE.

IRAQI FREEDOM. The Coalition invasion of Iraq in March 2003 was the first major conflict prosecuted largely through the exploitation of SIGINT, and in the first few hours of the campaign, Saddam Hussein's principal ground-defense command centers had been identified and neutralized. In the months prior to the offensive, a comprehensive electronic wiring diagram of the entire country had been constructed, and preparations had been made to sabotage the landline and fiber-optical cable infrastructure, leaving the military establishment wholly reliant on conventional radio transmissions for virtually all but the lowest echelon communications. The Coalition's demonstrably overwhelming superiority in electronic warfare persuaded others, including the Chinese Peoples' Liberation Army, to review their doctrine and invest heavily in SIGINT and **electronic countermeasures**.

IRAQI FREEDOM's initial success at eliminating the threat from ground defenses, which had been monitored and mapped since the imposition and enforcement of the United Nations' no-fly zone over country in 1991, was dependent on both external collection and the regime's use of equipment that had been compromised before delivery and installation. Accordingly, once air supremacy had been established, the Coalition was quite at liberty to select its targets in a uniquely risk-free environment, thereby transforming the nature of large-scale warfare.

IRELAND. During **World War II**, the Irish Republic's intelligence service, G-2, developed a signals-intelligence capability and constructed four intercept stations, at Dublin, Limerick, Cork, and Athlone. G-2's chief, Colonel Liam Archer, developed a useful relationship with Cecil Liddell, the head of **MI5**'s Irish Section, and agreed to share any useful intelligence with his British counterpart, despite Éire's supposed neutrality. The British encouraged Archer by supplying him with a consignment of some rare American National HRO radio receivers and suggested siting further stations close to the coast to monitor traffic that was thought to be exchanged between long-range Focke-Wulf Condor maritime reconnaissance aircraft and U-boats. Another priority was any signal broadcast from the German legation in Dublin, which was known to have retained an illicit transmitter with which to communicate with Berlin.

Although G-2 never invested in any **direction-finding** equipment, leaving the British to rely on an apparatus in Northern Ireland, the organization did attempt to monitor and trace clandestine broadcasts made by Abwehr agents and achieved considerable success in breaking a cipher hidden in secret writing by Dr. Hermann Goertz, an Abwehr spy languishing in an internment camp. While he believed he was in contact with Berlin, his correspondence was in fact being answered by Dr. Richard Hayes, a G-2 cryptographer in Dublin. *See also* WORLD WAR II.

IRTON MOOR. The **Admiralty Y Service** intercept station at **Irton Moor** in Yorkshire. *See also* DIRECTION FINDING (D/F); INTERCEPTION.

ISCICLE. The Office of Strategic Services codename for decrypts of Abwehr hand ciphers, known as **ISOS** by the British.

ISCOT. The **Government Communications Headquarters (GCHQ)** designation for Comintern wireless traffic intercepted during **World War II**, ISCOT originated at Alford House, overlooking London's Park Lane, where a small team of codebreakers, led by Bernard Scott, were isolated even from GCHQ's **Diplomatic Section**, which was also located in Mayfair, at Grosvenor Street. Given its political sensitivity, ISCOT was given very limited distribution to the relevant branches of **MI5** and the Secret Intelligence Service, and a detailed analysis of the traffic was completed by Jane Sissmore on behalf of Section IX. **Kim Philby**, confident that ISCOT contained nothing incriminating in respect to himself or the Communist Party of Great Britain, had assigned the task to her, "to keep Jane busy" and preoccupied with a topic where she could do him no harm. *See also* GOVERNMENT CODE AND CYPHER SCHOOL (GC&CS).

ISK. The **Government Code and Cypher School** designation of the Abwehr's **Enigma**-generated intercepted wireless traffic produced by a section at **Bletchley Park** led, until his death in 1942, by Dillwyn Knox. When selected officers of the U.S. Office of Strategic Services were indoctrinated into the source, it was codenamed **TRIANGLE**. *See also* ISOS; ULTRA.

ISOS. The **Government Code and Cypher School** designation of the hand ciphers employed by the Abwehr to communicate by radio with out-stations on channels where **Enigma** machine encryption was unavailable. Derived from "Intelligence Service Oliver Strachey" after the **Radio Security Service** cryptanalyst who solved the first Abwehr hand cipher intercepts, ISOS included a variety of wireless links, known as Roman-numbered groups, such as **Group I**, indicating a particular encrypted network. By the middle of the war, ISOS had become a generic term for decrypted enemy intelligence signals, irrespective of the source or encipherment method. By June 1941, ISOS was considered such a priority that 120 intercept positions were dedicated to the source, and the **Y Committee** then agreed that, if the traffic continued, they would be increased by a further 60 positions.

When selected officers of the Office of Strategic Services were indoctrinated into ISOS, it was referred to as **ISCICLE**.

ISOSCELES. The Office of Strategic Services designation for decrypts derived from the Abwehr's radio link between Vichy and Berlin.

ISPAL. The **Government Communications Headquarters** codename, actually an abbreviation of "Intelligence Service Palestine," for illicit Jewish Agency and Haganah encrypted wireless traffic broken by the **Diplomatic Section** in London in 1945.

ISRAELI SIGNALS INTELLIGENCE. *See* RAS-AL-GHALEB; UNIT 8200.

ITALIAN CIPHER BUREAU. Headed by Major Barbieri of the Servizio Informazioni Militare (SIM), the Italian Cipher Bureau concentrated on Allied and neutral diplomatic traffic, including the French, Romanian, Swiss, and Turkish ciphers, and routinely sponsored either the burglary of diplomatic premises to acquire copies of codebooks or telegrams that had been encrypted and transmitted. Before **World War II**, SIM employed two sources, the brothers Francesco and Secondo Constantini in the British embassy for this purpose, and in 1941 obtained a copy of the U.S. military attaché's **Black Code**.

In 1943, following the Italian surrender, Barbieri's organization, with a staff of 80, continued to work on diplomatic traffic from a base in Brescia under German sponsorship. At the end of the war, when Barbieri and his cryptographers had been captured and interrogated, SIM was encouraged to continue its activities and share its product with the British **Government Communications Headquarters (GCHQ)**. Under GCHQ's direction,

Barbieri turned his attention to Soviet military machine-cipher traffic, code-named TAPER.

IVY BELLS. The **National Security Agency (NSA)** codename for a submarine collection operation using the USS *Halibut*, *Seawolf*, and *Parche* that succeeded in placing recording devices over Soviet and Libyan underwater cables. The devices intercepted the traffic through an induction antenna that did not compromise the physical integrity of the cable. The program was betrayed to the Soviets by an NSA analyst, **Ronald Pelton**, in 1980, who sold the details of a pod in the Sea of Okhotsk near the naval base at Petropavlovsk, and the USS *Richard Russell* was deployed in 1992 to remove the last of the pods in the Barents Sea.

The operation, part of the Special Navy Collection Program, was high risk, being conducted inside territorial waters, and placed the submarine crews and NSA support teams in jeopardy when, every six weeks, the pod tapes were replaced. The raw material, acquired from the cable's multichannels, included some data that was unencrypted, apparently because the Soviets regarded the traffic as safe from hostile interception. *See also* NAVAL COLLECTION PLATFORMS.

J

J-19. The Allied designation of the Japanese diplomatic code employed during **World War II** and known in Tokyo as FUJI, J-19 was broken by American cryptographers at **Arlington Hall** in 1942.

JADE. The designation at **Arlington Hall** of a Soviet super-encipherment system applied in parallel by the **NKVD** to messages encrypted with codebooks and **one-time pads (OTP)**. Even though each text contained an **indicator** to tell the authorized recipient which page number had been selected, there was no chance of anyone else reconstructing the OTP unless they had access to the manufacturing process. In the Soviet trade traffic studied by American cryptographers in 1944, the indicator that told the cipher clerk which OTP page to use was the last two digits of the first five-figure group, and in one distinctive series of diplomatic telegrams, in a system designated ZDJ and codenamed JADE, the subtraction of the penultimate group from the last group was always 22222.

Quite rightly, Moscow enjoyed complete confidence in the arrangement that, when managed properly, meant that mere acquisition of one organization's codebook would not compromise the messages sent by another's. Furthermore, because each page of the OTP was, as the name implies, only employed once, the danger of potentially dangerous duplication was avoided. Predictable repetition, the very heart of all successful cryptanalytical attacks, was eliminated by employing OTPs, or at least in theory. To protect the OTPs themselves, the Soviets distributed them to designated cipher clerks in special pouches and kept them under constant armed guard.

Although Finnish cryptanalysts had enjoyed the huge advantage of access to what were to become known as the **Petsamo** documents, the **Signal Security Agency (SSA)** staff had none to begin with and, in the absence of any original plaintexts, had been obliged to use classic cryptanalytical techniques to reconstruct the NKVD's codebook. Much of this work was accomplished by **Meredith Gardner**, a linguist rather than a cryptanalyst, who exploited the unbelievable but undeniable fact that the Soviets had compromised the integrity of their own systems by reusing the same one-time pads.

Under normal circumstances, the numbers printed on one-time pads would be generated randomly, and single sheets, each unique, would be bound into books and used consecutively. However, eight months after Gene Grabeel had started on a **discrimination** task, one of the Arlington Hall cryptographers, Lieutenant Richard Hallock, a Signals Corps reservist who had been an archeologist at the University of Chicago, realized that many of the sheets used in different systems had been duplicated, and once the beginning of a page had been spotted, the rest of it could be predicted. Nevertheless, it was to be another six or eight months before he could, with Mary Jo Dunning, Dr. Burton Phillips, and Genevieve Feinstein (nee Grotjan), begin to extract the underlying numerical code groups. Following Hallock's discovery, prolonged study of JADE by Cecil Phillips and Genevieve Feinstein revealed an astonishing fact, that the JADE material had been enciphered using precisely the same OTPs as Moscow's trade traffic.

This extraordinary blunder took place in late 1942, at a time when the Soviet administration was under great pressure following the Nazi invasion. Whatever the still unexplained reason for the duplication, some 70,000 pages were duplicated and then inserted, often in different order, into one-time pads for distribution to cipher clerks, who doubtless never realized the fundamental flaw that accidentally had been introduced into the system. Altogether 35,000 duplicates were found, about half the estimated total, and probably represented the total production of three to six months. The earliest use was traced to May 1942, with the majority being used between July 1942 and July 1943. Hallock's discovery of the duplicate pages, which continued to appear in Soviet traffic until June 1948, allowed his colleagues to make impressive progress on the texts, of which there was an abundance between 1942 to 1944, a period of dramatic escalation in the volume of Soviet cable traffic overseas.

In retrospect, the most plausible explanation for the error is human failure. Someone, somewhere in the manufacturing chain, simply doubled the output of OTPs by running off extra copies of pages that were never intended to be duplicated. This astonishing act of negligence served to undermine the intrinsic security of the OTP and ultimately allowed British and American cryptographers to achieve what theoretically should have been impossible. Prolonged study by a team led by Cecil Phillips, which included Frank Lewis, Genevieve Feinstein, Lucille Campbell, and Frank Wanat, demonstrated that hundreds of OTPs used to recipher trade messages had been used for the same purpose by the NKVD. Lewis produced a method of aligning page numbers that exposed several hundred more duplicates in the trade traffic; Wanat, a navy chief from the **Naval Security Group**, spotted similar characteristics in the Havana–Moscow traffic, and Lucille Campbell had been alongside Cecil

Phillips when he noticed an excess of the number six in the first group of NKVD messages. These were vital faults that made the banal trade traffic so crucial and the NKVD texts so vulnerable.

The first stage of the cryptographic attack was to establish the extent of the Russian error, and this was achieved by transferring onto IBM punch cards the first five groups from 10,000 trade messages, on the basis that the first few words of any telegraphic message, regardless of origin, is likely to be stereotypic, containing the words "to" and "from" and perhaps a date and a potentially predictable serial number. The cards were sorted automatically by an IBM machine, and the comparison revealed seven matching groups. If the OTP numbers had been genuinely random, the odds of finding two matches was calculated to be about a hundred million to one. To find seven meant the Soviets had blundered and had allowed a tiny crack to be exploited. The crack enlarged when, having had the duplicates "dug out" by dozens of young women, among them Gene Grabeel and Katie McDonald, Gardner concentrated on the Soviet spell code, employed to encipher English words, and started to assign Russian meanings to the numeric code groups. This, effectively, was reconstructing by analysis a Soviet codebook, and after much effort, he found a message from New York in the JADE series dated 18 May 1944 that was entirely in English and, therefore, had been enciphered with the spell code:

> Your No. 2106, second paragraph:
> "If election were held today R [President Roosevelt] would probably obtain slender majority of popular vote but lose election due heavy concentration his vote in South where his big majorities count no more in final result that would simple 51 percent."

Although the groups containing the signature were unrecovered, Gardner had spotted that the code group for "t-i-o-n" had been repeated three times in quick succession, and this enabled him to read the rest, which must have been the answer to a request from Moscow for clarification of telegram number 2106. The content was obviously political, explaining how the president could lose in the Electoral College despite heavy Democratic support in the South, and not consular, and the use of English suggested a quote from an American political source. This was proof that JADE was diplomatic in nature and could be broken to reveal evidence of espionage.

The value of Gardner's work really became apparent when, on 31 July 1946, he began to read much of a message sent to the head of the NKVD's foreign intelligence branch, Lieutenant-General Pavel M. Fitin, codenamed VIKTOR, in Moscow on 10 August 1944, from the New York *rezident*, Stepan Apresyan, codenamed MAY, on the subject of coding procedures in Mexico.

No. 1142
Your No. 3502

1. In accordance with a decision of the HOUSE [Comintern] SISKIN and KARLOS [Christian Casanova] before departure from TYRE [New York] were instructed in the use of a reciphered code which they were to [46 groups unrecoverable] alphabet and in addition SISKIN was to write in English and KARLOS in Spanish. We cannot establish whether the pads for the codes were sent home since they were entered under the heading "SISKIN's Business" and "KARLOS's Business" [4 groups unrecovered] there are no inventories. From TYRE no enciphered letters were sent by [or to] SISKIN and KARLOS. Both codes ended up in STOCK's [Mikhail A. Shalyapin] hands.
2. For correspondence with ARTHUR a book in the Spanish language "Una Excursion a Los Indios Ranqueles" was used as a code.
3. For correspondence with ALEKSANDR the books "My Sister Eileen" and "Defense Will Not Win War" were used as codes.
4. For correspondence with MARGO [Margarita Neklen] the book "Opera" which was sent to you in accordance with your instructions, was used.
5. For correspondence with HARRY [Jacob Epstein], JUAN [Juan Gaytan] and ANITA books were used whose titles were not kept since they were destroyed after the departure of LUKA [Pavel Klarin] to the COUNTRYSIDE [Mexico].
6. In a sample cipher square TWAIN [Semyon Semyonov] trained [34 groups unrecoverable] ZUL, PAUL which were destroyed and for STAFFMAN which we are dispatching to you by the next regular post.

This was a hugely significant text because it was patently an NKVD message and proved beyond doubt that it was really possible to read NKVD communications, whereas hitherto there had been a strong suspicion that the collection of encrypted intercepts may have been trade telegrams of no intelligence interest whatever. This text mentioned numerous NKVD assets by their codenames, of whom Sam Semyonov, codenamed TWAIN, and Pavel Klarin, codenamed LUKA, were very familiar to the Federal Bureau of Investigation as members of the New York *rezidentura*, while ostensibly Mikhail Shalyapin (STOCK) was listed officially as a humble clerk. Neither ARTHUR, ALEKSANDR, ANITA, PAUL, nor STAFFMAN could be identified, but by May 1972, when the text was reissued for the last time, collateral evidence had named KARLOS as Christian Casanova Subercaseaux, JUAN as Juan Gaytan Gody, MARGO as Margarita Neklen Hansberger de Paul, and HARRY as Jacob Epstein. It subsequently emerged that the 10 August message was but one in a series, with references to KARLOS appearing in nine other VENONA signals, in a time span between May 1943 and December 1944.

On 13 December 1946, a further major breakthrough was made, with another telegram from Apresyan in New York, dated 25 July 1944, apparently based on a conversation held by one of his *rezidentura's* staff, codenamed SUK, with BEAR, a leading figure in the Republican Party, about Thomas Dewey's chances in the November elections. SUK's unidentified source correctly forecasted that President Franklin D. Roosevelt, codenamed CAPTAIN, would be reelected handsomely:

> To VIKTOR
> On 9th July SUK had a confidential chat with BEAR who expressed his opinion on a number of questions connected with the election campaign. While regarding KULAK as "a rising star" who "would probably become president" sooner or later, BEAR at the time expressed his conviction that CAPTAIN [President Roosevelt] would get "a huge majority" in the elections. As the chat took place before the convention, advise whether to telegraph more.

While the content of this **VENONA** message was not of any great significance, it was a complete text, but Gardner's third success, accomplished on 20 December 1946, had the very widest implications, for helpfully it contained plenty of non-Russian names that required to be spelled out in full, using a "spell table" to encode English letters, and proved the extent of Soviet interest in and knowledge of the wartime **Manhattan Project** two years earlier, in December 1944:

> Conclusion of Telegram No. 940
> Enumerates the following scientists who are working on the problem—Hans BETHE, Niels BOHR, Enrico FERMI, John NEWMAN, Bruno ROSSI, George KISTIAKOWSKI, Emilio SEGRE, G. I. TAYLOR, William PENNEY, Arthur COMPTON, Ernest LAWRENCE, Harold UREY, Hans STANARN, Edward TELLER, Percy BRIDGEMAN, Werner EISENBERG, STRASSENMAN [7 groups unrecoverable] our country addressed himself to NAPOLI and the latter, not wanting to listen to him, sent him to BECK [Sergei Kurnakov] as military commentator on the paper. On attempting to visit HELMSMAN [Earl Browder] he was not admitted to him by the latter's secretary. ANTON [Leonid Kvasnikov].

The fact that these 17 physicists were developing an atomic weapon was one of the biggest secrets of the war in December 1944, so the accurate list demonstrated the scale of the leak. Dr. George Kistiakowsky was the chief of the Explosives Division at Los Alamos; the American chemist Harold Urey headed the uranium production center in New Jersey; Lawrence directed uranium production at Berkeley by the electromagnetic method; Segrè was Fermi's protégé from Italy; Bethe directed the Theoretical Division at Los

Alamos, where Bohr joined him after his escape from Denmark, and where the Hungarians Edward Teller and John von Neumann were working on an implosion trigger at Los Alamos; Rossi, a close friend of Fermi's, had pioneered cosmic ray research; Percy Bridgeman was the Harvard expert on the measurement of high pressures; Sir Geoffrey Taylor was the Cambridge-educated meteorologist and leading British specialist on explosive hydrodynamics; Professor Compton was in charge of the Argonne Laboratories in Chicago and Oak Ridge, Tennessee, where uranium was refined by the spectrographic method; Hans Stanarn was, presumably, the Swiss physicist Hans Staub. Quite apart from the personalities involved, the sites where the Manhattan Project was undertaken were also considered top secret. Apart from the question of how the Soviets had acquired these names, the only real conundrum was the inclusion of Werner Heisenberg, who was at the time leading the Nazi's atomic research program in Germany, and Fritz Strassman, the chemist who had first identified transuranic elements and first split the nucleus of the atom. Neither was ever at Los Alamos, but doubtless their names had circulated among those working in the United States as likely competitors.

Evidently, the report, signed by the Amtorg engineer Leonid R. Kvasnikov, who had been based in New York between March 1943 and October 1945, meant that his source had approached Nicola Napoli, the manager of a Soviet film distributor for help, who had been embarrassed and had sent him to Sergei N. Kurnakov (codenamed BECK). Instead, he had called on the general secretary of the Communist Party of the United States of America, Earl Browder, who had refused to see him, doubtless anxious not to be so obviously compromised by espionage.

The unrecovered groups in ANTON's message were tantalizing, but the clues were followed up by additional work on the preceding telegram, which referred to a source codenamed MLAD, who was discovered to be Dr. Theodore Hall, the Harvard-educated physicist. This disclosure alone was sufficient to justify the most searching investigation into breaches of security at Los Alamos, but VENONA was to reveal much, much more about a joint **GRU**/NKVD operation codenamed ENORMOZ.

Although the discovery of the duplication of one-time pads looked like a breakthrough, there was a limit to the cryptographic success that it implied, and this was recognized at the outset by Meredith Gardner and his team. Almost an obsessive, and certainly an eccentric, Gardner was undoubtedly a linguistic genius. As one of the SSA team assigned to the Russian problem, he knew that progress would be very slow without the aid of a "crib" or shortcut, and this was the issue he raised with the Federal Bureau of Investigation's liaison officer, Special Agent Robert J. Lamphere.

JAPANESE SIGNALS INTELLIGENCE. Prior to **World War II**, the Japanese hardly invested in signals-intelligence research; although, the light cruiser *Yuhama* was deployed to monitor American radio traffic during fleet exercises conducted by the U.S. Navy off Hawaii in 1923. In 1932, a tanker, the SS *Erimo*, undertook much the same kind of collection mission, and from 1937, foreign wireless traffic was intercepted at Owada, just outside Tokyo. Later, two sites were established at Kita-tama in Tanashi and at the Nakano Communications School, and an overseas facility was built at Jaluit in the Marshall Islands.

Although the Japanese would prove highly proficient at the tactical **interception** of voice channels, their cryptographers were unable to read any American cipher and believed that the complexity of their own language would make their own systems secure. The Imperial Japanese Army relied on four-figure codes and super-encipherment using additive tables, which were changed regularly. Like the Imperial Navy, the army adopted geographic code areas so, if one code was compromised, the impact would be localized. The disadvantage of this procedure was the necessity to distribute thousands of new codebooks throughout a command whenever the loss of a ship or the capture of a position suggested an additive table had been captured.

On 29 January 1943, a pair of Bird-class minesweepers, HMNZS *Kiwi* and *Moa*, rammed the Japanese submarine *I-1* onto a reef in Kamimbo Bay, near the northwest tip of Guadalcanal, and recovered some 200,000 codebooks and additive tables, which were passed on to the **Central Bureau** in Melbourne. When the Japanese learned of the incident, planes were sent to bomb the submarine and destroy the crypto material, but they were too late. In similar incidents, a Japanese submarine was surprised on the surface at Salamaua in New Guinea as it made a delivery of codebooks, and another jettisoned a cargo of cipher material when it experienced engine failure between Truk and Rabaul. The production and distribution of new ciphers was the responsibility of the Imperial Japanese Navy's 10th Section of the 4th (Communications) Department, which printed codebooks at the Bunjido printing works in Yokohama. Altogether some 75 Japanese naval codes were broken by Allied cryptanalysts during **World War II**, including the stereotype weather codes in constant use across the Pacific. *See also* JN-25; JN-36; JN-40; LAUNG-YAUNG; OSHIMA, HIROSHI; PURPLE; WIRELESS EXPERIMENTAL STATION (WES).

JELLYFISH. The **Government Communications Headquarters** codename for line-of-sight radio links, the Michael-gerät and Rudolf-gerät, manufactured by Telefunken, which transmitted a low-power beam containing voice and teleprinter channels over distances of up to 100 kilometers. The traffic was hard to intercept as there was no ionospheric reflection, and it was protected by an extra encryption system introduced in February 1944. The principal JELLYFISH channels terminated at two controls, at Berlin for the Western Front, codenamed STRAUSSBERG, and at Königsberg, codenamed ANNA, for the Russian Front. The traffic was read by **Bletchley Park** from 1 June 1944 after the development of the Mark II **Colossus**, an improved version of the original computer incorporating 2,400 valves.

JENISSEJ. The Soviet codename for a signals-intelligence facility on the top of the Brocken, the highest peak in the Harz Mountains in East Germany, JENISSEJ was established by the Soviets in April 1947 and was taken over by the East German **Ministerium für Staatssicherheit** five years later and codenamed URIAN.

JN-25. Broken in 1939 by John Tiltman, JN-25 was the Imperial Japanese Fleet's principal cipher system. *See also* JAPANESE SIGNALS INTELLIGENCE; JN-36; JN-40.

JN-36. Japanese Imperial Navy wireless ciphers were given numeric designations by cryptographers at **Bletchley Park**, and when JN-36 and JN-37 were broken, they proved to be weather codes. *See also* JAPANESE SIGNALS INTELLIGENCE; JN-25; JN-40.

JN-40. The **Government Code and Cypher School** designation of the Japanese Merchant Navy cipher, broken in early 1942. *See also* JAPANESE SIGNALS INTELLIGENCE; JN-25; JN-36.

JU-JITSU. A Royal Air Force (RAF) codename for a series of clandestine signals-intelligence and photoreconnaissance overflights conducted from RAF Sculthorpe, Norfolk, by a newly formed Special Duties Flight between April and October 1952, JU-JITSU was repeated in April 1954. These missions were made by British aircrew, flying ostensibly British aircraft because, in June 1950, President Harry S. Truman had banned all incursions into Soviet airspace so as to avoid provoking the Kremlin. The restriction had been imposed when a U.S. Navy PB4Y-2 Privateer, a reconnaissance variant of the B-24 Liberator, on a mission from Wiesbaden to Copenhagen, had been shot down in April 1950 by Soviet La-11 fighters off Latvia with the loss of all 10 crew. Truman's ban lasted throughout the Korean War, so Strategic Air Command would not be able to make the overflights needed to monitor the Soviet air-defense radars, requiring the assignment to be passed to the RAF.

JU-JITSU aircrew, led by Squadron Leader John Crampton, a Canberra pilot from 101 Squadron, were trained on U.S. Air Force RC-45C Tornados at Barksdale Air Force base in Louisiana and then at Lockbourne Air Force base near Columbus, Ohio, and four of the aircraft were repainted into RAF livery and flown down the air corridor to Berlin to test Soviet reactions. The first JU-JITSU mission, authorized by Prime Minister Winston Churchill, was flown over around 30 targets in the Soviet Union at 36,000 feet on 17 April 1952 with a duration of 10 hours and was completed without incident, having been refueled in Danish airspace by KB-29 tankers, also from Sculthorpe. The flights were resumed on 28 April 1954, when three Tornados penetrated Soviet airspace at night to collect **radar intelligence**, with Crampton flying to Kiev before returning to a **North Atlantic Treaty Organization** base in Germany to be refueled. The other pair also completed their missions, one over the Baltic states and the other to Moscow, without incident.

Altogether, 33 RC-35Cs were built, and the planes at Sculthorpe were assigned to the 91st Strategic Reconnaissance Wing. Although the Tornado had proved vulnerable to MiG-15s in Korea, the Soviet fighters were not equipped with airborne radar, so the Tornados were relatively safe from **interception** at night when monitoring Soviet ground stations. *See also* WORLD WAR II.

K

KACZMARZYK, ADAM. A British **Secret Intelligence Service (SIS)** source in Poland, Adam Kaczmarzyk was a "walk-in" at the British embassy in Warsaw who identified himself as a cipher clerk working at the Warsaw Pact's headquarters willing to spy for the British. The local SIS station assigned Gerald Warner to handle the source, codenamed BENEFICIARY, who was motivated by money, which he spent on his girlfriend and a new car. He was compromised when he attempted to recruit a colleague and was arrested in August 1967 when SIS held a rendezvous with Kaczmarzyk's nominee. A member of the SIS station was expelled, along with his secretary, and BENEFICIARY was executed by a firing squad after a trial that lasted four days.

KAGNEW STATION. The **National Security Agency** intercept site in Ethiopia, Kagnew Station remained operational from 1943 until May 1977, when it was evacuated following a revolution in Addis Ababa.

KAMER 14. During **World War II**, a Dutch cryptographic organization worked on intercepted Japanese cipher traffic at the Bandung Technical College.

KAMISEYA. One of the U.S. **National Security Agency**'s largest overseas sites, Kamiseya was accommodated in the underground tunnels of a former Japanese **World War II** torpedo storage facility and processed traffic intercepted from China and the Soviet Union.

KEY. Allied cryptographers working on **Enigma** traffic during **World War II** referred to the individual machine settings as the key. The settings, or starting points, consisted of the device's four variables, being the selection of three rotors from the eight available; the order in which the three were inserted into the machine, the precise positions, from a choice of 26, for each of the rotors; and the wiring of the plugboard.

KOLBE, FRITZ. Born in Berlin in September 1900, Fritz Kolbe was an anti-Nazi member of the Reich's foreign ministry who traveled regularly to Switzerland as a diplomatic courier from August 1943. On his visit to Bern on 18 August, Kolbe made contact with Allen Dulles of the U.S. Office of Strategic Services (OSS) and began to supply him with copies of original foreign ministry telegrams. In his first batch were dispatches from the German ambassadors Franz von Papen in Ankara, Otto Abetz in Paris, and Konstantin von Neurath in Prague, and, in addition, Kolbe volunteered information about the Foreign Ministry's **one-time pad** cipher system, revealed that the German legation in Dublin contained a secret wireless transmitter that had not been declared to the Irish authorities, and claimed that some American military codes had been broken by German cryptanalysts.

Initially unconvinced that Kolbe was not an agent provocateur, Dulles received Kolbe on two further occasions during 1943, twice in 1944, and twice in 1945. Altogether, he supplied some 1,600 copies of correspondence exchanged between Berlin and German diplomatic missions in 30 different countries. Finally satisfied that Kolbe was authentic, Dulles codenamed him GEORGE WOOD and forwarded his material to Washington, DC, where it was judged authentic.

Although Kolbe's unexpected windfall might have been welcomed by the Allied signals-intelligence community, it actually caused dismay because exploitation of **Enigma**, **Geheimschreiber**, and **FLORADORA** decrypts had already made the GEORGE WOOD documents available, and the fear was that if Kolbe was caught attempting to smuggle telegrams into Switzerland his captors would quickly conclude that his previous deliveries would have made the Foreign Ministry's confidential communications channels vulnerable to cryptographic attack. Accordingly, Dulles received no encouragement to continue his relationship with Kolbe; although, his geographic isolation, surrounded by the Axis, prevented him from being given the true reason until he was able to travel to Paris to be briefed by OSS in August 1944. *See also* GERMAN SIGNALS INTELLIGENCE.

L

LATERAL. A transmitter communicating directly with another outstation in a radio net, instead of broadcasting through a central control or "star," is known to traffic analysts as a lateral. The term also included "sideways" traffic between neighbors where the usual routine was "forward" or to the "rear" in an order of battle. During **World War II**, German traffic was organized into "Netz" (network), where the outstations used their own individual callsigns and frequencies; "Kreis" (circle), with outstations sharing the same callsigns and frequencies; and "Stern" (star) systems, in which outstations could communicate laterally. The German headquarters network controlled eight armies and four Panzergruppen, the latter being organized into three groups, being Northern (Latvia), Central (North Poland), and Southern (Rumania).

LAUNGYAUNG. During the unsuccessful Japanese offensive codenamed Operation Z, which was launched in Burma in February 1944, with the objective of seizing Chittagong and invading India, the Japanese 55th Infantry Division, led by General Sakurai, overran the site of General Frank Messervy's 7th Indian Division headquarters, forcing his hasty retreat from Laungyaung. Although it was later claimed that all the secret documents and codebooks had been destroyed in time, the Japanese recovered a copy of Sakurai's complete battle plan, together with details of individual units, the names of their commanding officers, and their objectives. This information had been compiled from MONGOOSE and LEOPARD intercepts; although, when the official history of the campaign in Burma was published in 1968, it was alleged that compromising papers had been found in the possession of a dead Japanese officer.

Despite the Japanese discovery, Operation Z was defeated because of advance knowledge of the enemy's intentions, and the decisive battle proved a turning point for the war in Burma. *See also* JAPANESE SIGNALS INTELLIGENCE.

LEE, ANDREW D. Responsible for selling manuals relating to the **RHYOLITE** and **ARGUS** signals-intelligence satellite systems to the KGB *rezidentura* in Mexico City, Lee was a California drug dealer who obtained the classified material from his friend **Christopher Boyce**, who was employed by a defense contractor, TRW Inc. Lee was detained by police in Mexico City in January 1977 and then returned to the United States where in May he was sentenced to life imprisonment, having been convicted on eight counts of espionage.

***LIBERTY*, USS.** A U.S. Navy signals-intelligence collection platform, the USS *Liberty* was attacked by Israeli aircraft and gunboats in the Mediterranean off the coast of Sinai on 8 June 1967 with the loss of 34 crewmen killed.

Commissioned in December 1964 with the designation Auxiliary General Environmental Research (AGER), the *Liberty* was controlled by the **National Security Agency (NSA)** and monitored Egyptian air-force voice channels for any transmissions in Russian, two specialist linguists having joined the crew at Rota for that purpose. Commanded by William McGonagle, the clearly marked and flagged *Liberty* had been stationed in international waters 25 miles north of Al Arish when the Six Day War broke out. The *Liberty* was strafed and bombed with napalm by two pairs of Mirage and Super Mystère jets, before being attacked by three torpedo boats. Holed below the waterline, and with 171 wounded personnel, Commander McGonagle sailed his ship to Malta.

Israel later claimed that the attack had been a case of mistaken identity, insisting that the *Liberty* had been erroneously reported to be an Egyptian troopship, the *El Quseir*, and paid compensation of $13 million to the U.S. Navy, the survivors, and the dependents of those killed. Several subsequent investigations uncovered no evidence that the episode was anything other than a tragic accident; although, the U.S. Navy's official court of inquiry was conducted in conditions of such secrecy, and its subsequent report so highly classified, that there was considerable speculation about the possibility that the attack had been a deliberate attempt to prevent the NSA from collecting information about the Sinai campaign that might have been damaging to Israel's interests.

A comprehensive review conducted at the request of President Lyndon B. Johnson by Clark Clifford rejected the conspiracy theories by pointing out that the *Liberty* had no Hebrew linguists aboard and, therefore, had not the capability to monitor Israeli channels. He examined signals-intelligence and voice-channel intercepts, which revealed considerable confusion on the part of the Israelis engaged in the attack, and concluded that the incident had been "a flagrant act of gross negligence" but not "a deliberate act."

LONDON SIGNAL INTELLIGENCE BOARD (LSIB). Created in 1943, the LSIB acted as the chief authority for the supervision of all **British sig-**

nals-intelligence operations and consisted of representatives from all the interested parties, including the **Government Code and Cypher School** (later **Government Communications Headquarters**), the Foreign Office, the Admiralty, the War Office, and the Air Ministry. LSIB's role was to advise the British and Dominion governments on SIGINT policy, conduct negotiations with Allied partners, and enforce agreements entered into with other SIGINT authorities, such as the **BRUSA** and **UKUSA** Agreements.

LONG, LEO. A Cambridge graduate, Leo Long worked during **World War II** as an intelligence analyst in the War Office military-intelligence directorate branch MI14 studying **Enigma** intercepts to develop an accurate enemy order of battle. A university friend of Anthony Blunt, Long was recruited by him as a Soviet spy and relied on him as a conduit to pass decrypts and other highly secret signals-intelligence material to the **NKVD**.

After the war, Long transferred to the Intelligence Division of the British Control Commission for Germany and, having failed to join **MI5**, pursued a career in the movie business in London. Identified as one of his sources by Blunt in April 1964, Long was given an assurance, in return for a detailed confession, that he would not be prosecuted.

LONGFELLOW. The **Government Communications Headquarters** codename for the principal Soviet multirotor cipher machine employed by the Red Army until 1948 and read as part of the **BOURBON** project. *See also* COLERIDGE.

LOOKING GLASS. The **National Security Agency** codename for an airborne collection program based at RAF Mildenhall in Suffolk during the 1970s over the Baltic Sea, LOOKING GLASS flights were undertaken by RC-135s, the **electronic-intelligence** variant of the Boeing 707.

M

MAGIC. The American codename for decrypts recovered from Japanese diplomatic traffic encrypted on the **PURPLE** machine. *See also* UNITED STATES SIGNALS INTELLIGENCE.

MAGNUM. First launched from the Space Shuttle in 1985, the MAGNUM satellite system was a signals-intelligence collection platform constructed around a 160-foot parabolic dish. Partly funded by **Government Communications Headquarters**, MAGNUM succeeded the **RHYOLITE** system and replaced the British **ZIRCON** satellite that was scrapped on grounds of cost.

A second MAGNUM was launched on 23 November 1989, and a third was placed in orbit from the Space Shuttle on 14 November 1990. *See also* AIRBORNE COLLECTION PLATFORMS; BYEMAN; VORTEX.

MANHATTAN PROJECT. Clear proof that the Manhattan Project had become a target of Soviet espionage was revealed in December 1946 when a message in the New York **VENONA** traffic dated two years earlier, in December 1944, disclosed the names of 17 secret personalities working on the Anglo-American atomic-weapons program at Los Alamos. Over the following four years, **Arlington Hall** cryptographers uncovered further evidence suggesting that all five other major Manhattan Project sites in the United States and Canada, at Hanford, Oak Ridge, Berkeley, Chicago, and Chalk River, had suffered hostile penetration. Analysis of the VENONA texts indicated that an **NKVD** operation codenamed ENORMOZ and headed by Leonid Kvasnikov, based at an unusual, special *rezidentura* accommodated in the Amtorg building, had succeeded in acquiring some key sources inside those secret installations. When, in 1949, Klaus Fuchs was eventually identified as one of the principal leaks and helped lead the Federal Bureau of Investigation to Harry Gold, the remainder of the spy ring began to unravel. However, convictions obtained against the main players, including Julius Rosenberg, failed to prevent the Soviets from successfully testing a uranium bomb in August 1949.

MARSHALL, WILLIAM M. Employed as a radio operator by the Foreign Office's **Diplomatic Wireless Service (DWS)** at the organization's headquarters at Hanslope Park, William Marshall was arrested in London in April 1952 and convicted of breaches of the Official Secrets Act.

Formerly a Royal Signals soldier with service at Sarafand and Cairo, Marshall joined the DWS in November 1948 and was posted to Moscow. However, having been spotted by an off-duty **MI5** watcher meeting a KGB officer, Pavel Kuznetsov, who was working under diplomatic cover, Marshall was arrested as both men met in a park in Wandsworth. A search produced a scrap of paper with information about Hanslope Park, and Marshall was tried at the Old Bailey and sentenced to five years' imprisonment.

MARTIN, WILLIAM H. A **National Security Agency (NSA)** cryptographer, William H. Martin and his friend **Bernon F. Mitchell** disappeared while making an unauthorized visit to Cuba in June 1960.

A chess champion, Martin had served at a **Naval Security Group** facility in Alaska during his service in the U.S. Navy, and then had been transferred to **Kamiseya** in Japan. In 1954, he had switched to the **Army Security Agency** as a mathematician and, in February 1958, had joined the Communist Party of the United States of America. Both men participated in a press conference upon their arrival in Moscow and denounced overflights by U-2 and C-130A aircraft.

MASIRA. This strategically important Omani island in the Persian Gulf accommodated a Royal Air Force air base and a **Composite Signals Organisation** intercept facility, monitoring traffic in the region.

MASK. The encrypted clandestine traffic exchanged between Moscow and London from January 1934, transmitted by Stephen J. Wheeton, a veteran member of the Communist Party of Great Britain (CPGB), and the resulting transcriptions were circulated by the **Government Code and Cypher School (GC&CS)** on a very limited basis to **MI5**, the head of one Secret Intelligence Service section, and senior members of the cabinet. For more than two years, MASK represented a unique window into the most secret decisions and deliberations of the CPGB and eloquently demonstrated that, far from being independent, the party was nothing more than a very pliant instrument of the Communist International, directed by the Kremlin, and that the CPGB's long-serving general secretary, Harry Pollitt, received his instructions directly from Moscow. According to MASK, Pollitt sought Moscow's guidance on virtually all policy issues and even the most trivial aspects of administering the party. In a more sinister vein, the CPGB acted as a recruitment center for trainees to undergo indoctrination and espionage courses at the Lenin and Wilson Schools and vetted candidates nominated by the shadowy Soviet contacts, who disguised their identities behind codenames.

GC&CS's monitoring station at Denmark Hill, Camberwell, headed by Commander Harold Kenworthy, first began intercepting Wheeton's signals in February 1934, and he continued to transmit until April 1935, when he fell ill with tuberculosis and was replaced by another veteran Communist, William Morrison. The cipher used was studied by a veteran codebreaker, Leslie Lambert, and together with a seasoned colleague, Brigadier John Tiltman, he succeeded in reading the traffic until January 1937. Through the use of **direction-finding** equipment located at the army intercept station at **Fort Bridgewoods**, outside Chatham, at the Air Ministry W/T Section at Waddington, headed by Wing-Commander Lywood, and the Royal Navy's receiver at **Flowerdown**, near Winchester, GC&CS's technicians were able to show that Moscow was also communicating with a Comintern station in Vienna, using the call sign 3PD, and others in Shanghai, Paris, Athens, 3OS and 9RP in Prague. More were traced to Spain, Basel, Zurich, Copenhagen, and the United States. Altogether 939 messages from Moscow were read and 634 from London, making a total of 1,573 decrypts altogether. The length of the individual messages varied from a couple of lines to several paragraphs, but they revealed the scale of the subsidies paid by the Kremlin to the CPGB and to the *Daily Worker* and also identified by name clandestine couriers and the Comintern's secret policies concerning the 1936 general election and the civil war in Spain.

MATA HARI. At her espionage trial held in camera in Paris in July 1917 before a military tribunal, the evidence against Margaretha MacLeod, who worked as an exotic dancer under the stage name Mata Hari ("Eye of the Dawn" in Hindi), consisted of intercepted German intelligence messages, which appeared to identify her as a spy codenamed H-21, who was to be paid 15,000 pesetas. The signals, dated between 13 December 1916 and 8 March 1917, had been monitored from a French station on the Eiffel Tower before being decrypted by the Deuxième Bureau, and testimony from her interrogator, Pierre Bouchardon, proving that she had been in contact with two German attachés in Madrid, sealed her fate, and she was executed at Vincennes by a firing squad on 15 October 1917.

Born in August 1876 in Leeuwarden, Holland, Margaretha Zelle had married a middle-aged Dutch officer, Rudolf MacLeod, in 1895, and lived with him at Banjoe-Biroe in Java. Having divorced her husband, she traveled to Paris in 1903 to perform and, by 1912, had achieved considerable fame, appearing at Milan's La Scala in *Bacchus and Gambrini*. While in Berlin, her lover, the local police chief, Traugott von Jagow, is alleged to have recruited her as an agent and introduced her to Colonel Walter Nicolai, the head of the German intelligence service.

MAX. The **Radio Security Service (RSS)** codename for signals originating on the Russian Front purporting to be the transmissions of a German spy operating behind the Soviet lines. According to RSS analysts, the MAX traffic originated from areas along the line from Leningrad to Rostov and Kerch through the Caucasus from the north; from Novosibirsk to Batumi from Georgia, Azerbaijan, and Armenia; from Iran, Baghdad, and Basra as well as Kuibyshev and Astrakhan; and from the western side of the Caspian Sea.

The MAX traffic, closely associated with **MORITZ**, proved one of the enduring mysteries of **World War II** as it was never clear whether MAX was actually a double agent working under the control of the **NKVD**. If so, much of his information appeared to be authentic and seemed a heavy price to pay for establishing his bona fides with the enemy.

Extensive investigations at the end of the war conducted by several Allied intelligence agencies established that MAX had been handled by an Austrian Abwehr officer, Richard Kauder, codenamed KLATT, who had been based in Bucharest and then Sofia and claimed to be in contact with an extensive White Russian network headed by General Anton Turkul. Suspected of being a double agent, Kauder was arrested by the Gestapo in February 1945 but later liberated by the U.S. Army. Under interrogation by the U.S. Counter-Intelligence Corps, Kauder acknowledged that he had always known MAX and MORITZ had operated under Soviet control. During the quadripartite occupation of Vienna, Kauder was abducted by the **NKVD**, but his fate and his true loyalties remain a mystery.

MEASUREMENT AND SIGNATURE INTELLIGENCE (MASINT). A highly sophisticated, technical intelligence source, MASINT is the clandestine collection of data, often remotely through the use of sensors, that enables analysts to maintain surveillance on sensitive targets in denied areas. MASINT developed during the **Vietnam War** when javelin-like acoustic and seismic devices were dropped from aircraft into territory held by the Vietcong to record vehicle movements and tap into landlines. Codenamed IGLOO WHITE, one program involved the insertion of thousands of sensors in Laos to check on activity on the Ho Chi Minh trail.

In 1978, two atomic-powered MASINT sensors were found on Nanoa Diva in the Himalayas by the Indian authorities, who concluded that the apparatus had been designed to monitor Chinese nuclear tests, and another, which collected details of trucks driving in and out of a military installation, was recovered in 1985 from a hiding place buried in East Germany. A device codenamed CERVICAL RUB, disguised to look like a tree stump, was deployed to monitor aircraft movements at a nearby airfield. The information was then transmitted in a burst signal to a HERITAGE satellite via a small antenna that extended above the surface. The device's active life was entirely dependent on the power source, making the battery-operated versions disposable and very hard to detect.

MENWITH HILL. The largest **National Security Agency** facility outside the United States, RAF Menwith Hill, near Harrogate in Yorkshire, accommodated a detachment of the U.S. **Army Security Agency** in 1954 engaged in intercept operations on 545 acres of moorland covered with antenna arrays. Since then, the base has expanded and has a large number of radomes to service satellite ground links. In 1984, a major extension, codenamed STEEPLEBUSH, was added to the operations building to house new **satellite communications** equipment.

MEXICO CITY VENONA. A total of 81 **NKVD** telegrams were sent from Mexico City to Moscow, spanning December 1943 to November 1944, of which most were addressed to Pavel Fitin from the *rezident*, Lev Tarasov, but with nine sent by him to Lavrenti Beria. The other channel, from Moscow to Mexico, consisted of 127 texts, of which most were sent in Fitin's name, but Beria signed six. When studying the telegrams, the Federal Bureau of Investigation counterintelligence analysts were struck by their serial numbers. The NKVD had adopted a system of beginning each year afresh with the figure 001, and the three texts for December 1943 on the Mexico channel gave the serial 129 on 6 December and 181 on the last day of the month, suggesting that a total of about 50 had been sent to Moscow during the month. That in turn implied a considerable escalation if only 120 odd had been sent during the remainder of the year. On the reverse of the circuit, from Moscow to Mexico, the serials were 708 on 8 December 1944, 559 on 7 September 1945, and 269 on 22 May 1946, extrapolating a considerable decrease in the traffic at the end of the war.

MI1(b). Under Lord Kitchener's 1915 reorganization of the War Office, a new military-intelligence directorate included a signals-intelligence unit designated MI1(b), headed by Colonel Adrian Simpson with an initial staff of two German scholars. Simpson established a General Headquarters cipher bureau at Le Touquet to analyze material collected from intercepted enemy landline communications. Within a year, MI1(b) had built **direction-finding** stations at Leiston in Suffolk and Devizes in Wiltshire (with a control facility on the roof of the War Office in London), and Major Malcolm Hay, a noted historian and recently released prisoner of war, recruited into the Intelligence Corps a team of academics at 5 Cork Street, Mayfair, to process decrypts provided by a group of cryptographers designated **MI1(e)** and led by Brigadier Anderson. By 1917, 43 of MI1(b)'s 50 cryptanalysts were working on diplomatic traffic and were reading Argentine, Brazilian, Danish, Dutch, Italian, Japanese, Norwegian, Persian, Swedish, and Uruguayan communications, as well as the Vatican's diplomatic telegrams.

In 1919, MI1(b) and MI1(e) were amalgamated with **NID 25** to create the **Government Code and Cypher School**. By the end of hostilities, MI1(b) employed 70 personnel, a figure that later increased to 100 during the Versailles negotiations. In total, MI1(b) solved 52 diplomatic codes, 48 enemy trench codes, and some 700 individual cipher keys. *See also* BRITISH SIGNALS INTELLIGENCE.

MI1(e). Created in July 1916 and staffed with four cryptographers from MI1(b), the MI1(e) subsection was headed by Brigadier Anderson and **was** dedicated to monitoring **zeppelin** movements and was linked to a plotting room in Room 417 at the War Office. In 1917, MI1(e) was amalgamated into **NID 25**.

MI5. The British Security Service first began exploiting signals intelligence when, in March 1934, an agent inside the Communist Party of Great Britain provided the key, codenamed **MASK**, that gave **Government Code and Cypher School** cryptanalysts access to the clandestine wireless traffic exchanged with Moscow. Although this traffic dried up in 1937, the organization fully recognized the potential value of intercepted intelligence traffic, as opposed to the diplomatic traffic already in circulation as **BLACK JUMBOs**. During **World War II**, MI5, in conjunction with the **Radio Security Service**, came to rely heavily on **ISK** and **ISOS** to provide advance notice of enemy spies destined to undertake missions in the British Empire and, in the postwar era, exploited **VENONA** to identify traitors, among them Klaus Fuchs in 1949 and the X Group membership, including Ivor Montagu, J. B. S. Haldane, and the **GRU** spy codenamed BARON.

During the **Cold War**, MI5 mole hunters pursued Donald Maclean through leads provided by the VENONA decrypts but were only able to find compromising references to Ursula Kuczynski, Kim Philby, and Melita Norwood when it was too late to take action against them. The organization was also preoccupied with the protection of SIGINT secrets and investigated **John Cairncross** for having breached security at **Bletchley Park**, **Douglas Britten** (who was arrested at RAF Digby), and **Geoffrey Prime**. *See also* BRITISH SIGNALS INTELLIGENCE.

MI8. The British military-intelligence designation of the War Office branch responsible for signals intelligence, MI8 was created in 1914 and headed by Colonel Arthur Churchill with a staff of 25 clerks and 40 typists. It was divided into four subsections, being MI8(a), dealing with wireless policy; MI8(b), based at the **General Post Office**, handling commercial and trade cables; MI8(c) distributing intelligence derived from censorship; and MI8(d) liaising with the cable companies. Once **World War I** was declared, MI8 representatives were sent to the Marconi stations at Poldhu Point, at Mullion in Cornwall, and at Clifden in County Galway, to monitor the traffic, and eventually both sites came under Admiralty control, with Poldhu closing in August 1917.

In 1939, MI8 remained the War Office designation for the Military Intelligence Directorate's signals-intelligence organization, which would include the **War Office Y Group** and the **Radio Security Service**, MI8(c). During **World War II**, MI8 would retain direct control over three intercept sites, in the basement of Queen Anne's Mansions, designated Y-3 and linked to a satellite site in nearby Caxton Street, which later moved to Devonshire House, and Y-2 at **Harpenden**. *See also* BRITISH SIGNALS INTELLIGENCE.

MINARET. The **National Security Agency** codename for the targeted **interception** of telephone traffic involving domestic suspects, including civil-rights activists, political extremists, and antiwar protestors identified by the Federal Bureau of Investigation. Because interception did not occur within the continental United States but was conducted from British-controlled facilities, the program was not in breach of the ban on domestic interception as defined in 1934 by the Federal Communications Act. The names of some 1,600 U.S. citizens were placed on a watch list, which generated approximately 75,000 individual personal files. *See also* SHAMROCK.

MINISTERIUM FÜR STAATSSICHERHEIT (MfS). The East German Ministry of State Security included a significant signals-intelligence branch from 1953, when the Soviets allowed the new Communist state a degree of autonomy in the security sphere. Designated S/3, the MfS's *funk* unit fulfilled an internal monitoring role based at Gosen, in a Berlin suburb, and at Hohen-Luckow, with a staff that grew to 1,095 when, in 1973, it was amalgamated with an embryonic SIGINT-collection department designated Hauptabteilung III, which, from June 1971, had been run by Horst Männchen, a Stasi officer who had begun his career as an MfS radio operator. When the German Democratic Republic collapsed, the MfS's Third Main Directorate employed a staff of 2,361 and operated from 80 regional installations across the country with a headquarters concealed as the Institute for Technical Developments in the Ministry for Science and Technology building at Berlin-Wuhlheide. As well as monitoring domestic communications, the MfS ran telephone **interception** operations in West Berlin and reportedly tapped up to 2,300 lines serving U.S. facilities.

The MfS also conducted operations overseas from diplomatic missions and adopted military cover, even though the East German army, the Nationale Volksarmee, possessed its own SIGINT capability, the Verwaltung Aufklärung, with some 1,000 personnel and its own intercept sites, such as the installation codenamed URIAN atop the Brocken Mountain and a satellite ground station at Biesenthal. *See also* GERMAN SIGNALS INTELLIGENCE; JENISSEJ; ROYAL PALM.

MISAWA. One of the largest **National Security Agency** overseas facilities, located on the north coast of the Japanese island of Honshu, Misawa included an airfield and conducted signals-intelligence collection operations against the Soviet Union and China. *See also* AIR INTELLIGENCE AGENCY (AIA); SATELLITE COMMUNICATIONS.

MITCHELL, BERNON F. A **National Security Agency (NSA)** cryptographer, Bernon Mitchell and his friend **William Martin** disappeared while visiting Cuba in June 1960. They had both joined the NSA in July 1957, having served in the U.S. Navy together. After his naval service, Mitchell had graduated from Stanford University. After training, they had been assigned to the NSA's Office of Research and Development where, in September 1959, they learned that a U.S. Air Force C-130A, shot down over Armenia a year earlier, had been a **SUN VALLEY airborne collection platform**. Dismayed by their discovery, they had approached an Ohio congressman, Wayne Hays, but he had taken the matter no further.

Both men were alleged to be homosexuals and had visited Cuba in December 1959 together, where it was presumed that they had made contact with the Soviets, but the statement they left, in a bank in Laurel, Maryland, insisted they were acting independently. Once in Moscow, they denounced the NSA publicly, claiming that the organization had over "2,000 manual intercept positions" and was reading the wireless traffic of more than 40 countries, including "Italy, Turkey, France, Yugoslavia, the United Arab Republic, Indonesia and Uruguay."

After their arrival in the Soviet Union, Mitchell and Martin lived in Leningrad under the protection of the KGB, and both married Russian women, Galya and Valya, respectively. Mitchell, who had been born in San Francisco in 1929, worked on computers at the Bonch and Bruevich Electro-Mechanical Institute, and his wife, Galya Vladimirova, to whom he remained married for nearly 40 years, was his music teacher and a senior piano tutor at the Leningrad Conservatoire. In 1979, they applied to the KGB for permission to immigrate to the United States but were refused permission.

MONOPOLY. The codename of a tunnel dug from a townhouse under the Soviet embassy compound at Mount Alto in northwest Washington, DC, to collect technical intelligence from covert sensors and listening equipment. The operation was betrayed to the KGB by Robert Hanssen, a senior Federal Bureau of Investigation counterintelligence expert, prior to his arrest in February 2001.

MONTES, ANA B. The lead Defense Intelligence Agency (DIA) analyst on Cuban operations, Ana Montes had been recruited by the Cuban Dirección General de Inteligencia (DGI) before she even joined the DIA at Boling Air Force Base from the Department of Justice in September 1985. Of Puerto Rican descent and born to a U.S. Army doctor in February 1957, Montes spied for the DGI over a period of 16 years before she was arrested in September 2001 by the Federal Bureau of Investigation and charged with espionage. As part of her duties, she regularly attended highly classified interagency intelligence conferences concerning the Cuban target and was indoctrinated into the most sensitive **National Security Agency** programs, details of which she promptly passed to her DGI handlers. In a plea bargain in October 2001, Montes was sentenced to 25 years' imprisonment, to be followed by five years' probation, and is presently at Carswell, a federal prison in Fort Worth, Texas. A damage assessment completed by the U.S. Counterintelligence Executive remains classified.

MOON SHOT. Prior to the introduction of **satellite communications**, the **National Security Agency** experimented with a microwave system designed to link its **naval collection platforms** to receiving ground stations by bouncing a signal off the surface of the moon. Introduced on the USS *Oxford* in February 1964 and known as Technical Research Ship Special Communications, the channel avoided the risk of hostile **interception** as the microwave was a direct, line-of-sight transmission.

MORITZ. Closely associated with the **MAX** traffic, MORITZ was the **Radio Security Service (RSS)** codename for a clandestine transmitter ostensibly operating from Egypt, Libya, Palestine, Syria, Cairo, Iran, and Iraq. RSS analysts suspected both MAX and MORITZ were double agents under **NKVD** control, especially when they continued to broadcast after warnings about them had been sent to Moscow.

MORWENSTOW. Part of the **Composite Signals Organisation**, the satellite ground station at Morwenstow, near Bude in north Cornwall, is an intercept facility controlled by Great Britain's **Government Communications Headquarters**.

MUMBAI ATTACK. In November 2008, 10 heavily armed terrorists from Pakistan mounted a coordinated attack on hotels, cafes, a religious center, and the railway station in Mumbai, killing 172 people. During the attack, the Indian authorities disconnected local cell-phone communications and monitored satellite-phone conversations conducted by gunmen in the Taj Mahal and Oberoi Hotels, establishing that they were Kashmiri separatists being directed by Lashkar-e-Taiba, a notorious group of extremists banned in India and Pakistan. Interrogation of Mohammed Ajmal Kasab, the sole surviving terrorist, revealed that the Islamist militants had been equipped with GPS receivers and satphones in anticipation of the cell-phone network's suspension. Nevertheless, their reliance on the satellite system enabled their exchanges to be monitored and recorded, providing absolute proof of the terrorists' origins.

MUSTARD. The **War Office Y Group** designation of the **Enigma** channel used by its German counterpart, the **Chiffrierabteilung**, was codenamed MUSTARD and read daily continuously throughout the war from 27 June 1941.

N

NATIONAL RECONNAISSANCE OFFICE (NRO). Created in September 1961 to manage classified United States satellite programs, the NRO's existence was not publicly acknowledged until 1992, and its exact location in Westfield, Fairfax County, Virginia, is unpublicized. Hitherto, the NRO had been based mainly in California, with a small suite of offices in the Pentagon, but the construction of its ostentatious new headquarters building, with the attendant budgetary issues, attracted Congressional criticism and the early departure of its director. As well as controlling systems designed to collect imagery, the NRO supervised the **RHYOLITE**, **ARGUS**, and **MAGNUM** signals-intelligence platforms.

The NRO's first director was Joseph Charyk (1961–63), who was succeeded by Brockway McMillan, a Massachusetts Institute of Technology mathematician, who was dismissed in 1965. Other directors have been Alexander Flax (1965–69), John McLucas (1969–73), James W. Plummer (1973–76), Thomas Reed (1976–77), Hans Mark (1977–79), Robert Hermann (1979–81), Edward Aldridge (1981–88), Jimmie Hill (1988–89, 1993–94), Martin Faga (1989–93), Jeffrey Harris (1994–96), Keith Hall (1996–2001), Peter Teets (2002–5), and Donald Kerr (2005–present). In 1998, the NRO was amalgamated with the National Imagery and Mapping Agency. *See also* AIRBORNE COLLECTION PLATFORMS; CHALET; UNITED STATES SIGNALS INTELLIGENCE; VORTEX.

NATIONAL SECURITY AGENCY (NSA). Created in October 1952 from the **Armed Forces Security Agency**, which was itself the successor to the **Signal Security Agency** that had conducted **interception** operations during **World War II**, the NSA is based at Fort George G. Meade, Maryland, and is responsible for conducting signals-intelligence operations for the U.S. government.

The NSA was considered so secret, until its charter was published in 1984, that it was often referred to as "No Such Agency" or "Never Say Anything," even though it was the largest federal employer. The NSA runs its own intercept operations and processes raw intelligence collected by the U.S. **Naval**

Security Group, the Army Intelligence and Security Command, and the **Air Force Intelligence, Surveillance and Reconnaissance Agency**.

In addition to Fort Meade, the NSA operates from regional sites in the United States, at the Medina Annex on Lackland Air Force Base, San Antonio, and at Fort Gordon, Georgia. Overseas, NSA intercept sites shared facilities with the British and the Australians and managed independent facilities, either under semitransparent military or civilian communications cover, or from diplomatic premises. Because of intense secrecy surrounding the NSA, details of individual contributions to specific events are hard to identify, but the absence of a SIGINT-collection capability in Tehran is considered to have been a factor in the embassy's inability to monitor the Shah's communications and anticipate the collapse of his regime in 1979. Conversely, in April 1983, while reading the Iranian foreign-ministry traffic to and from Lebanon and Syria, the NSA found a reference to the payment of $25,000 linked to the recent devastating truck-bomb attack on the U.S. embassy in Beirut. The interception of President Hosni Mubarak's private telephone conversations disclosed sufficient information, in the form of 10 separate transcripts, in October 1985, to have the EgyptAir Boeing 737 flying the four *Achille Lauro* Palestinian terrorists from the al-Maza air base in Cairo to Algiers intercepted by four F-4s off a carrier, the USS *Saratoga*, over the Mediterranean and escorted to a **North Atlantic Treaty Organization** base in Sicily. Similarly, accurate decryption of 388 Libyan diplomatic messages in April 1986 confirmed that the bombing of the La Belle nightclub in West Berlin was a murderous atrocity committed by Libyan intelligence officers acting on direct instructions from their headquarters in Tripoli.

For understandable reasons concerning the NSA's sources and methods, such coups are neither celebrated nor even hinted at outside the confines of Fort Meade; although, in the German case, an exception was made, and President Ronald Reagan, during a television address broadcast to justify a U.S. Air Force raid on Tripoli and Benghazi, made explicit reference to three Libyan intercepts implicating Tripoli in the explosion, which had killed an American soldier and injured 50 others.

The NSA's internal structure is organized into five directorates: Operations; Technology and Systems; Information Systems Security; Plans, Policy, and Programs; and Support Services. The Directorate of Operations is subdivided into three groups, A Group conducts SIGINT operation across Western, Central, and Eastern Europe, concentrating on Russia and the former Soviet republics; B Group covers Asia, with a focus on North Korea, the People's Republic of China, and Vietnam; and G Group covers the rest of the world. C Group handles policy and resources; D Group liaises with the rest of the U.S. intelligence community through the Office of Director

NSA (DIRNSA); E Group deals with acquisition issues and engineering management; I Group supervises information security programs; J Group is responsible for legislative affairs; K Group is cryptologic research; L Group provides logistical support; M Group is administrative; N Group runs programs for determining mission priorities; P Group distributes finished intelligence; Q Group advises the DIRNSA on policy and plans; R Group has responsibility for research and engineering; S Group enforces standards and conducts security compliance inspections; T Group is the Telecommunications unit that manages the secure networks; U Group is the Office of General Counsel, overseeing legal issues; V Group enforces industrial security; W Group manages aerospace and satellite programs; and X Group controls participation in special-access programs.

Within the U.S. intelligence community, the NSA continues to be the dominant structure in terms of budget, personnel, and product—acquiring, processing, and distributing SIGINT to a limited group of policymakers and others authorized to receive it. NSA decrypts influenced military planners during the **Vietnam War** and played a significant role in locating a largely unseen enemy that had adopted guerrilla tactics launched from cross-border safe havens.

As the principal American SIGINT organization, the NSA was a priority target for hostile penetration, and numerous employees took the opportunity to sell classified information to the Soviets, among them **Jack Dunlap**, who spied between 1960 and his suicide in 1963; Robert Lipka, who later admitted to having passed material from September 1965 until his retirement two years later; **Ronald Pelton**, who sold documents to the KGB after his retirement from the NSA in 1979; and **James Hall III** of the **Army Security Agency**, who spied from 1982 until his entrapment in 1988.

NSA operations and capabilities are usually highly classified and generally consist of collection and cryptanalysis, but military action in Iraq and Afghanistan led to the development of offensive programs designed to disrupt an adversary's communications, which were deployed against the insurgencies. One project, codenamed JESTER, was bench-tested in combat conditions in Afghanistan by NSA teams, designated JACKAL, which sought to prove that electronic interference could play a key support role in a guerrilla conflict.

The Directors of NSA have been: Ralph Canine (1952–56), John Samford (1856–60), Lawrence Frost (1960–62), Gordon Blake (1962–65), Marshall Carter (1965–69), Noel Gayler (1969–73), Samuel Phillips (1972–73), Lew Allen (1972–77), Bobby Inman (1977–81), Lincoln Faurer (1981–85), William Odom (1985–88), William Studeman (1988–92), John McConnell (1992–96), Kenneth Minihan (1996–99), Michael Hayden (1999–2005), and

Keith Alexander (2005–present). *See also* AIR INTELLIGENCE AGENCY (AIA); ARGUS; CAPSTONE; COMPUTER NETWORK ATTACK (CNA); ECHELON; MINARET; RHYOLITE; SHAMROCK; UNITED STATES SIGNALS INTELLIGENCE; VENONA.

NATIONAL TECHNICAL MEANS. Within the United States intelligence community, the term "national technical means" is employed as a euphemism for satellite technology, usually meaning imagery- or signals-collection systems. *See also* NATIONAL RECONNAISSANCE OFFICE (NRO).

NAVAL COLLECTION PLATFORMS. Although most naval **submarines**, surface vessels, and aircraft can be converted to undertake signals-intelligence collection roles, either by the installation of self-contained suites with the necessary specialist **interception** equipment or by accommodating positions for intercept operators, some navies develop dedicated platforms, designated **AGI**s for Warsaw Pact trawlers and **AGTR**s for a class of U.S. Navy ships.

During the **Cold War**, the French Navy operated the 2,700-ton *Berry* in support of the **Groupement des Contrôles Radio-électriques** until its replacement by the *Argos*. The Royal Navy depended on T-class diesel-electric submarines, including HMS *Turpin*, *Totem*, and *Truncheon*, but a plan proposed by **Government Communications Headquarters** to build a dedicated nuclear-powered SIGINT vessel was scrapped following the capture of the USS ***Pueblo*** in 1968.

The Federal German Bundesmarine has also invested in surface collection platforms, commissioning the *Oste* in July 1998, followed by two other Type 423 ships, the *Oker* and the *Alster*. Officially designated as "fleet service ships," the trio are based at Eckernförde and attached to the 1st Submarine Squadron. They are quite unlike any other warships, with a single, very tall mast and an unknown number of antennae concealed inside a pair of large radomes behind the bridge. *See also* DeSOTO; *ENDURANCE*, HMS; HOLYSTONE; *LIBERTY*, USS; MOON SHOT.

NAVAL SECTION. Created in 1924 under the leadership of a barrister, William F. ("Nobby") Clarke, the Naval Section at the **Government Code and Cypher School** expanded to a staff of 24 upon the outbreak of **World War II** but within a year was employing more than 1,000 cryptanalysts and linguists.

NAVAL SECURITY GROUP (NSG). A component of the U.S. **National Security Agency**, the NSG managed ground intercept sites and other naval platforms, including warships and submarines. By 1989, it was estimated that the NSG ran 48 facilities in the United States and abroad, as well as an undisclosed number at sea. The NSG was founded in March 1935 and continued operations until September 2005, when it was absorbed into the Naval Network Warfare Command.

During the **Vietnam War**, the NSG manned intercept positions on U.S. 7th Fleet aircraft and surface vessels, concentrating on North Vietnam's VHF air-defense voice channels and the interdiction of boats, often disguised as fishing trawlers, which attempted to resupply the Vietcong from the sea. Known to the Vietnamese as Group 125, the unit's communications were intercepted in an operation that began in November 1962, but by July 1966, the attempts at infiltration appeared to have come to an end. However, the Group 125 net was heard again in November 1968 and in 1970 tactical intercept equipment. Known as FACTOR, the equipment was put aboard P-3s at Cam Ranh Bay for a very fruitful series of interdiction operations codenamed MARKET TIME.

NEW ZEALAND SIGNALS INTELLIGENCE. *See* COMBINED SIGNALS ORGANISATION (CSO); GOVERNMENT COMMUNICATIONS SECURITY BUREAU (GCSB).

NID 25. The Naval Intelligence Division's designation of the Admiralty's cryptographic branch better known as **Room 40**, NID 25 would be amalgamated with MI8 in 1921 to create the **Government Code and Cypher School**.

NKVD. The principal Soviet intelligence service during the period of the **VENONA** intercepts, 1940 to 1949, the Narodni Kommisariat Vnutrennih Dei was micromanaged from Moscow by General Pavel Fitin. His worldwide network of *rezidents* operated under diplomatic or consular cover, and their encrypted communications, exchanged over commercial cable systems, provided the traffic for the Anglo-American cryptographers. Before World War II, the NKVD had operated separately from its Soviet military-intelligence counterpart, the **GRU**, but Lavrenti Beria's decision to combine the two in an effort to concentrate on ENORMOZ would lead to the compromise of many of its networks. *See also* SOVIET/RUSSIAN SIGNALS INTELLIGENCE.

NORTH ATLANTIC TREATY ORGANIZATION (NATO). During the **Cold War**, NATO supplied a coordinated SIGINT-collection capability by ringing the Soviet Bloc with intercept sites stretching from Norway and Canada through Denmark, Holland, Germany, Italy, Greece, and Turkey to the United States. All either contributed directly to the SIGINT matrix or provided the necessary ground stations, **airborne collection platforms**, or naval facilities required to sustain 24-hour coverage of Warsaw Pact military activity. In the post–**Cold War** era, following the dissolution of the Warsaw Pact, NATO troops were deployed to Bosnia and Kosovo and then participated in campaigns in Afghanistan, Iraq, and Libya. In each of these theatres, SIGINT played an increasing role against unconventional forces, usually guerrilla irregulars who had mounted an insurgency but remained dependent on communications that were vulnerable to hostile interception. *See also* BUNDESNACHRICHTENDIENST (BND); CANADIAN SIGNALS INTELLIGENCE; DANISH DEFENSE INTELLIGENCE SERVICE (DIS); DUTCH SIGNALS INTELLIGENCE; NORUSA; UKUSA.

NORUSA. The signals-intelligence agreement signed with Oslo in October 1952, NORUSA gave the **National Security Agency (NSA)** access to Soviet signals intercepted at Tomaselv on Vadso Island, Kirkenes, Viksjofjell, and Fauske. Once collected, the traffic was sent to the NSA via an American communications relay at Gardermoen, on the outskirts of Oslo, not far from the Norwegian signals-intelligence headquarters at Saeter, or via a teleprinter to **Chicksands Priory** in England. In addition, the Norwegians ran two **direction-finding** stations at Heimdal outside Trondheim and at Randaberg, near Stavanger, with a further site at Nordstrand, near Oslo, which concentrated on the southern Soviet Union and the Warsaw Pact satellite countries. The significance of the Norwegian activity can be judged by the fact that the entire Norwegian Intelligence Service only ever employed 1,000 personnel, of whom more than half were declared to be engaged on "High Priority" work, which was a euphemism for SIGINT.

As well as operating a chain of coastal ground stations, the Norwegians used Royal Norwegian Navy submarines as collection platforms and created a commercial front company, Egerfangst, to deploy specially equipped trawlers in the Barents Sea, the first of which was the *Eger*, which began operations in 1956.

NORWEGIAN SIGNALS INTELLIGENCE. *See* NORUSA.

NURRUNGAR. Operational in 1969, the U.S. satellite-tracking station at Nurrungar was located close to the missile test range at Woomera in Southern Australia and closed in 1999. Partly because no public announcement was made by the Australian government concerning the Joint Defense Facility, Nurrungar, the site became the subject of political controversy and attracted some demonstrations, with protestors claiming that, because of the presence of the U.S. Air Force, the area had become a target for a surprise Soviet attack. *See also* DIEGO GARCIA; PINE GAP.

O

OFFICIAL HISTORIES. Many intelligence agencies seek to pass on to their successors such lessons as have been learned and, until recently, have resisted efforts to make these internal histories more generally available through a process of declassification. These same agencies have also sought to protect their sources and methods by censoring official histories, and early drafts of the Royal Navy's account of **World War I** were edited to remove references to intelligence appreciations and decrypts transmitted to British warships so as to avoid reigniting the controversy that had surrounded the battle of Jutland.

At the conclusion of **World War II**, **Government Communications Headquarters** authorized the preparation, in 12 volumes, of the work undertaken at **Bletchley Park**. The numbered typescripts, written by section heads and others selected shortly before they returned to civilian life, had very limited circulation and were intended for the archives to be a guide in any future conflict. The last volume, written by Wing Commander Eric Jones and Peter Calvocoressi, innocuously entitled *The German Police*, detailed what was known, and when, about the Nazi concentration camps. The entire series was eventually declassified in the United States by the **National Security Agency** in 1993.

In 1979, the British government agreed to the release of the first of five volumes of *British Intelligence in the Second World War*, written by a team of historians led by Professor Harry Hinsley. Although not exclusively devoted to signals intelligence, they described in considerable detail, and with great authority, the exploitation of signals by Great Britain's various intelligence services and its impact on the prosecution of the war.

The only other officially sponsored history of signals intelligence in the public domain is three volumes of *American Cryptology during the Cold War, 1945–1989* by Thomas R. Johnson, a heavily redacted version of which was declassified in 2007.

The continuing degree of sensitivity about disclosures relating to SIGINT may be gauged by the fact that Sir Lawrence Freedman's official history of the 1982 Falklands conflict experienced a delay of four years while various

Whitehall departments argued about the wisdom of making certain information public. When the two-volume account finally was released in 2006, it contained the very first official transcript of an enemy decrypt to be published by a British government since 1927.

ONE-TIME PAD (OTP). An encipherment system based on a table of randomly generated numbers that were only ever to be used once, the one-time pad appeared to offer impregnable security by denying cryptanalysts the opportunity to exploit patterns created by repetition. However, Allied cryptographers realized during **World War II** that German one-time pads depended upon machine-generated numbers, which meant they were not truly random. Having studied the German device, manufactured by the electronics firm Lorenz, it proved possible to replicate and predict the machine's product, thus enabling some German Foreign Ministry traffic, codenamed **FLORADORA**, to be solved and read.

Wartime scrutiny of Soviet cable traffic revealed that they also relied on OTPs, but through some error, individual pages within separate pads had been replicated, thereby undermining the system's intrinsic security. How this blunder occurred has never been explained, but the answer may lie in precisely how the Soviets generated their random numbers, for it is a cryptographic principle that it is almost impossible to truly replicate the random selection of numbers. Even a room full of women typing digits will eventually succumb to the rules of "psychological random," and interest in this heightened when T. A. (Tom) Wagoner made an astonishing discovery at **Arlington Hall** while attempting to solve German diplomatic one-time pads, copies of which had been illicitly acquired from a courier's bag. Extended cryptographic analysis showed that the sheets of five-figure numbers were far from random, and eventually a pattern emerged, which suggested that each page had been printed by a machine, with a single carbon copy, and stamped separately with a page number. Once this had been realized, the **Government Communications Headquarters** liaison officer, Brigadier John Tiltman, recognized the distinctive process of a single, commercially available Lorenz machine, and this breakthrough enabled the cryptographers to recreate their own versions of the one-time pads.

Buoyed by this remarkable achievement, attention was turned to the Soviet OTPs, and two suggestions have been made about how they were developed. One witness has described a room full of women calling out numbers, while the defector **Igor Gouzenko** has recalled a mechanical device, not unlike a lottery machine, in which numbered balls rolled into a frame. However, the best evidence came from Vladimir Petrov, who had actually worked in the room that printed the Soviet ciphers, and he remembered a machine with a flashing blue light, attached by cable to an inner sanctum to which he never had access. Such a device did exist in the West, turning electrical impulses generated on a filament into random numbers, and a lengthy but ultimately fruitless search was made by GCHQ in case a patent had been registered.

OP-20-G. Established in 1924, OP-20-G was the U.S. Office of Naval Intelligence designation for a small cryptographic organization intended to exploit a Japanese Imperial Navy cipher copied during the burglary of the consulate in New York two years earlier. In the absence of investment in independent cryptanalysis, OP-20-G would continue to depend on illicit entries, or **black bag jobs**, for its working papers and, until the fall of 1942, was based in the Navy Building on Constitution Avenue, Washington, DC.

Initially, OP-20-G, headed by Commander Joseph Wenger, had a staff of 655 and 75 officers, concentrated on **J-19** and then JN-25 Japanese naval traffic, which had been intercepted at three sites, at a station in Washington, DC, codenamed NEGAT, at Station CAST in Cavite, and at Station HYPO, located in the basement of the 11th Naval District's administration building at Pearl Harbor, where Lieutenant-Commander Joseph J. Rochefort, assisted by Thomas H. Dyer, reconstructed enough of JN-25 by early 1942 to read up to 30 percent of the enemy signals. Their task was eased by the use of IBM automatic calculators, card sorters, and musicians from the USS *California*'s band, who, following the loss of their ship, proved to be exceptional codebreakers. Later, additional intercept facilities were opened on Guam, Bainbridge Island, and at Puget Sound.

In February 1941, OP-20-G's Lieutenant Prescott Currier and Ensign Robert Weeks delivered a reconstructed **PURPLE** machine to **Bletchley Park** in return for the latest British breakthroughs on German **Enigma** ciphers, thereby cementing a transatlantic signals-intelligence relationship that would develop into a partnership to exploit Kriegsmarine **DOLPHIN** intercepts. A specialist subsection, designated OP-20-GM, was created at the old Mount Vernon Girls School on Nebraska Avenue to run American-manufactured bombes, which raced through the permutations of the daily key settings, and by mid-1944, some 96 of the devices had been installed to recover DOLPHIN and **SHARK**.

By 1943, OP-20-G had opened a branch in Hawaii under Fleet Radio Unit Pacific (FRUPAC) and later added the Radio Analysis Group Forward (RAGFOR) on Guam and the **Fleet Radio Unit Melbourne (FRUMEL)** in Australia. In addition, and in conditions of exceptional secrecy, OP-20-G began to study Soviet traffic, with 192 personnel assigned to a facility covering the Pacific on Bainbridge Island, Washington state. By the end of hostilities, OP-20-G had expanded to 6,955 personnel and 1,499 officers.

Later, the organization would also receive Soviet transmissions monitored from an **Army Security Agency** site on Skaggs Island, a vast 3,310-acre compound 25 miles northeast of San Francisco opened in May 1941, and would remain operational until 1993. *See also* UNITED STATES SIGNALS INTELLIGENCE.

OSHIMA, HIROSHI. The Japanese ambassador in Berlin, Baron Hiroshi Oshima's regular wartime reports to Tokyo contained important strategic information that was encrypted on a **PURPLE** machine, which had been solved by **William Friedman**. Originally appointed the Japanese military attaché in Berlin in 1933, after postings to Vienna and Budapest, Oshima maintained a close relationship with Adolf Hitler throughout the war and reported his frequent conversations with him almost verbatim in his lengthy messages to the Japanese foreign ministry. He received daily briefings on the progress of the Axis and undertook tours of inspection to see German fortifications for himself, and his candid observations proved exceptionally valuable to those authorized recipients of the **BLACK JUMBO** traffic. In August 1944, reports of his discussions with Hitler have the first clues to a surprise counterattack, which later occurred in the **Ardennes** in December.

Having been forced by Allied air raids on Berlin to move his embassy to Bad Gastein in Austria, Baron Hiroshi was detained by American troops at the end of hostilities, convicted of war crimes, and sentenced to life imprisonment. He was released in 1959 and died in 1975, aged 89. *See also* JAPANESE SIGNALS INTELLIGENCE; WORLD WAR II.

P

PALLADIUM. The codename for an ingenious transmitter designed to generate signals that could emulate the returns of an aircraft registering on the Soviet SPOON FEED target-acquisition radar system, PALLADIUM was designed to stimulate Soviet air-defense systems so their broadcasts, voice channels, and related traffic could be monitored.

PANDORA. The **Government Code and Cypher School** codename for the clandestine signals traffic encrypted on a **one-time pad** exchanged between Berlin and the German Legation in Dublin during **World War II**. When requested to do so, the German diplomats had surrendered a transmitter to the Irish authorities but retained another undeclared set, which was used only occasionally. *See also* FLORADORA.

PATCHETT, BRIAN. In July 1956, 25-year-old Corporal Brian Patchett of the Intelligence Corps defected to East Germany with his girlfriend. Formerly employed as an intercept operator at RAF Gatow in Berlin since the previous November, he possessed detailed knowledge of the deployment of British special signals units in West Germany. After his defection, he participated in a series of propaganda broadcasts in which he condemned British espionage and compromised much of the British SIGINT network in the British zone.

PELTON, RONALD. In November 1965, at the age of 24, Ronald Pelton was hired by the **National Security Agency** following seven years' service in the U.S. Air Force, with which he had served as a Russian linguist monitoring voice channels in Peshawar. Later, evidence emerged that Pelton had left the air force after some undefined disciplinary issues. He was then assigned to work in Great Britain, but when he returned to the United States, he experienced financial difficulties and declared himself bankrupt before he resigned in July 1979 to become a yacht broker in Annapolis, with the hope of supporting his wife and four children. His solution was to call the Soviet embassy in Washington, DC, to offer information and then visit the building, where he was interviewed by the KGB.

Following the defection of a KGB security officer, Vitaly Yurchenko in August 1985, who revealed the existence of an NSA "walk-in," the Federal Bureau of Investigation was able to trace a six-year-old tape of Pelton's telephone call and play it to NSA staff, who recognized his voice. He was then interviewed at the Annapolis Hilton and was arrested in November, just days before he was scheduled to fly to Geneva. Pelton confessed to having betrayed many NSA projects, including **IVY BELLS**, for a total of $35,000 paid over a period of five years. His knowledge of the NSA budget had given him access to the organization's most secret operations, at least seven of which he had betrayed to his KGB handlers, and he also identified the principal 60 Soviet communications channels subject to monitoring. Between 1980 and 1983, he met a KGB officer, Anatoly Slavnov, on numerous occasions and, in 1983, made the first of several trips to Vienna. He separated from his wife in the summer of 1985 and, on 25 November 1985, was arrested on six conspiracy and espionage charges. At his trial, he was represented by a former NSA deputy general counsel, and he pleaded not guilty, but on 16 December 1986, he was sentenced to three concurrent life terms, plus 10 years.

A subsequent damage assessment of Pelton's espionage drew attention to the fact that he had been able to join the NSA while already under a cloud, that his bankruptcy had gone unnoticed at the NSA, and that a classified U.S. Navy report compiled in 1981 about the loss of the IVY BELLS pod in the Sea of Okhotsk concluded that the operation probably had been betrayed by a traitor.

PETERSEN, JOSEPH S. In October 1954, Joseph Petersen, a **National Security Agency (NSA)** cryptanalyst and leader of the Russian section, was arrested and charged with passing information to the Netherlands embassy in Washington, DC. While working at **Arlington Hall** during **World War II**, Petersen, a graduate of Loyola University, had become friendly with a Dutch liaison officer, Colonel J. A. Verkuyl, and after the war, they had remained in contact. A routine check on letters mailed abroad by NSA personnel revealed that Petersen had disclosed details of work undertaken on the Hagelin B-211, the cipher system used by the Dutch foreign ministry. Petersen was sentenced to seven years' imprisonment and was released on parole in 1958.

PETROV, EVDOKIA. Married to the **NKVD** *rezident* in Canberra, who defected in April 1954, Evdokia Petrov was also an experienced NKVD officer and cipher clerk. Having been told that her husband had been abducted and murdered, Evdokia was escorted aboard a flight back to the Soviet Union when she was given the opportunity, during a refueling stop at Darwin, to speak over the telephone to her husband, Vladimir, then in the hands of the

Australian Security Intelligence Organisation (ASIO), and confirm that he was at liberty and in good health, and she was offered the chance to join him. She accepted and subsequently was debriefed by Western intelligence personnel, among them **MI5**'s George Leggett, who traveled to Australia after the couple had been resettled on their chicken farm, to interview her about Soviet encryption procedures; although, this aspect of their defection was omitted from their joint autobiography, *Empire of Fear*, which was released in 1956. Declining all invitations to travel abroad to be interviewed by liaison services, the Petrovs were only prepared to receive visitors under ASIO's supervision, usually at the Swan island facility near Melbourne.

PETSAMO CODEBOOK. The capture of a partially burned codebook by Finnish troops who briefly captured and occupied the small port of Petsamo during the Winter War provided Finnish cryptographers with access to a cipher system used by **NKVD** border troops. In 1944, **Calvin Hoover** of the local Office of Strategic Services (OSS) station was able to buy a copy of this document in Stockholm, and other versions were acquired later from the Finnish cryptanalysts who had traveled to Sweden on the ***Stella Polaris*** and settled there, of whom several subsequently immigrated to the United States, Colonel Paasonen and Colonel Hallamaa among them. In addition, a copy was retained by the Swedish **Försvarets Radioanstalt**, another by the French, apparently sold by Paasonen, who spoke the language and for a while lived in France, and yet another was bought by the Japanese military attaché, Lieutenant-General Makoto Onodera. At least one original, bought for 100,000 Crowns, wound up at **Government Communications Headquarters**, where the Stellist material was known as SOURCE 267, and special measures were taken to preserve the crumbling pages, which had been turned to ash at the edges, and a senior **BRIDE** cryptanalyst recalls handling a Soviet codebook with a bullet hole through the middle. A list of items sold by the Stellists, retrieved in Sweden, mentions a "five digit from 1941" sold to the British for 100,000 Crowns, a "strip code" purchased by the Americans, and a code used by Soviet armored formations bought by OSS *and* the Japanese.

PHILBY, KIM. Codenamed STANLEY in the **VENONA** texts dated September 1945, Kim Philby was the **NKVD**'s source inside the Secret Intelligence Service (SIS) on whom Moscow relied for information about the defection of the **GRU** cipher clerk in Ottawa, **Igor Gouzenko**. In his 1968 memoirs, *My Silent War*, Philby recalled a

> joint Anglo-American investigation of Soviet intelligence activity in the United States had yielded a strong suggestion that there had been a leakage from the British Embassy in Washington during the years 1944–45, and another from

> the atomic energy establishment in Los Alamos. I had no ideas about Los Alamos. But a swift check of the Foreign Office list left me in little doubt about the identity of the source in the British Embassy. My anxiety was tempered by relief, since I had been nagged for some months by a question put to me by my Soviet contact in Istanbul. He had asked me if I had any means of discovering what the British were doing in the case under investigation by the FBI—a case involving the British Embassy in Washington. At the time of asking there was nothing I could have done.

Philby served in Turkey from January 1947 to September 1949 and arrived in the United States to take up his liaison post in Washington, DC, in October 1949. Thus, Moscow had learned about the embassy investigation prior to Philby's departure from Istanbul, and the tip had come from **William Weisband** in February 1948. Equally clear is the fact that Weisband had no knowledge of how the Federal Bureau of Investigation (FBI) was pursuing the VENONA leads, and the NKVD was either partly or wholly reliant on Philby for help. This in turn implies that the NKVD had failed to penetrate the other Allied security agencies that had been indoctrinated into the secret, a view confirmed by Philby's official briefing into the background of the leak given by an SIS colleague, Maurice Oldfield, in London just before his departure. This news Philby relayed to his local handler, Yuri Modin, who, according to Philby, "after checking with headquarters . . . was left in no doubt that information from the FBI and my own referred to one and the same case."

In Modin's slightly confused version of this episode, he claims that "an American cipher expert from the **Army Security Agency** (later the **National Security Agency**), William Weisband, had just informed SIS that a brilliant analyst in his service named Meredith Gardner was attempting to decode telegraphic messages sent to the Lubyanka during the war by various KGB posts around the world. This caused a ripple of alarm at the Center. "We had no idea what the Americans might uncover, and this information hung over us like a sword of Damocles." Of course, Weisband had no contact with SIS, and his news must have gone to Moscow and then filtered back to Modin, but certainly Philby monitored the FBI's progress thereafter and was able to keep his contacts informed of impending moves against Klaus Fuchs and later Donald Maclean.

On the British side, just three embassy officials knew of the investigation: Philby's predecessor Peter Dwyer, who was about to retire from SIS and settle in Canada, his **MI5** counterpart, Geoffrey Paterson, and the embassy's security officer, Sir Robert Mackenzie. Although Philby attempted to warn of the net closing in on Fuchs, the message was delayed, and the scientist was interrogated and then arrested. In the HOMER case, Philby saw the FBI documents building up against Maclean and was able to alert the diplomat in sufficient time for him to escape to France.

Following his defection to Moscow in January 1963, Philby drafted his memoirs and gave a curiously sanitized account of his wartime work in SIS's Section V in which he undertook the role of a signals-intelligence analyst, studying **ISK** and **ISOS** intercepts for counterintelligence leads. Philby never referred to the enemy's use of the **Enigma** machine and presumably was persuaded by his KGB hosts not to go into any detail about the flaws in Soviet cipher systems that had been exploited by Anglo-American cryptographers.

Philby died in Moscow in May 1988, some 25 years after his defection, a chronic alcoholic and desperately disappointed that his Soviet hosts had neither trusted him nor exploited his encyclopedic knowledge of British wartime and postwar signals-intelligence operations. *See also* BRITISH SIGNALS INTELLIGENCE.

PHOU PHA THI. Located a dozen miles inside Laos, Phou Pha Thi was the location of a clandestine **National Security Agency** SIGINT-collection site on top of a remote mountain codenamed HEAVY GREEN, which enjoyed a reach into North Vietnam as part of the LIMA SITE program. In March 1968, it was partially evacuated of American personnel before it was stormed by North Vietnamese troops, who seized classified matériel and reportedly took two prisoners for interrogation.

PINCH. A clandestine operation conducted for the specific purpose of acquiring some cryptographic material that is judged likely to be of assistance to cryptanalysts is known as a "pinch," to distinguish between a plan intended to retrieve a prize and the fortuitous recovery of an unexpected windfall from, for example, jetsam. Pinches were only possible where there was a certainty that the required material would be available and that no adverse consequences would follow.

During **World War II**, a series of pinches were undertaken to assist the **Bletchley Park** codebreakers. Initially, the targets were unarmed enemy weather ships and the Kriegsmarine's Home Waters Enigma key known as *Heimisch*, later HYDRA, and broken briefly by the Allies in June 1941, following the capture of cryptographic material from the *München*. More documents, relevant to the **DOLPHIN** key, seized on 11 May 1940 from a patrol boat, the *Schiff-26*, enabled six days of DOLPHIN traffic to be read retrospectively for the previous month, from 22 to 27 April. In March 1941, a force of British commandos and Norwegian volunteers, codenamed Operation CLAYMORE, attacked the German-occupied Lofoten Islands, and a party from HMS *Somali* boarded an armed trawler, the *Krebs*, and recovered a box of Enigma rotors and the key list for February.

Consequently, DOLPHIN traffic for April succumbed, and then the messages from 1 to 8 August were read retrospectively. This was followed by breaking the daily keys for 18 and 19 September, and thereafter DOLPHIN was read with a delay of only 36 hours until the end of hostilities with the exception of a three-week period from 1 March 1943, when new bigram tables were introduced and required reconstruction by the cryptographers. In the interests of preserving the pinch's security, CLAYMORE's true purpose was not disclosed until 1974.

The code rooms of diplomatic missions have always been a target for pinches, and numerous attempts were made during World War II to make surreptitious entries into these premises. A plan by Office of Strategic Services personnel to burgle the Japanese naval attaché's office in Lisbon was canceled at the last moment because, unknown to the officers concerned, the cipher had been solved already and their well-intentioned opportunism would have compromised the break.

During the **Cold War**, the **Central Intelligence Agency** created a specialist unit, designated **Staff D**, to covertly obtain cryptographic data, while the Federal Bureau of Investigation's Tactical Operations Section performed similar duties domestically. *See also* TISLER, FRANTISEK.

PINE GAP. The U.S. **National Security Agency** facility located in central Australia, near Alice Springs, since 1966, Pine Gap is a large, secure satellite ground station technically under the control of the Australian **Defence Signals Directorate** and known as the Joint Defence Space Research Facility. Linked to a **RHYOLITE** satellite, which originally was placed in geostationary orbit in June 1970, with other ground stations at **Menwith Hill** and Buckley Air Force Base, Colorado, the site acquired some unwanted notoriety when the radical Labor Party leader, Prime Minister Gough Whitlam, threatened in 1975 not to renew the NSA's lease. He was subsequently dismissed by Governor-General Sir John Kerr, leading to claims that his hostility to the NSA had led to a conspiracy to remove him, a theory enhanced by assertions made by **Christopher Boyce**, a disaffected TRW employee convicted of espionage in 1977, who said he had seen secret messages on a highly classified **Central Intelligence Agency (CIA)** channel implicating the CIA station chief in Canberra in a plot to remove Whitlam.

POL AND 9WE. Detected at **Fort Bridgewoods** in 1939, the call signs POL and 9WE exchanged very distinctive encrypted traffic on a network controlled from Köthen that included Berlin, Lüneburg, Glogau, Neubrandenburg, Märkisch, Lübeck, and Friedland. The traffic, identified as Luftwaffe **Enigma** messages, was designated BROWN or the less common RED at **Bletchley Park**, and, by August 1940, additional transmitters had appeared at Bergen, Vannes, Morlaix, Cherbourg, and Den Helder, with a control at Boulogne, and handled a large volume of "KR" ("urgent") signals.

High volumes of RED came to be associated by traffic-analysis experts, who monitored the length and times of the transmissions, with imminent night air-raids on London, and until Christmas 1940, this represented Great Britain's principal early warning system.

POLISH CIPHER BUREAU. The Cipher Bureau of the Polish General Staff's Radio Intelligence Office, headed by Colonel Gwido Langer, began studying German-encrypted wireless traffic as soon as the Wehrmacht introduced the **Enigma** machine and collected the signals from a network of intercept stations located at Krzeslawice, Starograd, and in a suburb of Poznan.

In December 1932, Langer received from **Gustave Bertrand** a selection of documents that had been provided by his agent, **Hans-Thilo Schmidt**. By January 1933, Langer's organization had reconstructed the device's theoretic wiring diagram and a month later commissioned the AVA Radio Manufacturing Company to construct a replica. Within a year, AVA had built 15 copies and delivered them to the Cipher Bureau's secret cryptographic facility at Mokotov-Pyry, deep in the Kabackie Woods, south of Warsaw. There, the head of the Cipher Bureau's German Section, Major Maksymilian Ciezki, collaborated on the design of an electrical apparatus, known as the cyclometer, that moved two AVA Enigma rotors through every permutation of the machine's settings. This innovation led to the development of a bombe, so called because the inventors of the electromechanical device that incorporated six AVA Enigmas had been eating an ice-cream bombe when they had first discussed the idea, which could race through the 17,576 permutations in just two hours. A further improvement, created by Henryk Zygalski, was a set of perforated sheets, in a stack of 26 and marked with 51 holes, that enabled cryptanalysts to recover Wehrmacht Enigma keys.

News of this cryptographic progress was shared with the British and French at a secret meeting held in Paris in January 1939. However, as the political atmosphere in Europe deteriorated, a second meeting was called in Warsaw in July, which was attended by Bertrand representing the Service de Renseignements, Alastair Denniston and Dillwyn Knox from the **Government Code and Cypher School (GC&CS)**, and Humphrey Sandwith from the Admiralty. As a result, the British and the French were given an AVA replica Enigma each, which were dispatched to Paris in August and gratefully accepted by the local Secret Intelligence Service station commander, Wilfred Dunderdale.

Upon the outbreak of war, the Cipher Bureau was evacuated hastily to Romania and then was accommodated by Bertrand at the Chateau de Vignolles near Gretz-Armainvilliers, 30 miles northeast of Paris. Codenamed BRUNO, the site was linked to **Bletchley Park** by landline, but in 1940, the staff were evacuated again, first to Paris and then farther south to a Deuxième Bureau establishment, Les Fouzes, at Uzes in Provence. Meanwhile, the Poles went to England, where the cryptographers were posted to the Rothamsted Experimental Station in **Harpenden**, and the other staff were sent to a secure compound at Funny Neuk in Woldingham. When 17 Polish cryptographers had applied to come to England in October 1939, permission had been refused on advice from Bletchley Park, and although they were allowed to land in 1940, they remained isolated from GC&CS for the remainder of hostilities.

PRIME, GEOFFREY. Born in 1938 and educated at a Roman Catholic school in Stoke-on-Trent, Geoffrey Prime joined the Royal Air Force in 1956 for his two years' National Service and volunteered for the Russian course at RAF Crail. He was then posted to Kenya as a regular and, upon his return in April 1962, went on a course at the Joint Services Technical Language at RAF Tangmere. In May 1964, he was posted to RAF Gatow as an intercept operator monitoring Russian voice channels, and while there, he offered to spy for the Soviets in January 1968.

Although initially suspicious, perhaps because he claimed to be motivated by ideology, the KGB recruited him and urged him to apply for a transfer to **Government Communications Headquarters (GCHQ)**, which in September 1968 proved successful. There he worked as a Russian linguist in the London Processing Group (LPG) in Carlton Gardens but attended regular annual meetings with his KGB 3rd Chief Directorate handlers in Vienna, Ireland, Rome, and Cyprus. Between the rendezvous, he would fill dead drops with photographs taken on a Minox camera of classified GCHQ documents, listen for instructions on a shortwave radio, and encrypt messages on a **one-time pad**.

In 1976, when the LPG moved from its city of London offices in St. Dunstan's Hill to Cheltenham, he was posted to J30, a Soviets signals section in J Division. He received a promotion in June 1976 and moved to lead a group of transcribers in J25, where he was given a **BYEMAN** clearance, which gave him access to **RHYOLITE** and **CANYON** satellite data. However, he resigned without explanation in September 1977, soon after having married landlady Rhona Ratcliffe, a 33-year-old divorcee with three children, to work as a local taxi driver. Three years later, in April 1980, the KGB contacted him and invited him to stay in Vienna for three days and took him on a river cruise in Hungary, apparently in an attempt to persuade him to rejoin GCHQ. A further meeting took place in East Berlin in November 1981, but although he continued to supply old GCHQ documents he had copied before his resignation, they retained a high value and included details of the SAMBO project, which monitored Soviet low-frequency underwater transmissions.

Prime was arrested in April 1982 on pedophile charges. After he had been detained, his wife Rhona discovered espionage paraphernalia hidden in their home, and it was seized by **MI5**. Under interrogation, Prime confessed that he had spied for the KGB between January 1968 and his resignation in 1977, when he was a section head in the sensitive J Division. He was sentenced in November 1982 to 35 years' imprisonment and released on parole from Rochester prison in March 2001.

PROHIBITION. Between January 1920 and the repeal of the 21st Amendment in 1933, alcohol was banned in drinks in the United States and became a valued commodity that was smuggled into the country by highly organized rumrunners that were coordinated by radio. The U.S. Coast Guard responded by intercepting the wireless traffic and creating a cryptographic laboratory headed by Elizabeth Friedman to read the smugglers' encrypted messages in support of contraband interdiction operations. Altogether, some 22 destroyers, 46 patrol boats, and 28 cutters were fitted with transceivers and connected to shore stations in an effort to defeat the illicit trade.

PROXIMITY FUZE. A potentially war-winning device that detonated artillery shells upon command from a signal reflected from a target within a designated range, the proximity fuze was technology disclosed to the United States scientific community by Sir Henry Tizard in 1941 under strict security conditions. Initially, the apparatus was only fitted to antiaircraft artillery to prevent a round from falling intact into enemy hands. The fuze, designated the AN/CPQ-1, was manufactured in New York by Western Electric, where three Soviet spies, Joel Barr, Al Sarant, and Nathan Sussman, had access to the blueprints and appeared in the **VENONA** traffic. A fourth spy, Morton Sobell, also a member of Julius Rosenberg's network, stole details of the apparatus and, according to Aleksandr Feklisov Rosenberg, gave him an actual fuze as a Christmas present in December 1943. The technology was regarded as especially sensitive as quite simple countermeasures, such as a jamming signal emitted from a target aircraft, could easily neutralize it.

***PUEBLO*, USS.** In January 1968, the U.S. Navy AGER signals-intelligence collection platform *Pueblo*, on its first operational voyage, was captured by North Korean sailors while it cruised in international waters off the port of Wonsan. Commanded by Lloyd Bucher, the lightly armed *Pueblo* was a converted cargo ship, built in 1944, that had been commissioned the previous year as an auxiliary with a crew of 83 under the operational control of the **Naval Security Group (NSG)** at Kamiseya, Japan. When the ship was raked by 57 mm cannon-fire from North Korean gunboats, mortally wounding one member of the crew, Bucher obeyed an order to surrender, and the ship was boarded. The *Pueblo* was then sailed into Wonsan, where the classified compartments were emptied of documents, cipher machines, intercept equipment, and other electronics. According to KGB General Oleg Kalugin, an estimated 792 pounds of this equipment was subsequently delivered to the Soviet Union for scientific analysis.

Within hours, North Korean television was broadcasting seized secret documents with the codeword classifications TRINE and SAVIN. Despite protests from the **National Security Agency (NSA)**, the U.S. ambassador to the United Nations, Arthur Goldberg, was authorized by the White House to disclose intercepted North Korean radar tracking that proved the *Pueblo* had been well outside the 12-mile limit.

The survivors, including two U.S. Marine linguists who scarcely spoke any Korean, were repatriated on 23 December after 11 months of beatings and interrogation, but the ship remains in Nampo, where it has become a tourist attraction. Although Commander Bucher was recommended for prosecution by court-martial, the secretary of the navy stayed the proceedings, and he remained in the U.S. Navy until his retirement in 1973.

Following the *Pueblo* debacle, the House Armed Services Committee mounted an investigation chaired by Otis Pike, and in camera evidence was given on 10 March 1968 by the NSA's director, General Marshall Carter, but his secret testimony was later leaked to the media two days later by Pike. A damage assessment was completed after the crew had been debriefed in San Diego, and the NSA concluded that if the Soviets had been passed the *Pueblo* booty, then communications for some of 1967 and 1968 should be considered compromised. As a result of the fiasco, the entire AGER program was canceled in 1968. *See also* UNITED STATES SIGNALS INTELLIGENCE.

PURPLE. The **interception** and decryption of Japanese diplomatic traffic by the American Army Cipher Bureau, designated MI8, dated back to 1919, but the revelation by the State Department to the newly appointed secretary of state, Henry L. Stimson, that its **Black Chamber**, located in New York under the commercial cover of the Code Compilation Company, had been secretly reading foreign diplomatic telegrams, resulted in him ending the practice forthwith by withholding funding at the end of October 1929.

Despite Stimson's ban, the U.S. Army's small team of five cryptanalysts continued to work on what was termed "the Japanese problem" through the newly created **Signal Intelligence Service (SIS)** based at the Munitions Building in Washington, DC, in parallel with the U.S. Navy's Communications Security section designated **OP-20-G**. By 1935, after five years of toil, the SIS had achieved considerable success against RED, the "Type A" cipher machine used by Japanese naval attachés to encrypt messages that were handed over to commercial carriers for onward transmission to Tokyo.

As the Axis began to take shape, the huge volume of Japanese Foreign Ministry telegrams placed an increasing strain on the SIS's very limited resources, which boasted a staff of just seven, and an arrangement was made with OP-20-G to share the burden, with SIS handling traffic on even calendar days and the navy dealing with the odd days. This collaboration worked well, with SIS doubling its staff to 14, until March 1939, when the flow of decrypts suddenly ceased with the unexpected introduction of a more sophisticated replacement unit, the "Type B," codenamed PURPLE.

The PURPLE cipher generated by the Type B, of which only 25 models were built, resisted the decoders until 25 September 1940, when the first complete text succumbed and was distributed to the very limited group indoctrinated into the highly classified program. The breakthrough occurred after 20 months of pure cryptanalysis by a team led by **William Friedman**, who promptly underwent a nervous collapse. Friedman's subordinate, Frank B. Rowlett, had been able to reconstruct the PURPLE cipher, and thereby develop a replica Type B machine, by exploiting two flaws in Japanese procedures. The first was a common error, known as the stereotype, which consisted of the stilted diplomatic language which convention dictated should begin and end individual telegrams. The predictable repetition of the pedantic phrase "I have the honor to inform your excellency" allowed the experts to glimpse the construction of particular messages, as did the Japanese habit of reproducing word for word the content of U.S. State Department communiqués. Since the cryptanalysts, among them Genevieve Grotjan, worked with copies of the original texts, they had little difficulty in retracing the process by which the Japanese had transformed the plaintext into cipher. Similarly, a comparison between the embassy's diplomatic notes delivered by the embassy to the State Department and the original encrypted texts proved profitable when it was realized that, for cultural reasons, the Japanese staff would not dare alter one word of the content.

Once Friedman had grasped the principles upon which PURPLE was based, Lieutenant Leo Rosen was able to build an electromagnetic device constructed from telephone exchange relays, which duplicated the original machine, and by January 1941, a model had been delivered to OP-20-G in

Washington, DC, and another had been donated to **Bletchley Park**, which hitherto had concentrated on the Japanese Navy's hand ciphers. Another is alleged to have been sent to the **Far East Combined Bureau** in Singapore; although, there appears to be no record of its fate after the Japanese occupation.

The Allied window into the Japanese naval attaché circuits provided an invaluable source of intelligence, but it was an apparently routine seven-part cable, numbered 906, addressed to Colonel Saburo Hayashi in Berlin and Major Hirose in Helsinki, dated 6 October 1942 and translated early the following year, which proved to be crucial. The message, written in an "emergency system," perhaps based on a book, was just one of a mass of enciphered communications between Japan and Europe that was intercepted by Allied radio operators at the Vint Hill Farms Station, outside Warrenton, Virginia, and passed to a staff of 35 cryptographers and translators. It disclosed the results of a lengthy Japanese study of Soviet diplomatic and commercial traffic based on material exchanged between the embassies and consulates in Japan and Manchuria, and the Foreign Ministry in Kuibyshev. The Japanese had been interested in Soviet ciphers for years, at least dating back to 1921, when the Polish Deuxième Bureau had seconded one of its experts, Captain Kowalewski, who was himself of Russian origin, to modernize Tokyo's cryptographic procedures.

An American trawl through previous naval attaché telegrams revealed that the first report on Soviet cryptographic systems had been transmitted to Tokyo on 1 July 1941 from Berlin, Stockholm, Helsinki, and Hungary, and evidently since then, the Japanese had focused on the Moscow–Vladivostok link, as well as covering the diplomatic missions located at Seoul, Hakodate, and Dairen. According to the Japanese attachés, progress on the Soviet project had depended upon help provided in Helsinki by the Finnish authorities. Clearly, the Finns, exhausted by war with the Russians, had taken the strategic decision to cooperate with the Japanese, presumably in the hope of drawing Josef Stalin's attention to the Far East. One part of the collaboration was the exchange of cryptographic data, much of which had been gleaned from material recovered in the ruins of the Soviet consulate at the port of Lonahamari in Petsamo, in the extreme north of the country, which had been overrun by Finnish troops on 22 June 1941 before the **NKVD** staff could destroy their codebooks. The isolated building, that also housed the British, German, and Swedish consulates, overlooked the deep, iceless harbor, which was a valuable Finnish asset at the end of the narrow Petsamo corridor, providing valuable access to the Barents Sea.

The Soviet 104th and 52nd Divisions had occupied the port since the December 1941 offensive, which had resulted in a dictated peace the following March. The Lonahamari documents, four partially burned diplomatic codebooks, had allowed the Finns, who had established a listening post at Otaniemi, just to the west of Helsinki, to monitor Soviet and American diplomatic traffic and to read some Soviet messages, and their experience, shared with the Japanese, enabled Tokyo to break into it too. The four items were a diplomatic codebook designated Kod-26, the NKVD *Pobeda*-code, one for use by the GRU military intelligence service, and one by the Naval GRU. Studied in conjunction with the medium-grade military crypto items recovered on the battlefield, which included at least one NKVD Border Guards codebook, the material allowed the Finnish analysts to understand how the Soviets adapted military terminology in their systems, built code tables, and relied on a very straightforward mathematical formula to encode emergency signals. The Finns placed a heavy emphasis on cryptography, and Field-Marshal Carl Gustav Mannerheim's crucial victory against the Soviet motorized 44th Division at Suomussalmi, a turning point in the Winter War, was largely due to advance knowledge of the enemy's strategy gleaned from intercepted signals.

The collaboration between Finland and Japan was not entirely surprising. Although not a formal member of the Axis, and technically neutral for the first two years of the war, Finland participated in the assault on the Soviet Union the following year. Great Britain finally declared war on Finland on 6 December 1941, so by the time the 906 circular was intercepted, the Helsinki authorities had developed a strong relationship with the Tripartite countries. The fact that an NKVD codebook had fallen into Japanese hands and had been passed to the Germans was soon reported to Moscow by Harro Schulze-Boysen, an Air Ministry official in Berlin and GRU spy. His wife, Libertas, was having an affair with another member of his network, Horst Heilmann, who was an Abwehr cryptographer. Most likely, Heilmann came across the Soviet codebook, which had been a version used by the NKVD Border Guards, during the course of his work and alerted Schulze-Boysen, who relayed the news to his controller. However, once Moscow heard of it, there was nothing that could be done to retrieve the situation, and the Finns, the Germans, and the Japanese continued to concentrate on "the Russian problem" for the rest of the war.

Until the Finnish documents fell into American hands in 1945, there was uncertainty about how the Finnish cryptographers had achieved their success with the Soviet code, but the work described in the 906 circular, and parallel analyses of a further two-dozen Soviet military systems, encouraged the SIS's successor, the **Signal Security Agency (SSA)**, to do the same. The SSA, based at what had been SIS's wartime headquarters at **Arlington Hall** Station in northern Virginia, had inherited a large quantity of prewar Soviet intercepts, which had baffled the analysts, and, in June 1942, had agreed with the U.S. Navy and the Federal Bureau of Investigation to take responsibility for dealing with diplomatic problems.

Certainly the SSA had no shortage of raw material on which to work, for as well as acquiring the prewar collection of Soviet intercepts, the Agency had immediate access to all the current Soviet telegrams on the New York–Moscow–Washington, DC, circuits, which were sent by commercial carriers. Couriers from the embassy, consulate, and trade missions made daily deliveries to Western Union, and by law, the company was obliged to copy each telegram and supply it upon request to the U.S. Office of Censorship, then headed by General W. Preston Corderman, who, in 1943, commanded Arlington Hall. Thus, the SSA, which was to employ 7,000 men and (mainly) women at Arlington Hall, created a massive archive of thousands of enciphered texts, each consisting of varying quantities of five-figure groups. Apart from the recipient's cable address in Moscow and the identity of the paying originator, there was absolutely nothing to give a clue to the content; although, the five-figure system made it quite distinctive from the more usual six-figure convention or the four-figure employed by the Japanese.

News that the Japanese had achieved success inspired Colonel Carter W. Clarke, chief of the military-intelligence Special Branch, to create a new unit to start work on the accumulated collection of Soviet intercepts. Examination started on 1 February 1943, when a local school teacher of home economics, Gene Grabeel, who was one of a large number of civilians newly brought in as support staff, began the task of sorting the telegrams into categories based on the same distinct groups identified by the Finns. Miss Grabeel and her team, headed by Lieutenant Leonard M. Zubko, who had been at Arlington for several months, soon established from the Moscow cable address that more than half the texts were designated "Trade" and had originated with Amtorg and with the Purchasing Commission. These messages were designated colors, RED, GREEN, and LAVENDER, and the remaining categories were later found to be cipher systems employed by the NKVD, the Peoples' Commissariat for Foreign Affairs, the GRU, and the Naval GRU.

While the Foreign Ministry traffic was easily recognizable from the important consular centers they originated from (the GRU), Naval GRU was not identified until 1946, after the defection of the GRU cipher clerk Igor Gouzenko, and the NKVD material had to wait until 1948. However, progress under Zubko, a Slavic linguist who had lived in Russia, was slow, and the project was restarted by Captain Ferdinand Coudert, a son of one of the famous Coudert Frères international lawyers. Unfortunately, Coudert had no cryptanalytical skill, and he was replaced in November 1943 by Captain William B. S. Smith, a shrewd cryptanalyst who targeted the Russian diplomatic problem and set the scene for Oliver Kirby to take over in 1946. *See also* JAPANESE SIGNALS INTELLIGENCE.

PYRAMIDER. The **Central Intelligence Agency** codeword for a classified **satellite communications** system compromised by **Christopher Boyce** in 1976.

Q

Q CLEARANCE. The term applied to the security clearances required for access to **VENONA** material was known as "Q" and severely restricted to a small group of cryptographers and counterintelligence analysts and investigators.

QUICKSILVER. The wireless dimension to the FORTITUDE D-Day deception campaign was codenamed QUICKSILVER and consisted of vehicles containing broadcast equipment that generated signals that emulated the radio traffic usually associated with large military concentrations. The Allied objective was to provide enemy intercept operators with the evidence required to support assessments of the Allied order of battle and confirm the location of notional units, such as the nonexistent First U.S. Army Group.

R

RADAR INTELLIGENCE (RADINT). One component of the **electronic-intelligence** spectrum, RADINT is the term applied within the Anglo-American signals-intelligence community to information relating to the performance and apparatus of equipment designed to transmit and receive radar signals.

RADINT-collection operations include the deployment of **airborne collection platforms** in ferret missions to test reactions to incursions of sovereign airspace and provoke the transmission of hostile radar signals and other traffic for recording and analysis. *See also* CUBAN MISSILE CRISIS; RAS-AL-GHALEB.

RADIO FINGERPRINTING (RFP). Designed to counter difficulties encountered during **World War II** by the British with **call-sign analysis**, radio fingerprinting was developed as a technique to identify particular target traffic. During **World War I**, intercept operators had reported an ability to recognize characteristic "fists," being the speed, timbre of note, and other individual peculiarities associated with a single person. The foundation of RFP was a comparison in 1918 of photographs taken of the traffic, and this was later refined using modified sound-recording equipment purchased from Gaumont cinemas, which captured the sound of the radio traffic on a standard 17.5-millimeter film track with a string galvanometer. By relying on an oscillograph with a flat frequency response as a modulator, light from a low-voltage sound-recording lamp was focused by a complex optical system onto a galvanometer mirror. A tiny reflection of this light, measuring 1,000th of an inch, was then directed as an intense rectangle by a separate optical system onto sensitized film. By passing an electrical current through the oscillograph, the mirror is diffracted, causing the beam of light to more or less cover the track. Meanwhile, the film is moved at right angles to the movement of the light, exposing the film along its length at a standard speed of 18 inches a second.

When developed, the film offers a complete trace of the transmitted signal and, within a few minutes, provides a reasonably accurate, easily comparable visual record of a particular operator. This proved invaluable in the **traffic**

analysis process in identifying a transmitter using more than one call-sign or several different transmitters using the same call-sign or isolating a specific transmitter within a large net. The technique remained very secret because of the potential ramifications, especially in the clandestine field of agent impersonation.

RADIO INTELLIGENCE DIVISION (RID). On 1 July 1940, the Federal Communications Commission (FCC) received a secret allocation of $1.6 million from the White House and appointed George E. Sterling to create the Radio Intelligence Division, initially from seven sites in the United States and based on the British **Radio Security Service**, to monitor illicit transmissions. Known as the National Defense Communications Section, it quickly built an eighth monitoring station in Laredo, Texas, and arranged through the army to use facilities at a further 30 military posts.

Although the RID had not possessed the means to read the intercepted traffic, it was able to do so from June 1941 with advice from Paul A. Napier, the head of the cryptology section of the Federal Bureau of Investigation's (FBI) technical laboratory. For example, the RID had been monitoring a transmitter in Chile, with the call sign FMK, since April 1941, but had not succeeded in breaking its code. However, with the FBI's help, the RID was able to routinely intercept and read almost every Abwehr message transmitted from Latin America throughout the rest of the war. At certain times of the year, and at certain times of the day, it was almost impossible to exchange shortwave signals between Latin America and central Europe, so messages were relayed through a transmitter in New York, which was under the FBI's control.

To coordinate its activities with other agencies, including the FBI, the FCC, late in the autumn of 1940, created the Defense Communications Board, with representatives from the U.S. State Department, coast guard, and the army and navy, and gained approval from the Treasury in January 1941 to expand the intercept network by an additional 5 primary stations, with support from a further 60 so as to provide comprehensive monitoring 24 hours a day. With the FBI's help, the RID became highly proficient at reading the enemy's traffic, to the point that some technicians even suspected the Abwehr might have been deliberately providing the material to divert their attention from "more sinister operations of a professional nature." They were bemused by the frequent lapses in the most fundamental security procedures, such as the occasions when call signs PYL in Chile and LIR in Brazil even mentioned in clear text the titles of the novels their ciphers were based on.

The RID's **interception** operations were not restricted to North America, for one of its **direction-finding (D/F)** experts from Laredo, Robert Linx, was posted to Rio de Janeiro in late March 1941 for a temporary assignment that was to last four years, and John DeBardeleben was sent to Chile. With Linx's help, the Brazilian authorities developed their own radio direction-finding system, located at six coastal sites as well as Sao Paulo, Belo Horizonte, and Porto Alegre, which proved very effective and provided the basis of the evidence used to initiate the mass arrests of August 1942 that silenced all five clandestine transmitters. DeBardeleben's experience in Chile was one of rather less cooperation, and he was obliged to work independently of the pro-Axis authorities, with help from two FBI Special Intelligence Service Special Agents, and operate his D/F equipment while the trio traveled the country masquerading as miners to monitor the PYL transmitter, which was first traced to Cerro Alegre and then, after it went off the air until October 1942, to a farm outside Quilpué, near Valparaiso, and directed by Major Ludwig von Bohlen, codenamed BACH, who was the German air attaché accredited to Santiago. When the PYL site was raided by the Chilean Dirección General de Información e Inteligencia, sufficient espionage paraphernalia was recovered to compromise the other German networks in the region.

RADIO SECURITY SERVICE (RSS). Based at Arkley View in Barnet, North London, the Radio Security Service was created in late 1939 under the leadership of Colonel J. P. G. Worlledge to monitor illicit broadcasts from the British Isles. Lord Sandhurst, of the Radio Society of Great Britain, assembled amateur license holders, known as Voluntary Interceptors (VI), to scan the airwaves for suspicious transmissions. By the end of the year, no clandestine signals had been detected by Jimmy Sandhurst's team, but a small group of cryptographers, led by Oliver Strachey, had solved several of the Abwehr's hand ciphers, designated **ISOS**, and this breakthrough led to access to some Abwehr **Enigma** keys, subsequently codenamed **ISK**. By August 1941, RSS staff were taking 23,000 complete logs a month, with up to 10,000 sheets of intercepts a day. The small RSS analytical unit, designated the Radio Analysis Bureau (RAB), had been created by Walter Gill and consisted of Hugh Trevor-Roper, Stuart Hampshire, and Gilbert Ryle. In addition to the VIs, who, as cover, were enrolled into the Special Branch of the Royal Observer Corps, RSS personnel worked at static sites at **Wymondham**, Gilnahirk, St. Erth, Forfar, and Thurso.

In January 1942, Sandhurst joined **Special Communications Unit 1**, passing his command to Jack Hester, and the following year, RSS, which had become a virtual adjunct of **Bletchley Park**, came under the nominal control of the Secret Intelligence Service's communications branch, Section VIII, replacing Worlledge and Gill with Professor F. J. M. Stratton and Kenneth Morton Evans as his deputy. The RAB was then renamed the Radio Intelligence Service (RIS), and in 1946, the entire organization was absorbed into **Government Communications Headquarters**. *See also* BRITISH SIGNALS INTELLIGENCE.

RADIO SILENCE. Although intended as a security measure to prevent hostile interception, the imposition of radio silence can be interpreted by skilled traffic analysts as a significant break in a pattern of behavior, and maybe the prelude to an important event, such as the start of a surprise offensive. In the early days of **traffic analysis (T/A)**, when War Office operators based at **Fort Bridgewoods** monitored German transmissions, there was very little military traffic to cover, and there was no perceptible change in activity prior to the invasion in April 1940 of Denmark and Norway. The first clue consisted of the identification of mobile Sturmabteilung stations southwest of Aachen on 10 May, followed by more in the vicinity of Maastricht over the following days, and the establishment of a military transmitter in Amsterdam on 11 May linked to Paderborn, Lippstadt, and a site near Bremen. On 15 May, Chatham caught the clear-text signal on the same network reporting that General Kurt Student had been wounded.

A T/A study conducted after the surprise offensive provided no clues and no evidence of low-level **Enigma** traffic; although, a major Luftwaffe network controlled from Dortmund was seen to move to a location between Brussels and Liege on 1 June and then to Paris on 24 June. Altogether, 1,500 messages were intercepted from this network, but none could have provided advance notice of the enemy's strategy.

Traffic analysts first concluded that the **BROWN** traffic, from 10 groups of stations, was potentially very important when it was suspended between 28 July and 4 August 1939, a period during which it was known the Luftwaffe was conducting large-scale exercises. The traffic, traced to the Kiel–Dusseldorf–Berlin triangle in northwest Germany, was then collected in sufficient quantities for the cryptographers to attack it.

Radio silence imposed before the **Ardennes offensive** in December 1944 was later considered to have been misinterpreted, and, in October 1973, on the eve of Yom Kippur, Israeli **Unit 8200** analysts failed to spot the significance of an unexpected cessation of Syrian traffic. *See also* WORLD WAR II.

RAFTER. British codename for an ingenious apparatus designed to identify the frequencies to which radio transceivers had been tuned. **MI5** technicians discovered by accident that receivers located inside the Soviet embassy in London radiated a weak signal that could be detected and monitored, and the objective was to use the same technique to find other equipment tuned to the same wavelengths. The procedure was described accurately and in some detail by Peter Wright in his notorious memoirs *Spycatcher* in 1986.

RAMASUN STATION. The principal **National Security Agency (NSA)** facility in Southeast Asia during the **Vietnam War**, Ramasun Station was built in 1965 on the U.S. Air Force base at Udorn in northern Thailand and included **direction-finding (D/F)** and signals-**interception** antennae. Staffed by more than 1,000 **Army Security Agency (ASA)** and **Air Force Security Service** personnel, the site supported American forces engaged in combat operations across the region.

RAS-AL-GHALEB. In December 1969, Israeli commandos raided an Egyptian air-defense radar site at Ras-al-Ghaleb on Green Island in the Red Sea and removed the entire communications caravan and radar antenna on a pair of Sikorsky helicopters to Israel, while another lifted the four tons of Soviet radar equipment. Codenamed ROOSTER-53, the operation was conducted at night and provided technical analysts in Israel and the United States with an opportunity to study the electronics and develop suitable countermeasures. *See also* RADAR INTELLIGENCE (RADINT).

RED. The Luftwaffe's principal **Enigma** key was known at **Bletchley Park** as RED and contained high-level operational data. By the end of 1942, some 500 RED messages were being intercepted daily.

REPRIEVE. Codename applied within the West's intelligence community for clandestine intercept equipment installed in **diplomatic premises**.

REYNOLDS, S. WESLEY. In August 1948, Special Agent, S. Wesley Reynolds of the Federal Bureau of Investigation was asked by **Arlington Hall** to compile a list of Soviet codenames that the bureau had encountered in the hope that they might appear in the **VENONA** texts. The Reynolds list amounted to some 200 names, gleaned from technical surveillance, informants, and defector testimony, but only a few of the cover names appeared in the traffic. However, for those that did, it represented useful collateral, and in October 1948, Reynolds joined Robert Lamphere as a permanent link with the bureau to exploit further leads.

RHOMBIC ANTENNAE. The distinctive diamond-shaped aerials characteristically associated with signals-**interception** facilities, the rhombic antenna is usually rigged from four masts at a height of up to 24 meters above ground and is capable of monitoring very weak signals emanating from the direction of the diamond point.

RHYOLITE. A **National Security Agency** signals-intelligence satellite system, manufactured by TRW at Redondo Beach, California, the first RHYOLITE was launched in June 1969 from Cape Canaveral and transmitted signals to ground stations equipped with huge dishes at **Pine Gap**, **Menwith Hill**, and Buckley Air Force Base, Colorado. Three RHYOLITES were successfully placed in a very high geosynchronous orbit 24,000 miles above the earth, and, in 1975, the system was replaced by an enhanced version, initially known as "Advanced RHYOLITE" or "AR," the **ARGUS**.

A concept conceived by the deputy director of science and technology at the **Central Intelligence Agency (CIA)**, Albert D. ("Bud") Wheelon, the RHYOLITE project was headed by one of his engineers, Lloyd K. Lauderdale, who achieved success after Wheelon had left the CIA in 1966, after just four years in his post. The task of separating the invaluable **telemetry** data from other, less-important collected traffic was given to an interagency group, the Telemetry and Beacon Analysis Committee (TEBAC), which, over 18 months, recommended **discrimination** procedures that ensured the system produced the most impressive results and served to revolutionize signals-intelligence collection and processing.

In May 1977, a second generation, codenamed **AQUACADE**, was brought into service, and there were three further launches. The satellites were designed to collect Soviet communications and proved so successful that the Soviets began to encrypt the telemetry of their missile tests. However, both systems were compromised in 1976 by a TRW employee, **Christopher Boyce**. *See also* AIR INTELLIGENCE AGENCY (AIA); LEE, ANDREW D.

RIVET JOINT. Throughout much of the **Cold War**, U.S. Air Force **airborne collection platforms** routinely conducted SIGINT operations by flying missions along the periphery of hostile airspace to record target transmissions. Designated RIVET JOINT, these flights were undertaken by specially equipped RC-135 aircraft based at Offutt Air Force Base, near Omaha, Nebraska; Eielson Air Force Base in Alaska; Hellenikon Air Force Base near Athens in Greece; Kadena in Okinawa, Japan; Howard Air Force Base in Panama; and from RAF Mildenhall in England. Those flights over the Black Sea were designated CHROME CAP.

ROCKET. Soviet codename for an apparatus installed in the KGB's *rezidentura* in New York that monitored the radio communications of the surveillance teams deployed by the Federal Bureau of Investigation's local field office. *See also* IMPULSE; ZENIT.

ROCKEX. The successor of the British wartime **Typex** cipher machine, the Rockex was an improved version, which gradually replaced the Typex. The system was compromised in August 1967 when the British embassy in

Beijing was ransacked and the communications vault looted by Chinese purporting to be demonstrators.

ROOM 40. Established at the Admiralty upon the outbreak of war in August 1914, Room 40 was headed by Sir **Alfred Ewing** and studied German encrypted wireless traffic that had been intercepted at Hunstanton and identified by Marconi **direction-finding** facilities at Lowestoft, York, Murcar, and Lerwick. Room 40 remained operational until it was absorbed into the Naval Intelligence Division in 1917.

Ewing's cryptanalysts were assisted by the acquisition of several German codebooks recovered from ships, submarines, and **zeppelins**, and from the German consulate in Bushehr, which, in April 1915, produced copies 3512 and 89734 of a diplomatic code in current use. Another code, designated 13040, which proved the basis of the message known as the **Zimmermann Telegram**, was reconstructed by Nigel de Grey.

According to Admiralty records, Room 40 handled some 37,000 naval intercepts and distributed 54,000 individual messages between October 1914 and June 1918. In addition, a total of 30,000 encrypted diplomatic telegrams were studied, of which 90 percent were read. *See also* BRITISH SIGNALS INTELLIGENCE; MI1(b); MI1(e).

ROYAL AIR FORCE (RAF). The signals-intelligence branch of the Royal Air Force, established at RAF Waddington in Lincolnshire shortly before **World War II**, expanded to include a chain of **AI-1(e)** voice channel intercept stations and AI-4 signals-intelligence analytical centers at West Kingsdown in Kent, Cheadle in Staffordshire, and Montrose in Scotland.

Overseas, RAF Y was conducted by 380 Wireless Unit at the Chateau Beraud, at Draria, just outside Algiers; 351 Wireless Unit in Gibraltar; and 371 Wireless Unit in Freetown, Sierra Leone. Larger intercept facilities, designated 276 and 329 Wings, were responsible for managing a very large organization that included No. 2 at Amriyah, Alexandria; No. 3 at Maqab in Habbaniyah; No. 6 at Nerab near Aleppo; No. 9 on Kos; No. 10 at Siggiewi on Malta; No. 23 at Cap Bon, Tunisia; and No. 25 at Calabria; After the invasion of Italy, other sites were constructed at Conversano and Bitonto.

Airborne **interception** was provided by 192 Squadron, based initially at RAF Watton in Leicestershire, with a detachment at Maison Blanche just outside Algiers, but when 192 Squadron was disbanded, Watton become the **Central Signals Establishment**. In September 1948, a Lancaster and Lincoln B.2 converted for signals-interception duties were added and deployed to Habbaniyah in Iraq for flights along the Soviet border until May 1959, when the RAF moved to Cyprus. During the **Cold War**, the organization operated from bases at the Wireless Intelligence Development Unit at RAF Wyton in Cambridgeshire, under 162 Squadron cover, and from RAF Digby in Lincolnshire.

In July 1951, 192 Squadron was re-formed with Avro Lincolns and four Washingtons (the RAF version of the B-29), and in March 1953, one of the Lincolns was shot down by MiG-15s in East Germany, just outside the Hamburg air corridor, with the loss of six of the seven aircrew. On the same day, MiGs attacked another Lincoln near Kassel.

In February 1953, the squadron was equipped with Canberras carrying a third, extra member of the crew, a Russian-speaking intercept operator and **electronic-intelligence**-collection missions were flown on behalf of **Government Communications Headquarters** to and from Bodø in Norway and from Akrotiri. In anticipation of the Anglo-French invasion of the Suez Canal zone in October 1956, aircrew based in **Cyprus** undertook ferret missions to discover that Egyptian radar operators routinely switched off their equipment soon after midday and, in 1955, was credited with the discovery that MiG fighters had been fitted with airborne radar. The squadron had a forward operating base in Germany at Wünsdorf and later at Wildenrath and, in August 1958, was designated No. 51 Squadron. In April 1957, the Washingtons were replaced by the first of three Mark 2 Comets with a range of 3,060 miles, a cruising altitude of 40,000 feet, and provision for 10 intercept operators and 12 other passengers. Flights from Wyton included missions codenamed HEM, PAT, and TIBET. When deployed to Akrotiri, the Comet operations in the Mediterranean were codenamed DAMAGE, DOLVIN, and DUSTER. The Comets were withdrawn in 1974 and replaced by the Nimrod R-1.

Following the collapse of the Soviet Bloc in 1992, the RAF's SIGINT capability underwent a major review, but the three Nimrod R-1s, which had been introduced into service in 1974, actually a variant of the converted Comets deployed as maritime patrol aircraft, survived the cuts but were moved in 1995 from RAF Wyton to RAF Waddington. As part of their duties, they were deployed to detect and collect traffic from Russian vessels watching the transits of the Royal Navy's ballistic missile submarines, and, in May 1995, one of the planes was obliged to ditch in the Moray Firth. *See also* AIRBORNE COLLECTION PLATFORMS; BRITISH SIGNALS INTELLIGENCE; JU-JITSU.

ROYAL PALM. The codename of a joint East German and Cuban project to monitor signals to and from the U.S. Navy base at Guantanamo Bay in 1982, ROYAL PALM was sponsored by the **Ministerium für Staatssicherheit** and began with a team of a dozen or so technicians led by an English and Russian linguist, Major Fritz Gregor, who surveyed the site and then arranged for a mobile SIGINT suite to be delivered from Rostock by a freighter at the end of January 1984. The plan called for Guantanamo's communications traffic to be intercepted for two months, but the operation continued until August 1984, when the equipment was returned to East Germany.

RUGBY. The large **General Post Office (GPO)** high-power transmitters located at Rugby were employed during **World War II** to jam enemy communications. In May 1940, a landline connected the **War Office Y Group** site at **Fort Bridgewoods** to Rugby, and as soon as the intercept operators found the frequencies being used by Stuka pilots engaged in ground-attack operations, the GPO jammed them by broadcasting readings from the bible with 25 kilowatts of power. The interference proved highly successful at disrupting the Luftwaffe's campaign.

RUPERT. In his memoirs, *The Man behind the Rosenbergs*, first published in Paris in 1999 and then in New York in 2001, the **NKVD** case officer Alexander Feklisov revealed that, in July 1944, he had been assigned the task of running an American source codenamed RUPERT; although, he would not meet him until February 1945, when the agent, "fluent in several oriental languages, including Japanese," disclosed that he had been transferred to the U.S. Army's **Signal Intelligence Service (SIS)**. RUPERT, who claimed to have spent the last few months in Hawaii, revealed that the SIS had broken the diplomatic cipher used by the Japanese to communicate with Moscow. He also said that 70 percent of an Amtorg message had been solved and that considerable resources were being devoted to attacking Soviet systems. Feklisov held only one further meeting with RUPERT, whom he handed over to a colleague named Tikhov to run. RUPERT was identified in an internal memorandum compiled by Anatoly Gorsky in 1948 as Vladimir V. Sveshnikov. *See also* SOVIET/RUSSIAN SIGNALS INTELLIGENCE.

RUSSIAN SIGNALS INTELLIGENCE. *See* SOVIET/RUSSIAN SIGNALS INTELLIGENCE.

S

SAMBO. A joint Allied project based on **CLASSIC BULLSEYE** technology to identify and monitor low-frequency transmissions from submerged Soviet submarines while on patrol, SAMBO could pinpoint a target while on patrol. The entire system was compromised in 1980 by **Geoffrey Prime**, the mole inside **Government Communications Headquarters**, who passed the details to his KGB 3rd Chief Division contact. *See also* CLASSIC WIZARD.

SAND ISLAND. Located close to Midway in the Pacific Ocean, Sand Island was the site for a major **National Security Agency** intercept site designed to monitor Soviet submarine and shipping communications.

SANDRIDGE. Built in September 1939 by the **General Post Office (GPO)**, the direction-finding and intercept station at **Sandridge**, near St. Albans in Hertfordshire, was linked by landline to identical GPO facilities at Brora and **Cupar**. Initially intended to monitor and identify illicit, clandestine wireless traffic, Sandridge was absorbed into the **War Office**'s **Y Group** during **World War II** but afterward reverted to Home Office control.

SARAFAND. The principal British signals-intelligence intercept site in the Middle East, Sarafand was closed down in 1947 as the British mandate in Palestine drew to a close. Operating under cover as part of the Empire Chain wireless relay system, the Sarafand site included some cryptographic and **traffic-analysis** facilities and, in 1930, broke the Soviet Black Sea Fleet's call-sign code.

SATELLITE COMMUNICATIONS. As soon as the first generation of intercontinental communications satellites were established in 1965, by the placing into orbit in 1964 of the International Telecommunications Satellite Organization's (**INTELSAT**) Early Bird, by its subscriber countries, which grew to number some cooperating 144 governments, the five **UKUSA** agencies built parallel ground stations conveniently close to the INTELSAT facilities. Thus, **Morwenstow**, Sugar Grove, and Yakima covered the INTELSAT links to Goonhilly, Etam, and Carlton, respectively. As subsequent generations of telecommunications satellites were developed, the UKUSA agencies expanded to extend their interception capabilities by building new sites at **Pine Gap**, Australia; **Hong Kong** (which moved to Geraldton); Sabana Seca in Puerto Rico; Waihopai in New Zealand; Misawa in Japan; **Menwith Hill** in England; and **Bad Aibling** in Germany.

Independently of INTELSAT, other groups of countries created their own regional satellite communications links, among them INTERSPUTNIK in 1971, ARABSAT in 1976, EUTELSAT in 1977, INMARSAT in 1979, PANAMSAT in 1988, and PALAPA in 1995. Also, individual countries have strengthened their own internal communications by deploying satellites, among them France's TELECOM, Italy's ITALSAT, Israel's AMOS, and Spain's HISPASAT.

Additionally, various countries have developed their own exclusive military satellite-communications systems, with Great Britain operating SKYNET, France's SYRACUSE, the United States' MILSTAR, the Italian SICRAL, and the Russian MOLNIYA. *See also* AIR INTELLIGENCE AGENCY (AIA); MOON SHOT; PYRAMIDER.

SAUTERNE. The **Government Communications Headquarters** codename for Red Army traffic in the Far East intercepted and read during the postwar **BOURBON** project.

SCHMIDT, HANS-THILO. A corrupt German cipher clerk employed by the Reichswehr's **Chiffrierstelle**, Hans-Thilo Schmidt sold cipher-machine manuals and other classified crypto material to **Gustave Bertrand** of the French Service de Renseignements.

Born into an aristocratic family in Berlin in May 1888, Schmidt was decorated during **World War I** and afterward was given a job as a civilian clerk by his elder brother, Rudolf, who had remained in the army and had been appointed to head the Chiffrierstelle. Later, at the age of 52, Lieutenant-General Schmidt would command the 39th Panzer Korps.

However, by June 1931, Schmidt had adopted a lifestyle and a mistress he could not afford, so in July, he approached the French legation in Berne with an offer to sell classified crypto material. The Service de Renseignements accepted with alacrity, and Bertrand, accompanied by Rodolphe Lemoine, met the spy, codenamed ASCHÉ ("H. E."), at the Grand Hotel in Verviers, Belgium. At this rendezvous in November 1931, Bertrand received more than 100 technical documents relating to the **Enigma**, which he shared with the British Secret Intelligence Service station commander in Paris, Wilfred Dunderdale, and with the **Polish Cipher Bureau** in Warsaw.

For their next meeting, in Verviers, Bertrand took a Leica camera and photographed the army Enigma keys for November and December. On 8 May 1932, ASCHÉ produced pictures of the Enigma's plugboard settings for that month and, at a subsequent meeting at the Hotel Adlon in Berlin, approved a method of paying him through the French embassy with a diplomat delivering cash to a poste restante in the Tiergarten. At the end of October 1932, they met again at the Hotel d'Angleterre in Liège, and ASCHÉ handed over the Enigma key settings for November and December 1932. On 31 March 1933, ASCHÉ summoned Lemoine to the Adlon to reveal a plan to create the **Forschungsamt**, which was established on 10 April. He supplied more Enigma material to Bertrand and Guy Schlesser at the Hotel Davidsbaude in Splindlermühle on 16 September, having hiked over the border from Bavaria, including the daily keys for September and October.

At a series of meetings held over five days in January 1934, at Mürren in Switzerland, ASCHÉ revealed that he had been appointed to a liaison role with the Forschungsamt and supplied the January and February keys together with a list of the organization's secret radio call-signs and a 30-page internal manual. On 20 April, he met Schlesser at the Tiergarten Taverne and, the following day, passed him documents in the Adlon's bar, among them the keys for April and May. Then, in Montreux in July, he gave the June and July keys and revealed that the Forschungsamt had moved its headquarters into the Schillerstrasse and was monitoring the French embassy's landline communications. His 11th meeting took place in Zurich in January 1935, followed by one on 18 October in Bern, on 24 January 1936 at the Hotel Euler in Basel, and on 24 April in Lucerne, where he gave Bertrand the keys for April and May 1936. On 17 October in Basel, he met Major Perruche, the fourth Frenchman to know his true identity.

However, on 6 November 1937, he made an emergency rendezvous at the Charlottenburg railway station with the *Le Journal* correspondent Georges Blun to warn of Adolf Hitler's plan to absorb Austria and Czechoslovakia. Blun had hurried with the news to the French embassy where a diplomat named Dejean, had sent an encrypted cable to the Quai d'Orsay, unaware the message would be intercepted at Glienecke and read by the Forschungsamt and the Gestapo. This was confirmed by ASCHÉ when he next met Bertrand and Lemoine at the Hotel Euler on 18 November when he passed on the Enigma keys for November and December. On 26 January 1938 in Bern, he reported to Schlesser that the Forschungsamt's Dr. Kurzbach had failed to identify the culprit responsible for the leak, despite greater surveillance on the embassy, and met him again in Basel on 21 March.

That same month, Lemoine, while on a mission in Cologne, was taken into custody and questioned. Originally born in Berlin as Rudolf Stallman,

Lemoine had adopted his French wife's surname. He did not betray ASCHÉ but told Paul Paillole that he had bluffed the Gestapo by offering to supply details of French diplomatic codes used to communicate with Prague, Rome, and Budapest, which he knew had been compromised. Paillole accepted Lemoine's version but directed that he should not meet ASCHÉ again and should not leave France.

On 23 May, ASCHÉ gave Bertrand and Perruche the Enigma keys for May and June and in August took a vacation at La Madeleine in Paris, revealing to the two French officers that he was to be transferred to the Forschungsamt's intercept site at Templin. Further meetings took place in Basel on 29 January 1939, attended by Henri Navarre, and then again in Basel in early July, with a final rendezvous in the Hotel Eden in Lugano on 10 March 1940.

Following the Nazi occupation of France, Lemoine was arrested at his home in Saillagouse, in the Languedoc, and under interrogation in March 1943, he compromised Schmidt, who was himself taken into custody. Schmidt committed suicide in prison in September 1943, and his brother was cashiered, while Lemoine survived his incarceration and died in August 1946.

SEKTION 114. Between 1950 and 1955, the Federal German foreign ministry sponsored a cryptographic organization, designated Sektion 114, which was headed by Adolf Paschke and concentrated on East European cipher systems. Sektion 114 was later absorbed into the **Bundesnachrichtendienst**.

SERVICE 5. *See* GROUPEMENT DES CONTRÔLES RADIO-ÉLECTRIQUES (GCR).

SHAMROCK. The **Army Security Agency** codename for an agreement reached in 1945 with the three biggest U.S.-based international commercial carriers, RCA Global, ITT World Communications, and Western Union, to supply the U.S. government with copies of all overseas cables. *See also* MINARET.

SHARK. The **Government Communications Headquarters** codename for the Kriegsmarine's TRITON **Enigma** key reserved for the use of U-boats, introduced in October 1941 and broken by **Bletchley Park** on 13 December 1942, after a four-rotor Enigma was recovered from *U-559*, a German submarine captured in October. The switch to the new machine was not completed until 1 February 1942.

Like the other three Enigma rotors, the fourth, codenamed BETA, had 26 starting positions, but it was rigid in that it did not move in sequence with the others and was not interchangeable with them. For the Hut 9 cryptographers, anticipating that the current 17,576 permutations might be multiplied by a factor of 234 and not by a catastrophic 3,024, proved highly advantageous.

The introduction of SHARK effectively terminated the U-boat intercepts from January 1942 until the end of the year, a period that saw a dramatic increase in shipping losses, which ranged from 60 to 108 ships a month. SHARK was broken on 13 December 1942 following the capture of cryptographic material from the *U-110* in October, but this success only lasted until 10 March 1943, when a new short weather-signal book was introduced. However, with help from the material recovered from the *U-559* and the *U-205*, the intercepts resumed a week later, on 19 March, when traffic for the previous three days was read. Additionally, the Kriegsmarine was found to be repeating messages, originally transmitted in the short weather code, in the *Werftschlüssel*, the easily broken dockyard hand cipher, thus making the cryptanalytical task much easier.

On 1 July 1943, SHARK became inaccessible briefly when the Kriegsmarine introduced two new measures, a new fourth rotor, codenamed GAMMA, to replace BETA, and a new reflector, codenamed CAESAR, to replace the conventional one known as BRUNO. These modifications were defeated on 19 August when it became possible to read retrospectively SHARK for all the rest of the month, using broken **DOLPHIN** texts that had been rebroadcast on SHARK as "cribs," and from 12 September, it was read every day until the end of hostilities.

SHEYMOV, VIKTOR. A defector from the KGB's Sixteenth Directorate, Viktor Sheymov was exfiltrated to Finland in 1980 after a boating accident had been staged to suggest that he and his wife and daughter had drowned.

A Muscovite born in 1946 and a graduate of Moscow's Technical University, Viktor Sheymov joined the KGB in 1971 at the age of 25, having been recommended by friends of his father, a doctor. After training, he was assigned to the elite Eighth Chief Directorate, responsible for all the KGB's communications. Initially, he worked as a cipher clerk and then was posted temporarily to the First Chief Directorate's new headquarters in Yasenevo to assist in the installation of a new mainframe computer. Having gained a reputation as a computer expert, Sheymov became the Eighth Chief Directorate's troubleshooter, checking on the security systems protecting KGB facilities across the globe. In particular, he scored well in Warsaw where the local *rezident* had fallen foul of the ambassador over a dispute concerning the size of the *referentura*. Sheymov overruled the ambassador and ensured the KGB's prestige was untarnished.

Late in the autumn of 1975, Sheymov was promoted to the Third Department, dealing exclusively with security systems, and there acquired an unrivalled knowledge of the precautions taken by the KGB to prevent hostile **interception** of its machine ciphers. However, the more he learned about the KGB, the less satisfaction he found in his job, and he was suspicious about the circumstances of a friend's death in which he suspected the KGB had played a part. He was particularly disillusioned by the scale of the KGB's penetration of the church and, as a minor party official within the KGB, he realized through the required study of Communism that the leadership had betrayed the party: "What I've learned so far has shocked me deeply. It's clear that ever since the Revolution the leaders of the Communist Party have deviated more and more from the original Party goals. Now all we have is a bunch of bureaucrats who have usurped all the power and are doing nothing for the good of the country."

Sheymov obtained his wife's consent to defect, and he established contact with the **Central Intelligence Agency** in November 1979 while on a visit to Warsaw. He was smuggled out by car with his wife (Olga) and their five-year-old daughter, Elena, in May 1980 to Finland. Their disappearance went unreported in the West for 10 years while Sheymov underwent a lengthy debriefing. His autobiography, *Tower of Secrets*, was published in 1993.

SIGNALBUCH DER KAISERLICHEN MARINE **(SKM).** On 17 October 1914, the Russian naval attaché in London, Admiral Wolkoff, delivered a copy to the Admiralty of the SKM, which had been recovered from the German light cruiser *Magdeburg*. It was the principal German naval codebook, and it remained in use until May 1917. One of the first intercepts to be read indicated that a German auxiliary had been dispatched to a particular area to search for some missing torpedo boats. Admiral Oliver then sent a cruiser to the designated spot and intercepted the enemy ship, thus proving the validity of the decrypts. Further study of the SKM by Paymaster Charles Rotter of the Naval Intelligence Division (NID) revealed two keys, one of which proved to be the personal cipher of the commander in chief.

In May 1917, the SKM was replaced by the *Flottenfunkspruchbuch* (FFB), a two-part codebook that was introduced simultaneously with a change of call signs and other procedures, including a map-grid system. According to a contemporary report on **Room 40**'s cryptographers, "most of the call-signs could be guessed immediately, at least in so far as the type of vessel or area to which they belonged. A few, however, baffled analysis for some considerable time." *See also* WORLD WAR I.

SIGNAL INTELLIGENCE SERVICE (SIS). Created in 1929, the Signal Intelligence Service was based at the Munitions Building in Washington,

DC, and originally consisted of **William Friedman**, Frank Rowlett, Solomon Kullback, Abraham Sinkov, and three clerks operating with an annual budget of $17,400. As well as devising ciphers for the U.S. War Department, SIS cryptanalysts studied the encrypted wireless of other countries and, with the help of battlefield seizures in the Pacific theater, achieved considerable success in reconstructing various Japanese codes.

The SIS's intercepts from two stations, at Battery Cove, Virginia, and Fort Monmouth, New Jersey, were enhanced with copies of telegrams supplied by commercial carriers. By 1938, the Signal Corps' intercept capability had increased by the construction of new intercept facilities at Fort Sam Houston in Texas and at Fort Scott in the Presidio, San Francisco. Overseas, SIS collected traffic from Quarry Heights in the Panama Canal Zone, Fort Shafter on Oahu, and Fort McKinley in the Philippines. In March 1939, Fort Hancock in New York was added to the network, with Fort Hunt, Virginia, contributing the following October.

The SIS was absorbed into the Army **Signal Security Agency** in 1947, and then renamed the **Armed Forces Security Agency** in May 1949. According to **Alexander Feklisov**, from 1945, the **NKVD** ran a spy inside the SIS, codenamed **RUPERT**. *See also* UNITED STATES SIGNALS INTELLIGENCE.

SIGNAL SECURITY AGENCY (SSA). Created in 1944, the Signal Security Agency took over the staff and assets of the U.S. Army **Signal Intelligence Service**, which included 900 intercept sites across the globe, and was absorbed into an amalgamated multiservice organization, the **Armed Forces Security Agency**, in May 1949.

SILVER. A technical operation conducted by the British Secret Intelligence Service, SILVER was a tunnel dug under a street in Vienna to tap the telephone and teleprinter cables of the Soviet *Kommandatura*, accommodated in the old Imperial Hotel. Altogether, 81 speech circuits had been tapped, producing a daily total of 162 reels of tape, each with a duration of two and a half hours. The operation proved so successful that plans were made, codenamed GOLD, for a similar enterprise in Berlin. *See also* BERLIN TUNNEL.

SIRDAR. The Allied codeword employed during **World War II** to indicate special intelligence derived from India. *See also* ULTRA.

SKIP ZONE. Depending on the prevailing atmospheric conditions, signal **interception** from ground stations can only take place at locations where a target transmission wave bounces off the ionosphere and is reflected to the earth's surface. The distance between a signal's origin and the point of interception, where no traffic can be detected, is known as the skip zone. *See also* VIETNAM WAR.

SOCIAL NETWORKING. The use of the Internet to communicate provided SIGINT agencies with new channels to monitor; although, it was not until the "Arab Spring" of December 2010 that YouTube, Twitter, and Facebook acquired a strategic significance with the media being used to exchange information beyond the authoritarian control of the local security apparatus of totalitarian regimes. Whereas previous attempts to organize antigovernment demonstrations in Iran in 2009 had been suppressed by the expedient of disconnecting Internet servers and cell-phone towers, social networks were utilized to coordinate demonstrators, first in Tunisia and then successively in Egypt, Bahrain, Libya, and Yemen in January 2011, followed by unrest in Syria in March. Advice, ranging from methods of relying on international dial-up services to circumvent restrictions on local servers to the application of vinegar and onions in facemasks to neutralize the effects of tear gas, was delivered to protestors, making their impact much more effective. In contrast, previously ubiquitous local security agencies found themselves wrong footed and ill equipped to keep up with the pace of instantaneous mass communication employed by technologically adept opponents.

External analysis of text-messaging volume became one of the accepted benchmark indicators of urban discontent; although, the source could not provide actionable intelligence, either for relevant *mukhabarat* or for foreign observers engaged in **interception**.

SONGBIRD. During the **Vietnam War**, U.S. **Air Force Security Service** intercept operators monitored enemy wireless traffic to identify the location of downed American aircrew. Codenamed SONGBIRD, this traffic was swiftly exploited and often enabled search-and-rescue missions to be mounted before the Vietcong or North Vietnamese ground forces could react.

SOSUS. The acronym for Sound Surveillance System, SOSUS was developed in 1954 as a passive underwater sonar apparatus constructed on the Ramey Air Force Base in Puerto Rico with technology developed by American Telegraph and Telephone in partnership with Western Electric. This was followed by installations at Sandy Hook, New Jersey, and on Eleuthera in the Bahamas, which were designed to monitor the sound of Soviet submarines transiting through unavoidable geographical choke-points, such as the Straits of Gibraltar.

The hydrophones, some as large as a double-decker bus, were constructed by Bell Laboratories and recorded signals on the ocean floor in the 5- to

150-hertz bandwidth. Altogether, 18 acoustic-detection stations equipped with digital spectrum analyzers were located in the Pacific, Atlantic, and Mediterranean at a cost of $51 million during the first decade of development. SOSUS was able to detect Soviet submarines at very long distances, analyze the noise they emanated, eliminate the ambient sound, and identify individual boats by their particular, characteristic acoustic signatures.

The extent and efficacy of SOSUS was betrayed to the Soviets by **John Walker** in 1968, but the Soviets found it hard to develop countermeasures apart from attempting to mask the sound of submarines by having them rendezvous with a Soviet Bloc merchantman and transit the area directly below the ship's propellers in an effort to conceal the noise of its own screw.

During the **Cold War**, SOSUS terminals were maintained on Bear Island, Norway; Tudor Hill in Bermuda; and Argentia, Canada, and at **Diego Garcia**, Greenland, Hawaii, Iceland, Italy, Japan, Norway, the Philippines, Scotland, Turkey, and RAF Brawdy in Pembrokeshire, Wales.

SOVIET/RUSSIAN SIGNALS INTELLIGENCE. Having developed a highly reputable collection and decryption bureau during the Czarist era, Russia's principal cryptographers, including Felix and Ernst Fetterlein, fled abroad during the Bolshevik Revolution and peddled their knowledge to eager British and French cryptographic organizations. Thereafter, such evidence as there is suggests that Soviet signals intelligence was never considered a priority, and a small joint GRU/NKVD SIGINT unit pioneered by Gleb Bokii was dismantled after his execution during a purge in 1937. Breaks into foreign diplomatic ciphers were achieved through the recruitment by the **NKVD** of code-room staff, among them Tyler Kent and Henry Antheil of the U.S. embassy, who had both passed sensitive cipher material to their Soviet contacts. Similarly, the recruitment of Captain John King, a Foreign Office code clerk, and the Constantini brothers, who were both employed at the British embassy in Rome, supports the impression that the NKVD's SIGINT effort was directed more at "**pinches**" than the application of any particular cryptanalytical skills. Certainly, Soviet cipher systems proved vulnerable to the **Government Code and Cypher School**, which had gained access to some of the Kremlin's diplomatic channels prior to the public statements of August 1920 and May 1927 and to the Comintern's clandestine wireless traffic between February 1934 and January 1937.

During **World War II**, the NKVD's apparently fruitful attack on German Enigma ciphers were made possible by original **ISK**, **ISOS**, and **ULTRA** summaries supplied by Anthony Blunt, Leo Long, John Cairncross, and Kim Philby, while a GRU source, codenamed BARON, passed fresh decrypts to his Soviet handlers directly from **Bletchley Park** from early 1941. With the benefit of these cribs, Soviet cryptographers achieved some limited success with encrypted Axis wireless traffic.

In the **Cold War** era, the KGB's 16th Directorate is known, according to Victor Sheymov, to have supplied the Kremlin with a huge quantity of raw SIGINT, leaving it to the politburo membership to undertake its own interpretation. Sheymov, who defected in 1980 with nine years' experience, was the first KGB officer to give an inside account of the organization. Previous defectors with knowledge of Soviet cipher systems, including the GRU's **Igor Gouzenko** and the NKVD's **Evdokia Petrov**, had been cipher clerks within highly compartmented departments who had learned little of Soviet SIGINT beyond the existence of clandestine collection facilities within their own *referenturas*, which reputedly were engaged mainly on tactical counterintelligence tasks in support of local operations. Nevertheless, the global nature of the Soviet collection network, based on diplomatic premises, airborne and naval platforms, and a significant investment in SIGINT satellites, combined with a network of ground intercept bases located in Cuba, Syria, Ethiopia, and Vietnam, implies an effective capability. Additionally, the KGB's capability must have been greatly enhanced by information offered by "walk-ins," such as **Bernon Mitchell**, **William Martin**, **Victor Hamilton**, **Ronald Pelton**, Robert Lipka, **Jack Dunlap**, **John Walker**, **James Hall III**, **Brian Patchett**, and **Geoffrey Prime**, and by study of materiél recovered in 1968, first from the USS ***Pueblo*** in January, and then two months later from **Phou Pha Thi** in Laos.

Following the Soviet collapse, the 16th Directorate was absorbed into FAPSI, an ostensibly independent civilian SIGINT organization, which inherited the KGB's assets. Having withdrawn from the intercept site at Cam Ranh Bay, Vietnam, in May 2002, the principal remaining overseas Russian SIGINT facility was at the naval base at Tartus in Syria. *See also* CURZON NOTE; IMPULSE; MASK; ZENIT.

SPECIAL COMMUNICATIONS UNIT (SCU). The signals organization responsible for maintaining a wireless link between the Secret Intelligence Service (SIS) and individual **Special Liaison Units** in the relevant theaters were known as Special Communications Units. SCU 1, also referred to as "Main Line," was based at Whaddon Hall, SIS's Section IX headquarters, which was supported by local stations at Nash, Weald, and Windy Ridge. SCU 2 was the transmitter at Ravenscroft, directly opposite Arkley View in Barnet. SCU 3 was located at Hanslope Park, and SCU 4 was based at Dingli under Middle East Relay Service cover. Later in World War II, additional SCUs were established at Algiers, Beirut, Malta, Naples, Eindhoven, Bad Salzuflen, Bad Oeynhausen, Bad Godesberg, and Brisbane. SCU 11 went to

Palestine, with a satellite station at Habbaniyah, and there were further links to Delhi, Colombo, and Melbourne. The entire organization was supervised by Group Captain Fred Winterbotham, formerly head of the Secret Intelligence Service's air branch, Section IV. *See also* CENTRAL BUREAU; SPECIAL WIRELESS GROUP (SWG); WIRELESS EXPERIMENTAL STATION (WES).

SPECIAL INTELLIGENCE. Within the Allied intelligence community, the term "special intelligence" conventionally has been employed as a euphemism for signals intelligence. However, in January 1946, Anglo-American cryptographers agreed on a lexicon that narrowed the meaning to intelligence derived from cryptanalytical sources and designated a prefix, CREAM, to distinguish it. Content with a political, rather than military, significance was codenamed MAUVE. *See also* COMMUNICATIONS INTELLIGENCE (COMINT).

SPECIAL LIAISON OFFICER (SLO). Employed by Richard Gambier-Parry's Section IX of the British Secret Intelligence Service (SIS), SLOs were responsible during **World War II** for the handling of "special intelligence" and ensuring it was never seen by unauthorized personnel. SLOs operated from **Special Liaison Units** attached to overseas commands and were in direct radio contact with Section IX's headquarters at Whaddon Hall, Buckinghamshire. *See also* BONIFACE; ULTRA.

SPECIAL LIAISON UNIT (SLU). To distribute signals-intelligence decrypts securely, the communications branch, Section IX, of the British Secret Intelligence Service (SIS) introduced a channel staffed by indoctrinated personnel in 1940 and opened a radio link between Whaddon Hall's transmitter at Windy Ridge in Buckinghamshire and the SLU attached to Lord Gort's British Expeditionary Force headquarters, where a pair of experienced SIS officers, Humphrey Plowden and Robert Gore-Brown, later to be joined by Squadron Leader F. W. "Tubby" Long, based at the Advanced Air Striking Force headquarters at Meaux, handled the summaries and was responsible for personally decoding the messages sent in a **one-time pad** cipher system.

As **World War II** progressed, more SLUs were attached to senior commanders, in the Mediterranean theater at Athens, Cairo, Alexandria, Malta, and 8th Army Headquarters. Then, in March 1942, Benghazi was added, followed by Beirut, Tunis, Bizerte, Sicily, and Gibraltar. In 1943, Salerno, Taranto, and Bari were included and then Bastia, La Marsa, and Vasto. Finally, Morocco, Verona, Leghorn, Naples, Florence, and Ancona were indoctrinated.

The senior commanders were permitted to examine what was termed "special intelligence" only in the presence of the **special liaison officer**, an SIS officer charged with remaining in direct control over the material at all times and being responsible for its secure destruction. *See also* BONIFACE; BRITISH SIGNALS INTELLIGENCE; ULTRA.

SPECIAL WEATHER. The technique of intercepting signals containing local weather reports, in reference to flying meteorological missions, is known as "special weather," and a reliable source will obviate the need for mounting telltale flights prior to conducting air raids. During the **Vietnam War**, distinctive weather-reconnaissance flights were correctly interpreted as advance notice of an imminent enemy operation, so the **Air Force Security Service** resorted to an **interception** program to supply the U.S. 7th Air Force with the required data without risking alerting North Vietnam's air defenses.

SPECIAL WIRELESS GROUP (SWG). Following the evacuation of No. 2 GHQ Signals from France in May 1940, the Royal Corps of Signals created a specialist intercept capability, the Special Wireless Group, and trained intercept operators at the regiment's depots at Trowbridge and Bridgewater and at a requisitioned holiday camp at Prestatyn in North Wales.

Initially, SWG 1 was posted to Rothamsted Manor in **Harpenden**, a facility designated Y-1, and a subsection, Y-3, was established in the basement of Queen Anne's Mansions near the Secret Intelligence Service headquarters in Broadway, Westminster, with another on the roof of the Passport Office in Petty France. Elsewhere in England, Special Wireless Sections (SWS) were formed, with 3 SWS mobilized at Rottingdean; 104 at Maidstone; 108 in Cheveley, near Newmarket; 110 at Frant in Kent; and 108, 116, and 118 went to France immediately after D-Day. SWS **direction finding** led directly on 10 June 1944 to an air raid on the headquarters of Panzergruppe West at La Caine, near Thury-Harcourt, in which 17 senior German staff officers were killed, among them the chief of staff, which seriously disrupted the enemy's armored operations in Normandy.

A second unit, SWG 2, was posted to Sarafand to reinforce No. 2 Wireless Company, Royal Signals, and, by March 1941, had established an intercept facility in the King Farouk Museum at Heliopolis, next to the station manned by RAF Nos. 50 and 53 Wireless Units. 102 SWS was posted to Gibraltar in January 1941, and an SWS was established at Dingli in Malta in March 1941, having been delivered by submarines, HMS *Regent* and *Thresher*; 3 SWS went to Port Tewfik; 51 to Bizerte; 101 to Constantine; 103 to El Coefia; 105 to Derna; 107 and 114 went to Oran and Algiers; 113 to Tripoli; while 106 and 110 participated in the invasion of Sicily; 200 operated under Eastern

Wireless Sub-Centre cover at Barrackpore, India. SWG 3 was formed in Italy in July 1944. No. 2 Special Wireless Detachment went to Accra in West Africa; 151 SWS and 153 SWS ran seven stations from Dingli, including Beirut, Mosul, Kirkuk, and Tehran.

After training in Heliopolis in April 1941, several SWS were posted to a large site at Dobi Road, New Delhi, as 5 Wireless Detachment to provide support for the **Wireless Experimental Station**. In addition, SWG 3 was established at Ballygunge in Calcutta, 112 and 115 SWS at Bombay, 202 SWS at Uknia, and 203 SWS at Chittagong.

The only SWS to fall into enemy hands was 105 SWS, which arrived at Tobruk from England in January 1941 but, on 27 June, having intercepted Erwin Rommel's plan to attack Mersa Matruh, found itself surrounded at El Daba. It survived undetected behind enemy lines for three days but was then captured by a reconnaissance column of the 21st Panzer Division. Fifty prisoners were interrogated at Benghazi and then shipped to Bari for captivity at Fara Sabina before ending the war in Germany at Stalag VII-A. Captain J. B. ("Bingo") Mellor, 105's commanding officer, ensured the destruction of his unit's secret papers and evaded capture but was heavily criticized for allowing valuable radio equipment to be recovered by the Germans. *See also* BRITISH SIGNALS INTELLIGENCE; WAR OFFICE Y GROUP (WOYG).

SPECIAL UNITED KINGDOM LIAISON OFFICER (SUKLO). Under the terms of the **BRUSA** and **UKUSA** Agreements, a British representative of **Government Communications Headquarters** was posted to the British embassy in Washington, DC, to handle SIGINT issues. The first SUKLO, who fulfilled the role until his retirement in 1954, was John H. Tiltman, who was replaced by Patrick Marr-Johnson.

SPECIAL UNITED STATES LIAISON OFFICER (SUSLO). Under the terms of the **BRUSA** and **UKUSA** Agreements, an American representative of the U.S. Army-Navy Communication Intelligence Board was attached to the U.S. embassy in London to handle SIGINT issues, reciprocating with the **special United Kingdom liaison officer (SUKLO)**. *See also* GOVERNMENT COMMUNICATIONS HEADQUARTERS (GCHQ); NATIONAL SECURITY AGENCY (NSA).

STAFF D. The section of the **Central Intelligence Agency (CIA)** that concentrated on the acquisition of Soviet cryptographic material was designated Staff D and, headed for more than 12 years by Carleton Swift, undertook clandestine operations, usually against cipher rooms located inside Soviet diplomatic premises, to eavesdrop on communications and learn more about coding procedures. In 1978, Staff D was absorbed into a joint CIA and **National Security Agency** unit, the Special Collection Service, which operated small, two- or three-man Special Collection Sites concealed in U.S. diplomatic premises in target countries. Known as Special Collection Elements, these units were equipped with the latest, most sophisticated eavesdropping technology, including laser equipment that could pick up conversations by bouncing an invisible beam off a window, and were concentrated in embassies in Eastern Europe and the Middle East.

STATION X. *See* BLETCHLEY PARK.

STELLA POLARIS. At the end of hostilities between Finland and the Soviet Union on 17 September 1944, the Finns feared that, as had happened in Romania, the country would be occupied by the Soviets, and a group of nearly 800 intelligence officers and their families, led by the chief of military intelligence, Colonel Aladár Paasonen, and the head of the signals-intelligence section, Colonel Reino Hallamaa, participated in Operation STELLA POLARIS, the removal of Finland's cryptographic archive to Stockholm.

Paasonen, still regarded as the father of Finnish cryptography, was a brilliant linguist who had been taught most European languages by his Hungarian mother. Although he had studied Soviet cipher systems in Germany, Poland, and the Baltic and had made the first purchases of intercept equipment in Italy in the 1920s, the principal figure in STELLA POLARIS was Hallamaa, who was convinced that Finland was about to become a Soviet surrogate, and to some extent, he was proved correct. The Communist Party, together with the Social Democrats, held the majority in the first postwar government, and the Security Police fell into its hands.

As well as achieving considerable success with Soviet ciphers, Paasonen's department had recruited and run dozens of Soviet sources, and he knew that both they and their case officers would be doomed if the **NKVD** got its hands on his files. In addition, Paasonen also ran a commando battalion, designated ErP4, which operated behind Soviet lines, whose membership was also at considerable risk. Accordingly, with the tacit approval of the Finish President, Field Marshal Gustaf Mannerheim, and his military chief of staff, Erik Heinrichs, Hallamaa had invited his Swedish counterpart, Major Carl Petersen, the director of the defense staff's intelligence branch, known as the C-Bureau, to open negotiations at his villa in June 1944 at Sökö, just outside Helsinki. His objective was to gain Swedish consent to his plan to reestablish his entire organization in Sweden and continue operations against the Soviets.

For their part, the Swedes knew that if Finland fell to the Soviets, their own territory would then become the front line, and few then doubted that Josef

Stalin's ambitions included the whole of Scandinavia. Petersen was also aware that as well as possessing useful assets across the Soviet frontier, the Finnish codebreakers had accomplished far more than the Swedish SIGINT branch, known as the Swedish National Defence Radio Institute (FRA), then headed by Commander Torgil Thorén. Relations between the FRA, which had been created in 1942, and the Finns, had been sufficiently close to allow Thorén to second one of his subordinates, Captain Ake Rossby, to Helsinki during part of the Russian-Finnish War, and he was again chosen to negotiate with Hallamaa. However, Thorén himself was kept deliberately isolated from the STELLA POLARIS plans because he had become distrustful of the Finns following an incident earlier in the war when the Finnish military attaché in Stockholm, Colonel Stewen, tipped off the Abwehr to the success the Swedes had achieved in intercepting and decrypting teletype traffic on the German Oslo–Berlin landline, which happened to pass through the capital's telegraph exchange.

The extent to which the talks conducted by Hallamaa, Petersen, and Rossby were officially authorized remains a matter of speculation, with both governments insisting, probably to assuage Moscow, that they were an unauthorized private venture. Nevertheless, Mannerheim and the chief of the Finnish Defense Staff, General A. E. Heinrichs, knew of the scheme, as did the chief of the Swedish Defense Staff, Major-General C. A. Ehrensvärd. Whatever the truth, which is still politically delicate, Hallamaa assigned two subordinates, Lieutenant-Captain Heikki Aulio and Lieutenant Veikko Virkkunen, to plan STELLA POLARIS, and they completed the preparations at the end of August 1944. A secret meeting was held by Hallamaa at his headquarters at Mikkeli to brief the heads of his five intelligence units, Lieutenant-Commander Pekka Visa, his brother Captain Paavo Visa, and Captains Mauri Hartikainen, Aarre Tunkelo, and Bror Erkki Sten Pale, who were instructed to gather their equipment and families at the ports of Närpes, near Kaskö and Nystad, ready to embark on the *Seagull* and two other coastal steamers for the voyage to Sweden.

To ensure a proper reception, Virkkunen was transferred to the Finnish legation in Stockholm as assistant military attaché with instructions to represent the C-Bureau and make contact with his Axis counterparts, the Japanese General Makoto Onodera and the Abwehr's Büro Wagner. Elsewhere, the other two STELLA POLARIS components were put into action, with Colonel Jussi Sovio leading the ErP4 commandos to Liminka, conveniently close to the Swedish frontier, and the ErP4 vehicles, trailers packed with SIGINT equipment moving to Tornea, with orders to drive over the border to Haparanda at a moment's notice.

The first group of intelligence personnel left Närpes for Härnösand on 20 September 1944 aboard the *Maininki* and *Georg*, followed by the others on the *Osmo*, which left the next day for Gävle. Upon arrival, most of the "refugees" were accommodated at a displaced-persons camp while their archives and equipment were taken by ship in more than 700 crates to a secret Swedish Army store in a cave outside Härnösand and then moved, not to the FRA's headquarters on the island of Lövön, but to a separate FRA facility in Stockholm at Karlaplan 4. Finally, the cipher documents were moved to the cellars of the Hotel Aston, where, over a period, they were photographed by Pale and put onto microfiche using equipment supplied by the Finnish press attaché Heikki Brotherus. The film was then sent to Colonel Uljas Käkönen in Helsinki for processing, and when Pale finally left Stockholm in March 1945, he was carrying 900 meters of undeveloped film, which was later returned to his C-Bureau successor in Stockholm, Lieutenant Pentti Hartikainen.

In November 1944, once the "Stellists" were safely established in Sweden, a delegation of senior Finnish officers, led by the chief of staff and including Lieutenant-General Harald Öhquist, Chief of Military Finance Colonel Runar Bäckström, Colonel Martin Stewen, the Finnish military attaché, Colonel Hans von Essen, Major Emil Lautkari, and Captain Erkki Pale, arrived in Stockholm to negotiate the sale of all the crypto equipment to the FRA director, Commander Torgil Thorén, for 252,875 crowns. The deal, which was completed on 10 January 1945 in Stockholm's Grand Hotel, included the boxes of Soviet ciphers and codes and was backdated to 1 July 1944, the day before the armistice, and the name of a Swedish manufacturer was placed on the actual bill of sale, together with that of Colonel Johan Staahlstrom. According to General Öhquist, the Soviet material was crucial to the sale, which the Swedes would not conclude without it, and a separate annex was completed that gave the Swedes control over the archival material for 50 years. This was signed by the general on behalf of the Finnish Defence Minister, General Rudolf Walden, but the exact nature of the "property owned by the Finnish Defense Forces" sold was not specified so as to ensure that the transaction was a strictly private one between Finnish officials and Swedish individuals.

Altogether, 100 boxes of Soviet cryptographic material were transferred by Captain Pale in March 1945 to a mansion run by the FRA, and Major Emil Lautkari delivered the other archives to an FRA flat. In addition, Lautkari supervised the construction by the Stellist refugees of a radio facility at Lidongö, outside Stockholm, which was used to open a radio channel with Colonel Paasonen in Helsinki. There, Colonel Uljas Käkönen and Lieutenant Tauno Kylmänoja developed what amounted to a "stay-behind" network in anticipation of a Soviet takeover. Another line of communication was maintained with the military radio station at Rauhaniemi, in the extreme east of the country, through the Finnish legation in Stockholm, where Lieutenant

Pentti Hartikainen and a secretary, Anneli Martola, had access to the diplomatic bag. According to a British Secret Intelligence Service report from the local station commander, Rex Bosley, "all the 'Stellists' in Sweden received a regular weekly salary in kronor and those who worked in Pale's bureau received a monthly salary varying from 400 to 1,000 kronor."

The precise fate of the STELLA POLARIS archives remains a mystery, for the Swedes insisted that 29 boxes of the files, which were traced to the manor house of Rottnerös, owned by an industrialist, Major Svente Pählson, and 136 cases to Horningsholm, the ancestral home of a royal equerry, Count Carl Bonde, were removed and burned in November 1960. Both men were respected figures with considerable interests in Finland and appear to have acted in a semiofficial capacity, storing the STELLA POLARIS files with the deniable knowledge of the Swedish authorities. Major Pählson, whose estate included a large pulp mill, was a former director of the Defense Staff's security section, and he held strong views about the need for Scandinavian unity. His death in 1959, and the reluctance of his son to be further involved in keeping the files in the basement, may have prompted their destruction, which was carried out at an incineration plant at Lövsta by Captain Pehr Swartz, acting on instructions of the Swedish commander in chief, Lieutenant-General C. A. Ehrensvärd.

Doubt also surrounds what happened to the money paid for the Soviet material; although, Colonel Hallamaa is known to have joined French intelligence with Paasonen after the war and to have bought a substantial estate at Churriana in Spain, where he died in August 1979. The affair became public knowledge in November 1946 when an embarrassed Finnish government issued an official but highly disingenuous communiqué, explaining the entire episode, which Rex Bosley relayed to London:

> In the summer of 1944 the Finnish High Command asked the Swedish military authorities whether certain wireless equipment and personnel could be transferred to Sweden in the event of a military occupation of Finland. It was at the outset made clear by the Swedes that the establishment of a Finnish intelligence unit in Sweden could in no circumstances be permitted, but that there would be no objection in principle to the transfer of equipment and personnel, though the latter would be liable to internment if still in military service. The matter was dropped for the time being, but shortly before the Finnish armistice, the Finns expressed a desire to ship equipment and personnel to Sweden. Authority was given for their reception at Härnösand and Gävle, where they arrived in considerably larger numbers than the Swedes had contemplated, including women and children. They were treated as ordinary refugees, for such persons as had held military rank were already demobilized, and the majority had already returned to Finland. As regards the equipment, part of this was sold by the Finnish State to the Swedish State and negotiations in respect thereof were carried out by duly

> authorized representatives on either side. Payment had been made to a properly accredited Finnish authority and such equipment as was not purchased was returned to Finland. A number of packing cases containing archives had arrived with the equipment; these were not taken over by the Swedish authorities, but deposited with private individuals by an authorized Finnish representative.

Of course, this version was far from the full story, which the Finnish Security Police only learned as some of the Stellists returned to Helsinki and were arrested. Erkki Pale underwent 20 months' imprisonment and interrogation and eventually was confronted with a carbon copy of the STELLA POLARIS sale contract. As well as being infuriated by the circumvention, the Finnish Communists suspected, wrongly as it turned out, that the almost defunct German intelligence service had played a role in the affair, a claim made initially in the Communist *Ny Dag*. Certainly, the Germans had trained and equipped a resistance network in Finland in early 1945 in anticipation of the widely expected Soviet occupation, but the members of the "Büro Cellarius" had been rounded up in 1946. Named after a certain Captain Cellarius, it had been led by a prominent Nazi, Major Alarich Bross, who had evacuated his Sonderkommando Nord to Swinemünde in Germany and ran a radio station at Sockenbacka, managed by his subordinate Thoralf Kyrre. Even after the Finnish armistice, Bross continued to infiltrate saboteurs across the Baltic by plane and by U-boat, and it was only when the organization had been broken up that the Security Police were satisfied that the Sonderkommando Nord had not masterminded STELLA POLARIS; although, suspicion remained that the millions of rubles accumulated by the Büro Cellarius to finance an anti-Soviet guerrilla organization, which subsequently disappeared, may have been used to pay for STELLA POLARIS.

ST. HELENA. The British colony of St. Helena in the Atlantic, almost midway between Brazil and Angola, was the site of a large **Government Communications Headquarters** signals-intercept station, which monitored traffic from Africa and was closed down after the Rhodesian crisis had been resolved in 1980.

ST/POLLY. The **Central Intelligence Agency** Far East Division cryptonym for airborne signals-intelligence-collection missions flown during the 1950s by P-2V aircraft from Hsinchu on Taiwan.

SUBMARINES. Because the loss of a submarine, unlike a surface vessel, is likely to be sudden, without opportunity for the crew to jettison any crypto equipment, submarine wrecks, where accessible, have always been visited by divers seeking to salvage confidential papers. Such diving operations were conducted throughout both world wars, and on the rare occasions a submarine was recovered intact, usually when scuttling charges failed to detonate, attempts were made to conceal the coup, by isolating the surviving witnesses from other captured crew. During **World War I**, records indicate that cryptographic material was recovered from the *U-8*, sunk in March 1915. During **World War II**, cipher documents or hardware (or both) were retrieved from the *U-33* in February 1940, *U-26* in July 1940, *U-110* in May 1941, *U-559* in October 1942, *U-205* in February 1943, *U-175* in April 1943, *U-744* in March 1944, *U-505* in June 1944, *U-570* in August 1944, *U-1204* in December 1944, *U-260* in March 1945, and *U-1024* in April 1945.

During the **Cold War**, **North Atlantic Treaty Organization** submarines routinely undertook signals-intelligence-collection missions inside Soviet territorial waters. Following the accidental sinking in the Pacific of the Soviet Golf II diesel-electric *K-129* in March 1968, the **Central Intelligence Agency** undertook an ambitious scheme, codenamed AZORIAN, with the objective of retrieving various items, including Red Banner Fleet crypto and communications equipment. *See also* HOLYSTONE; IVY BELLS; NAVAL COLLECTION PLATFORMS; SAND ISLAND; SHARK.

SUN VALLEY. The **National Security Agency** codename for the signals-intelligence variant of the Hercules C-130 transport aircraft, SUN VALLEY missions were flown during the **Cold War** by the C-130 A-II and the C-130 B-II of the 7406th Combat Support Squadron from Rhein-Main. Most routes took the planes along the periphery of Soviet airspace from Norway to Turkey but occasionally experienced harassment. On 2 September 1958, a flight from Incirlik was shot down near Sasnashen by four MiG-17 *Frescos*, with the loss of all 17 crew, when it flew over southern Armenia. On 21 March 1973, another mission, from Hellinikon, Greece, was attacked by Libyan Mirages over the Mediterranean. *See also* AIRBORNE COLLECTION PLATFORMS.

SWELL. The Allied codename employed during **World War II** to indicate special intelligence derived from the Middle East. *See also* ULTRA.

T

TANNENBERG, BATTLE OF. During the Battle of Tannenberg in August 1914, the Russian First and Second Armies had almost encircled General Paul von Hindenburg's Eighth Army and were about to close the last escape route when a German signals-intelligence technician, monitoring Russian telephone communications, learned of a gap in the enemy's pincer movement. Hindenburg exploited the opportunity and was dubbed "The Lion of Tannenberg"; although, he never publicly acknowledged the source of his crucial intelligence. Since the Russian forces enjoyed overwhelming superiority over their adversary, the victory was highly significant and enabled the Kaiser to pursue a war on two fronts. Altogether, the inferior German force, which suffered less than 20,000 casualties, killed or wounded 78,000 Russian troops and captured 92,000 troops and 500 artillery pieces, effectively destroying the Second Army. *See also* WORLD WAR I.

TARGET INTELLIGENCE COMMITTEE (TICOM). In the belief that the Nazis would methodically destroy anything of use to the Allies, a view confirmed in January 1945 when a Wehrmacht signals-intelligence unit was found in Leopoldsburg, Belgium, totally destroyed, TICOM was created as an Anglo-American operation to seize the enemy's cryptographic assets. The local **Special Liaison Unit**, which searched what remained of the Leopoldsburg base, reported that the fleeing Germans had even left a note of welcome for the arriving troops—written in English.

News of any interesting seizure by Allied troops was passed to TICOM headquarters, based in the Paris office of the senior Supreme Headquarters Allied Expeditionary Force (SHAEF) signals officer, Colonel Bicher, who had prepared ambitious plans to dispatch teams to take control of enemy sites and had even drawn up a contingency scheme to drop commandos into the path of the Russians to deny them access to vital crypto equipment. One such installation, at Burgscheidungen, was of interest to TICOM 6 because of the recent arrival at the castle of a group of Reich Foreign Ministry officials bearing the entire records of the **Balkanabteilung** cryptographic unit that had been dedicated to "solving the Russian problem," which meant the **interception** and decryption of Soviet signals.

On 21 May 1945, another TICOM team, led by Howard Campaigne, was taken by some German prisoners of war to a hiding place in a basement under an Oberkommando der Wehrmacht (OKW) communications center at Rosenheim, near the Austrian frontier, where German technicians had intercepted Soviet military wireless traffic. The TICOM men were shown a still operational, fully equipped intercept facility, complete with a machine for demodulating teletype signals, which they promptly dismantled and shipped to **Bletchley Park**. Altogether, seven and a half tons of hardware were carried to England in a convoy of captured German trucks escorted by Selmer Norland and his British counterpart, Edward Rushworth.

TELECHECK. *See* CHECK.

TELEMETRY. The signals transmitted from missiles in flight to ground stations are known as telemetry and provide vital information, sometimes abbreviated to TELINT, relating to the weapon's performance, including rocket thrust, motor behavior, guidance system, engine shutdown, and warhead separation. Study by **National Security Agency (NSA)** analysts revealed that Soviet telemetry was contained in an AM frequency channel consisting of up to 48 different unencrypted components, and this data became available upon the construction of an intercept facility at Sinop on Turkey's Black Sea coast, which could monitor missile tests conducted from Kapustin Yar.

A member of the **North Atlantic Treaty Organization**, Turkey had been an active participant in the Allied signals-intelligence coverage of the Soviet Union since the NSA's site at Karamursel had been opened in 1957. Success at Sinop led to the construction of similar facilities at Astara, Behshahr, Kabkan, Khorramabad, and Shirabad in northern Iran to watch the Tyuratam range.

Assisted by technical information supplied by Colonel Oleg Penkovsky, a **GRU** officer who had access to highly classified details of Soviet ballistic-missile systems prior to his arrest in October 1962, and the construction of an intercept site in Qitai in the People's Republic of China, NSA scientists were able to divine all there was to know about Soviet intercontinental ballistic missiles (ICBM). The development of the **RHYOLITE** satellite, first launched in June 1969, was a further attempt to acquire more TELINT, but disclosure of the system's capabilities by **Christopher Boyce** led the Soviets to encrypt their telemetry, a countermeasure that later would be reversed by the Strategic Arms Limitation Talks. In 1992, Colonel Alexander Simakov, a Russian telemetry expert who had supervised every major Soviet missile from 1984, defected to the British and tied up many of the loose ends from the **Cold War**.

TELINT. *See* TELEMETRY.

TEMPEST. The codename for unavoidable electromagnetic radiation emitted from electronic equipment, including computers, cipher machines, and other potentially sensitive equipment located in accommodation otherwise considered secure.

TISLER, FRANTISEK. Codenamed ARAGO by the Federal Bureau of Investigation (FBI), Frantisek Tisler was a Czech cipher clerk based at the embassy in Washington, DC, who routinely allowed FBI Special Ops teams into the embassy code room over a period of two years until his defection in 1958. *See also* PINCH; STAFF D.

TORII. The **National Security Agency** intercept site located near Sobe on Okinawa, Torii provided collection facilities just 300 miles east of mainland China during the **Cold War** and consisted of an antenna field covering three square miles and a large **direction-finding** station.

TRAFFIC ANALYSIS (T/A). One of the principal signals-intelligence disciplines, along with **call-sign analysis** and **discrim**, T/A is the science of extracting useful information from intercepted traffic, irrespective of whether the actual content is intelligible. Study of the frequency allocation, wavelength, transmitting characteristics, preambles, procedure signals, operator chat, group linkages, regularity, time, and duration of the signal may indicate whether the channel is a familiar one, if the traffic is acknowledged by a correspondent, the identity of the operator, the operator's location, or the routine or otherwise nature of the signal. Combined with **direction-finding** information, T/A can suggest an unexpected change in routine, a breach of **radio silence**, or an error such as the repetition of a message sent in a higher-grade cipher that will potentially be of immense value to cryptographers.

Thus, even when a cipher or code remains unbroken, there can be considerable advantage available from T/A and definitely a contribution to the construction of a diagrammatic picture of a target communications network. The pioneering work in this specialist field was undertaken at **Fort Bridgewoods**, Chatham, during the winters of 1933 to 1935, when Russian wireless networks came under scrutiny, and success with them led to an equally fruitful study of Italian traffic originating in Spain and Abyssinia between 1935 and 1937. Thereafter, T/A came of age during **World War II** when Axis nets were the subject of 24-hour coverage across the globe, with some T/A being undertaken at individual intercept sites before the intercepts were passed to intelligence centers for decryption.

During World War II, T/A provided cryptographers with crucial clues that led to breakthroughs, the most notable examples being the solving of VULTURE, the development in early 1945 of the entire Wehrmacht general-staff wireless net, and the first clues to the **BARNYARD** system in the spring of 1945.

At the end of World War II, Anglo-American cryptanalysts agreed on a common lexicon and designated IVORY as the codename for intelligence drawn solely from T/A.

TRIANGLE. The Office of Strategic Services codename for decrypts derived from communications traffic encrypted with the Abwehr's **Enigma** machine cipher key and known to the British as **ISK**. *See also* ISOS.

TUNNY. The wartime **Government Communications Headquarters (GCHQ)** generic codename for intercepted German high-speed teleprinter traffic generated on the 10-rotor **Geheimschreiber** machine. Manufactured by Siemens, the machine was an online automatic-encipherment system that produced a punched tape at 66 words per minute that fed directly into a wireless transmitter using the international Baudot telegraph code. In May 1940, a Swedish cryptanalyst, Arne Beurling, accessed the German landline that was routed through Stockholm and solved the cipher. The **Försvarets Radioanstalt** then built a decryption machine to transform the traffic into plaintext, but GCHQ was unable to do the same until the **Colossus** was introduced in December 1943. A further refinement was the Schlüsselzusatz 40, made by Lorenz, but this too was defeated at **Bletchley Park**, allowing top-level communications exchanged between important military headquarters to be read.

TYPEX. The British version of the German **Enigma** cipher machine, used throughout **World War II** in all theaters, was the Typex, which, though not so portable, offered greater security, an attribute recognized by the Axis, which declined to devote resources to a task that was considered a lost cause after several devices, captured intact at Dunkirk, were examined by enemy experts. After the war, the Typex was gradually replaced by an improved apparatus, the Rockex.

U

UKUSA. The acronym for the signals-intelligence agreement made in March 1953 to replace the **BRUSA** agreement, UKUSA added Australia, Canada, and New Zealand to a partnership covering the **interception** and exchange of external communications. The terms of the treaty were declassified in June 2010 and provided the foundation for the pooling of collection facilities among the participants and the sharing of the product. It also noted the transfer of the **National Security Agency**'s interception facility in **Hong Kong** to Okinawa.

The UKUSA document was negotiated by Clive Loehnis, Arthur Bonsall, and the **special United Kingdom liaison officer** in Washington, DC, John Tiltman.

ULTRA. The generic codename introduced by the **Government Code and Cypher School** in March 1942 to replace **BONIFACE** for signals-intelligence summaries derived from cryptographic work on German **Enigma** and **Geheimschreiber** communications, ULTRA came to be the codeword classification for most German intercepts that required special handling.

To supervise the distribution of ULTRA, **MI8** and the Secret Intelligence Service created a network of **Special Liaison Units** to ensure the decrypt summaries were seen by the indoctrinated military commanders, and Edward Crankshaw was sent to Moscow to pass suitably sanitized versions to the Soviets. *See also* BLETCHLEY PARK; ISK; ISOS; RADIO SECURITY SERVICE (RSS); SCHMIDT, HANS-THILO; SPECIAL COMMUNICATIONS UNIT (SCU).

UNIT 8200. The Israeli Defense Force's (IDF) principal signals-intelligence organization, Unit 8200 was originally designated Unit 515 and then renamed 848 before acquiring its current title after the Yom Kippur War, but it has always been known as the Central Warning Unit and continues to occupy a large military compound at Herzliya. Intercept facilities were located atop the Har Avital on the Golan Heights and at Un-Hasiba in the Sinai.

Under the command of Joel Ben-Porat, the organization gained a reputation during the June 1967 War for identifying the call signs of the Syrian,

Jordanian, and Egyptian armed forces and reading much of those countries' encrypted military and diplomatic traffic. However, prior to the Yom Kippur conflict of 6 October 1973, Unit 8200 failed to interpret the unexpected imposition of **radio silence** along the front lines on the Golan Heights and on the Suez Canal and was unable to give advance warning of an imminent Arab attack. Unit 8200 also monitored the en clair radio and landline messages exchanged by multinational United Nations observers in the Sinai Desert with their headquarters in Tiberius, reporting the appearance of Soviet advisers in civilian clothes along the canal and the transformation of the Syrian posture on the Golan from a defensive to an offensive stance with the unexpected deployment from Homs of the 47th Infantry Division and the dispersal of Sukhoi-17 aircraft.

Following the conflict, the Israeli government conducted a postcombat review, the Agranat Commission, headed by the president of the Supreme Court, Shimon Agranat, and an unpublished annex to his *Interim Report* severely criticized Unit 8200 for a massive breakdown and a reluctance to analyze the evidence available and blamed military tacticians for placing too heavy a reliance on signals intelligence to provide early warning of a surprise attack. According to Agranat, strategists had developed "the concept," which included a belief that Egypt would not risk engaging in another war until its air force could match the performance of IDF aircraft. Another contributing fact was supposedly the "cry-wolf syndrome," which had developed following two false alarms from Unit 8200 since 1972, which had resulted in very expensive mobilizations.

UNITED STATES SIGNALS INTELLIGENCE. Signals **interception** by the United States originated with the War Department during **World War I**, but in recognition of the value of SIGINT, a civilian **Black Chamber** was established by the State Department in New York until it was closed in October 1929.

During the interwar years, the country's SIGINT was fragmented, with the U.S. Army, U.S. Navy, and U.S. Coast Guard conducting uncoordinated independent operations. It was not until May 1941 that the navy passed control of its chain of 11 **direction-finding** stations to the coast guard, so as to allow other sites to concentrate on Japanese channels, but it was the surprise attack on Pearl Harbor six months later that demonstrated the need to centralize all SIGINT activity, and the **Signal Intelligence Service (SIS)** made work on Japanese naval and diplomatic traffic a priority. Interception was conducted from Cheltenham, Maryland; Winter Harbor, Maine; Chatham, Maine; Jupiter, Florida; and Amagansett on Long Island, New York.

Japanese diplomatic messages to the embassy in Washington, DC, sent from Tokyo by radio to two commercial stations at San Francisco, RCA's RHP at Marshall and ITT Mackay's KFS at Half Moon Bay, were then relayed by landline to their destination. This traffic was routinely intercepted by WVY, a U.S. Army facility established in the Presidio in 1931 by a SIGINT pioneer, Colonel Joseph Mauborgne.

During **World War II**, the U.S. Navy's **OP-20-G** applied its considerable resources to the development of automated processes to solve cryptographic challenges, and success with the **PURPLE** machine cipher led to the distribution of **MAGIC** decrypts, which, like the **ULTRA** material produced by Anglo-American cryptanalysts at **Bletchley Park**, proved strategically significant. By the end of hostilities, much of the Axis wireless networks had been penetrated, with devastating consequences for the enemy, and work had begun on Soviet military and intelligence systems, among them **BOURBON** and **VENONA**.

Wartime investment in personnel and SIGINT technology achieved a global reach with partnerships formed with Australia, Canada, New Zealand, and the British **Government Communications Headquarters** and ensured joint collection and exploitation with the principal target being Warsaw Pact communications. **North Atlantic Treaty Organization** analysts calculated that a Soviet mobilization would give three weeks' warning of any military aggression in Europe, and some highly imaginative, clandestine cable-taps were developed by the **Central Intelligence Agency** and the **National Security Agency** to act as advance "trip-wires." However, other collection platforms, both naval and airborne, proved vulnerable to adversaries and led to the loss of reconnaissance aircraft, aircrew, and some naval personnel.

Although SIGINT failed to alert Washington, DC, to the impending attack on South Korea in 1950, the **Vietnam War** was the first conflict to be entirely integrated. Technical advances made through combat pressure led to increasingly sophisticated equipment and the development of SIGINT satellites and drones. *See also* AIRBORNE COLLECTION PLATFORMS; BERLIN TUNNEL; BRUSA; FRIEDMAN, WILLIAM; IVY BELLS; *LIBERTY*, USS; NAVAL COLLECTION PLATFORMS; *PUEBLO*, USS; SHARK; UKUSA; YARDLEY, HERBERT O.

V

VASSILIEV NOTEBOOKS. Formerly a KGB officer, Alexander Vassiliev resigned in 1990 to become a journalist but, six years later, accepted an invitation from the SVR to collaborate with an American historian, Allen Weinstein, in the preparation of *The Haunted Wood: Soviet Espionage in America—The Stalin Era*, based on the KGB's archives covering the 1930s and 1940s. During his research, Vassiliev compiled eight handwritten notebooks and a collection of loose papers amounting to 1,115 pages, which he had sent from Moscow after he had moved to London in 1996. These documents, subsequently translated into English, lodged with the Library of Congress, and placed on the Internet, amount to a unique insight into previously secret files that disclose the identities of 61 spies who appeared in the **VENONA** traffic and refer to more than 50 other sources who had cooperated with the **NKVD**.

VEGETABLE KEYS. The Luftwaffe's **Enigma** keys designated CELERY and LEEK were categorized at **Bletchley Park** as VEGETABLE keys.

VENONA. The arbitrarily chosen codeword given in 1961 to the greatest secret of the Cold War, VENONA represented a collection of more than 2,000 partly decrypted Soviet secret messages and, as an authentic glimpse into the clandestine activities of the **NKVD** and the **GRU**, was more highly prized than any other comparable asset. VENONA was limited in nature, covering communications to and from Moscow between 1940 and 1948, and represented only a tiny fraction of the traffic that was exchanged between individual diplomatic missions abroad and what was referred to as "the Center." Nevertheless, cryptographers in Great Britain and the United States continued to study the material until 1977, some 34 years after the U.S. Army's **Signal Security Agency** began work on what was then termed "the Russian problem." In the end, 2,200 individual messages were translated, but judging from their serial numbers, this was only a small proportion of the total number of Soviet communications, amounting to just under half of the NKVD's New York–Moscow circuit in 1944, 15 percent of the 1943 traffic (about 200 out of 1,300), and only 23 of the estimated 1,300 telegrams sent in 1942, which represents only 1.8 percent of the total. As for the Naval GRU, half was recovered from the 1943 Washington–Moscow channel, but none for any other year. In 1940, for instance, the New York GRU is calculated to have sent 992 messages to Moscow, with 335 from the NKVD.

VENONA was rightly regarded by the very few counterintelligence experts indoctrinated into the program as the holy grail, a vast database of incredibly secret information that provided clues to the identities of thousands of Soviet spies across the globe. Thousands of cryptographic hours and millions of dollars were devoted to accessing an espionage mother lode, which proved to be the catalyst for hundreds of investigations across the globe, some successful, many less so. When combined with other intelligence information, known as collateral, the bare bones of a VENONA text could be enhanced by the use of footnotes. Often the collateral took the form of surveillance records, and it was physical observation that helped the Federal Bureau of Investigation (FBI) to work out the senior personalities operating under legal cover in Washington, DC, San Francisco, and elsewhere. For instance, it was noted that one of the doormen at the Soviet consulate, who was known to participate in countersurveillance operations, showed unusual deference to a particular, but relatively junior, vice consul. In fact, Stepan Apresyan was the local NKVD *rezident*, a figure with a greater status than even the consul general, and his wife Aleksandra was also a senior NKVD officer, codenamed ZOYA.

As a direct consequence of hints gleaned from the often fragmented texts, Ethel and Julius Rosenberg died in the electric chair, Alger Hiss and Harry Dexter White were accused of being NKVD agents, Klaus Fuchs was imprisoned, Donald Maclean defected, and Ian Milne, an Australian diplomat, fled to Czechoslovakia.

The work on VENONA was considered so secret that the **Central Intelligence Agency (CIA)** was only told of its existence in 1952. Each time a significant new word was decrypted in a text, or additional "collateral" had been acquired, the individual text would be recirculated to the handful of security officials who had been allowed to learn about what amounted to the most important code-breaking exercise since **ULTRA**. While testimony from a handful of Soviet defectors, such as Walter Krivitsky, Victor Kravchenko, **Igor Gouzenko**, and Vladimir Petrov, added flesh to VENONA's bones, more collateral came from disaffected members of the Soviet apparatus in the United States. Whittaker Chambers, Hede Massing, and Elizabeth Bentley achieved considerable notoriety for their denunciations of former comrades, and their Congressional evidence was seized on by politicians of all persuasions. The left condemned an unprincipled witch hunt of loyal civil servants, the right demanding a purge of traitors who had penetrated deep into the administration. Alger Hiss proclaimed his innocence, and Laurance

Duggan committed suicide. But it is VENONA that offers a glimpse of the unvarnished truth.

From the American perspective, VENONA offered a fascinating insight into the way the Soviets ran their intelligence operations, but while it had an interesting historical significance, there was always doubt about the practicality of devoting so many resources to what some perceived as an exercise that was unlikely to uncover further active Soviet networks. This unusual view gained greater circulation after the announcement of the first Soviet nuclear test, prompting the suggestion that as the secrets of the **Manhattan Project** had been stolen, there was nothing further to be betrayed. Certainly, the Americans took a rather more sanguine line on the dangers of hostile penetration of the government bureaucracy and, in 1954, reduced their commitment to **BRIDE**, a reflection of the difficulty of keeping people working on the same challenging problem for years on end. Certainly, the loss of Oliver Kirby and Cecil Phillips was a setback, as was the retirement from the FBI of Robert Lamphere in July 1955. In contrast, the British at **Government Communications Headquarters (GCHQ)** remained convinced there was plenty of scope for progress, and this turned out to be true when, in 1957, the naval attaché channel between Washington and Moscow succumbed. GCHQ was also always conscious of the NKVD and GRU spies mentioned in the early London traffic and was particularly concerned about the possibility of a traitor inside GCHQ itself.

The cause of this long-term anxiety, which was fully justified, centered on a single VENONA text dated May 1941, but broken in 1962, that quite clearly was a raw ULTRA intercept from **Bletchley Park**, listing the railway stations across the Ukraine that the German armored divisions intended to utilize in the forthcoming BARBAROSSA offensive. The certainty that there was a spy inside or close to Bletchley Park in May 1941 was traumatic for two reasons. Firstly, the most obvious candidate for a traitor at Bletchley was **John Cairncross**, who became an espionage suspect in 1951 and was the subject of a lengthy Security Service investigation. While he was eventually to confess to having passed his NKVD contacts thousands of ULTRA decrypts, he had not arrived at Bletchley for a further year. And secondly, had that person received promotion to a position where he or she could influence GCHQ's recruitment and arrange for another spy to be inserted? When the BRIDE team realized the implications, Nigel de Grey, one of the century's great cryptographers, who had solved German problems in **World War I**, gathered the staff in the **Eastcote** lounge and delivered a chilling lecture on the probability of a Soviet spy at large in the building.

This conundrum was eventually resolved when, in April 1964, Anthony Blunt named Leo Long as his source in MI14, the German order-of-battle section in the War Office, which was on the distribution list for ULTRA. Long had extracted the Russian railway decrypt and had handed it to his Soviet contact, an admission that relieved GCHQ from the specter of penetration dating back to May 1941.

Both Kim Philby and Anthony Blunt knew about GCHQ's work on BRIDE, but there was very little that either could do to prevent the progress made jointly at Eastcote and **Arlington Hall**. With his more direct access, particularly while he was SIS's liaison officer in Washington, DC, from 1949 onward, Philby could monitor events and keep his fellow conspirators informed, but it was too late for the Soviets to repair their flawed cipher systems. In fact, there is some evidence that the NKVD learned as early as 1944 that BRIDE had been initiated, for in April that year, a directive, intended to enhance the system's security (while achieving exactly the reverse), was suddenly circulated to all *rezidents*. Unusually, the intended change was notified by message and not in a couriered instruction, as might have been expected, and became operational at very short notice, on 1 May. The new arrangement eliminated the identification in clear of the relevant **one-time pad**'s page number, known as the "indicator," and substituted a procedure in which the first five-digit group of the page was used in clear, to become the indicator, leaving the recipherment to begin with the *second* group on the page. This supposed improvement only changed the starting point by one group, and actually made the task of the cryptanalyst easier as it dispensed with the time-consuming necessity of matching page numbers to particular sections of 60 numeric groups. Instead, the attack could be concentrated on comparing the new indicators on vulnerable texts to those already thought to have been duplicated.

When word eventually emerged of the existence of a cryptographic goldmine, pressure was applied to the American authorities to release the data that had a historic value, for instance the material that demonstrated the guilt of both Ethel and Julius Rosenberg or implicated Alger Hiss. Until the collapse of the Soviet Bloc, this seemed unlikely, but in 1996, the decision was taken to declassify the entire collection. While this was an anathema to some of those who had toiled over VENONA for decades and had sworn to go to their graves without revealing anything of their work, there were actually few practical reasons for continuing the ban on publication beyond the general principle, often espoused by insiders, that anything said publicly about cryptanalytical methods is damaging. However, with the end of the **Cold War**, there was no necessity to fear the KGB, so the **National Security Agency** collaborated with the CIA to make public what none had dared hope would ever be seen outside the most closely guarded government vaults.

The VENONA documents offer a breathtaking insight into Soviet clandestine activities over a period of eight years, but because only the final texts have been released, it is impossible to trace how a particular text developed, perhaps from an initial few meaningless words into a message that was comprehensible. If the final versions had been available from the outset, history would have been quite different. The Soviets would have made much slower progress in constructing an atomic bomb, and the more notorious spies, such as Guy Burgess, Kim Philby, and Anthony Blunt, would have been rendered impotent many years earlier.

The only restriction placed on the release concerns ongoing investigations and the provisions of the U.S. Privacy Act, which protects private citizens from unwelcome disclosures from official files. This has led to deletions, with sensitive names being blacked out in the American releases. Similarly, the British government demanded some redactions; although, the method adopted is more difficult to detect, with the excisions appearing as inexplicable gaps in either the VENONA texts or in the accompanying footnotes. Whereas the American policy appears to have given a measure of protection to the living, being those suspected Soviet sources who were never positively identified or confronted with the allegations, their British partners seem to have adopted political embarrassment as their principal criterion for eliminating sensitive names. The only other deliberate excision in the declassified documents is the consistent removal throughout of all references to the first date of circulation. Each VENONA is marked with the last, and therefore most recent distribution, but it is impossible to determine precisely when the first break in a particular message was achieved or to chart the subsequent progress of the cryptographers.

VENONA inflicted irreparable damage on Soviet espionage but also took its toll on the men and women cryptographers who wrestled with the problem over 35 years. Some had worked on the material for a relatively short period, Cecil Phillips having moved on in 1950 to a career in automation and management, but others, like Gene Grabeel, devoted their entire careers to breaking out just a few groups at a time, teaching and inspiring successive generations of VENONA cryptographers. Mildred Hayes, who spent 28 years on VENONA at the NSA, was given the painful task of closing the operation down and consigning the messages and worksheets to the archives. Others did not survive so well, a few developing muscular difficulties in their hands because of the repetitive nature of the work, sliding pieces of paper up and down lists of unrecovered code groups. Nervous breakdowns occurred, a linguistic genius took to the bottle, another committed suicide, and one woman became so obsessive that she abandoned personal hygiene. *See also*

BRITISH SIGNALS INTELLIGENCE; SOVIET/RUSSIAN SIGNALS INTELLIGENCE; UNITED STATES SIGNALS INTELLIGENCE.

***VERKEHRSBUCH* (VB).** On 3 December 1914, a copy of the German *Verkehrsbuch* was delivered to the Admiralty after it had been recovered from the sea near the Dutch island of Texel by a British trawler based at Lowestoft. The same package, thought to have been thrown overboard from a German vessel during a naval engagement in November, was found to contain a copy of the *Handelschiffsverkehrsbuch*, the **Admiralstab**'s method of communicating with the German merchant marine. Various versions of the VB were found to be used by German military attachés posted abroad, by the German foreign ministry, the Reich colonial office, and the Kriegsnachrichtenwesen. Eventually, the VB was phased out and replaced by a series of Nordo codes. *See also* MI1(b); MI1(e); PINCH; ROOM 40.

VIETNAM WAR. The first **National Security Agency (NSA)** personnel in Vietnam arrived for work at a hangar on the Tan Son Nhut air base in April 1961 and lived in the Majestic Hotel in Saigon as civilians. Designated the 3rd Radio Research Unit (RRU), and occasionally USM-9J, the detachment with a strength of around 100 was subordinate to USM-9 in the Philippines and collected traffic from sites in Saigon and Da Nang, with a third facility established soon afterward in the Mekong Delta at Can Tho.

Initially, the NSA's strategy consisted of SABRETOOTH, which was assistance to the Special Security Technical Branch (SSTB), the local South Vietnamese SIGINT unit, and WHITEBIRCH, a mobile **Army Security Agency (ASA)** unit working from vans that carried ancient AN/TRD-4 high-frequency **direction-finding** (HFDF) and Morse intercept equipment. The Vietnamese signals-intelligence infrastructure consisted of some old **World War II** vintage equipment inherited from the French and some AN/PRD-1 sets for HFDF donated in 1961 by the U.S. **Central Intelligence Agency** and located at Pleiku, Da Nang, and Ban Me Thuot. For technical reasons, WHITEBIRCH proved impractical, and a Vietcong (VC) ambush on 31 December 1961 resulted in the death of nine Vietnamese technicians and James T. Davis, an ASA adviser, later described by President Lyndon B. Johnson as the first American soldier to die in the conflict. Thereafter, the 3rd RRU compound became known as Davis Station, and the ASA adopted the airborne D/F techniques pioneered by the French in the war against the Vietminh.

The single-engined U-6 was converted into an airborne HFDF platform and, from March 1962, proved very successful and led in January 1963 to the identification of a Vietcong concentration at Ap Bac, which became the first major land battle of the war. Airborne collection would later expand, with

QUEEN BEE CHARLIE, followed by the deployment of a pair of RC-130s flying daily missions from Bangkok, codenamed QUEEN BEE DELTA, an operation enhanced by **Naval Security Group** flights with the EC-121 and the EA-3B over the Gulf of Tonkin. By 1967, four more RC-130s were transferred to Da Nang from Europe, and the larger RC-135s were assigned from the Strategic Reconnaissance Wing to pick up Hanoi's traffic. By June 1966, the **Air Force Security Service (AFSS)** had also embraced airborne D/F with PHYLLIS ANN, some 47 EC-47s based at Da Nang, Pleiku, and Nha Trang, designated the 699th AFSS.

When RC-130s flew voice-intercept missions in October 1967, they scored a considerable coup by tapping into a logistics channel at Vinh, an assembly area north of the Demilitarized Zone, which allowed the AFSS to predict the infiltration of regulars down the Ho Chi Minh trail. This invaluable source became known as the "Vinh Window," but the volume of low-level voice intercept material threatened to overwhelm the NSA's limited language skills, so an emergency program, codenamed DANCER, was instituted to recruit and train additional linguists from SABRETOOTH units to transcribe and interpret the tapes. Finally, even greater catchment was required, and SHORTHAND was introduced to hire suitable South Vietnam government officials.

In 1963, a division of labor was agreed with NSA managing the larger sites, such as Phu Bai, and leaving the ASA to run the mobile facilities. Phu Bai was then developed into a major cryptographic center, the largest of its kind in the world with 100 radio intercept positions, and linked by a secure underwater cable, codenamed WETWASH, from Nha Trang to the 9th RRU at Clark Air Force Base in the Philippines.

It was not until 1964 that the AFSS was deployed in any numbers to Vietnam, when a radar station was constructed atop Monkey Mountain, near Da Nang. Then, the following year, Ramasun Station was built by the ASA at Udorn in northern Thailand and staffed partly by AFSS cryptographers, but the AFSS's attempts at airborne collection, including a plane packed with sophisticated computers, codenamed HAWKEYE, generally failed. Nevertheless, the American SIGINT capability had developed sufficiently to be able to provide the first reports at the end of 1964 of the North Vietnamese 325th Infantry Division at Dong Hoi moving south into Quang Tri, representing a major escalation in the conflict with the intervention of regulars in support of the Vietcong. On 6 February 1965, Pleiku was attacked with rockets, killing 8 and wounding 108 Americans, so the retaliation took the form of an air raid on Dong Hoi, and a month later, U.S. Marines landed at Da Nang, marking the first deployment of ground troops in the war. Their first major engagement, Operation STARLIGHT, entrapped a large Vietcong force at Chu Lai, which had been pinpointed in August by NSA intercept operators.

This victory, over an enemy that had intended to launch a surprise attack on Da Nang, illustrated the advantage of close cooperation between the NSA and the combat troops seeking to engage a largely unseen adversary. It was followed by close attention paid to elements of the 325th Division, which had encircled a U.S. Special Forces base at Plei Me. Two regiments of regulars, the 52nd and 33rd Infantry, were identified and then harassed as they withdrew up the Ia Drang Valley until they were cornered at Chu Pong and suffered up to 60 percent casualties before fleeing into Cambodia. Evidently, the thought of airborne signals **interception** never occurred to the North Vietnamese regulars who blamed spies for their defeat.

By 1967, the North Vietnamese Army (NVA) and the Vietcong had become dependent on an extensive but technically primitive radio communications network that proved vulnerable to NSA traffic analysis to the extent that, according to the NSA's official history, "the SIGINT system predicted every major VC or NVA offensive. This included date, point of attack and units involved."

American bombing raids over North Vietnam, conducted daily from March 1965, initially went unchallenged because Hanoi had only a few MiG-17s and MiG-21s, but gradually a network of 150 radar stations was built, connected to 40 reporting centers, which were linked to SA-2 batteries. This air-defense system generated manual Morse communications intercepted by the AFSS in a program codenamed HAMMOCK and resulted in warnings circulated to aircraft on KW-26 secure circuits. It was just such a HAMMOCK message that resulted in the first confirmed destruction of a MiG on 27 April 1966.

To counter the threat from the SAMs, SILVER DAWN airborne collection missions were flown to identify the FAN SONG radar sites and associated missile batteries, and once the coordinates had been obtained, the information was passed to the 7th Air Force for an IRON HAND SA-2 suppression operation. Later, HAMMOCK would be improved by IRON HORSE, an automated system that alerted pilots to imminent danger in seconds rather than minutes after a threat had been detected. When the North Vietnamese responded by using their FAN SONG radars at low power, to reduce the risk of interception, a BIG LOOK apparatus was fitted to the EC-121 and was still capable of interception and D/F of the reduced transmission. A further refinement, WEE LOOK, was put aboard the EA-3B to act as a threat emitter.

The major turning point in the war was the Tet Offensive of January 1968, the culmination of four months of unprecedented Vietcong activity and a dramatic change in their tactics with a series of major attacks launched with apparent and uncharacteristic disregard for casualties. However, although there was plenty of intelligence, from SIGINT, prisoner-of-war interrogation, and scrutiny of captured documents, that a massive assault was imminent, the most likely target seemed to be the border provinces rather than the urban areas. When two NVA divisions were discovered to be moving south toward Khe Sanh, the objective seemed obvious, but tactical SIGINT gave the edge to the U.S. Marines who came under siege on 21 January in a battle that lasted until April and left 10,000 NVA dead for 500 American casualties. The main, coordinated attack took place on "N-Day" over the Tet holiday on 30 January and concentrated on several cities, including Hue, Nha Trang, and Quang Tri, with NVA troops entering Saigon and reaching the presidential palace and the U.S. embassy before being repulsed. A post-action investigation for the White House concluded that "communications intelligence was able to provide clear warning that attacks, probably on a larger scale than ever before, were in the offing" but noted that "the indicators, however, were not sufficient to predict the exact timing of the attack."

The NSA did not regard Tet as a failure of intelligence but rather one of interpretation. The organization's facilities were concentrated in the central and northern provinces because that was where the main threat had always been perceived, and assets in the south were limited because, for technical reasons, intercept conditions there were historically poor. The dense, humid jungle dissipated the ground wave to such an extent that D/F work much beyond five miles of a transmission was useless, the **skip zone** was up to 90 miles long, and the sky wave was almost vertical, making interception next to impossible. In those circumstances, it was not surprising that the NSA had opted to concentrate its resources in more productive areas. With limited assets in Vietnam's southern provinces, it was inevitable that there would be less SIGINT product from the south, which turned out to be the VC and NVA focus.

Although the Tet Offensive resulted in 30,000 NVA dead and the VC being decimated, the White House and the U.S. military command looked disorganized and close to panic as newsreels showed that the enemy had penetrated to the heart of Saigon, even if only briefly, and the political will in Washington, DC, to continue simply ebbed away. At the end of March, President Lyndon B. Johnson announced that he had suspended the bombing of North Vietnam, that W. Averell Harriman had been appointed to lead a peace delegation, and that the president would not seek reelection.

The new American strategy was "Vietnamization," which required the National Security Agency to train and mentor the Special Security Technical Branch, which, at the time, had a strength of 1,000 operating three intercept sites (at Saigon, Can Tho, and Da Nang), four U-6 **airborne collection platforms**, and four D/F stations. The task of developing the SSTB over three years was shared by ASA operators, who helped establish a new intercept site at Pleiku, and by the AFSS, which donated 20 EC-47 aircraft. Another project, codenamed LACEBACK, indoctrinated the Vietnamese into communications security and provided crypto equipment, including the KL-9 cipher machine, the NESTOR voice scrambler, and some old World War II vintage M-209s. However, the 1970 Vietnamization plan was intended to reduce American combat troops, and the NSA anticipated a reduction from 8,500 personnel in theater to 6,654 by 1973, even if some 250,000 U.S. troops were to be withdrawn over the same period, leaving a total of 95,000 American soldiers, of whom only 6,000 would be engaged on combat duties.

At the end of April 1970, President Richard Nixon authorized an incursion into the sanctuary area over the Cambodian border from Tay Ninh province, and a major objective was the NVA and VC's combined theater headquarters, known as Central Office South Vietnam (COSV), the shifting location of which had been monitored by D/F operators. Carpet bombing by B-52 air raids had failed to destroy COSV, so ground troops crossed the frontier but failed to find the headquarters, which was evacuated to central Cambodia. Although a major logistics center, known as "the city," was obliterated, and the NVA had a narrow escape, the overall incursion was considered a strategic failure.

In October 1971, another incursion, this time into Laos, was planned based on SIGINT. Codenamed LAM BON 719, the objective was to seize the Ho Chi Minh trail over the Laotian frontier and prevent NVA troops and supplies from being infiltrated into South Vietnam. Although Vietnamese forces reached Tehepone, the NVA had withdrawn from the area and was massing for a counterattack. A retreat then turned into a rout, and the lesson was that whatever the quality of intelligence, there was no substitute for a determination by ground troops to fight and win. Nevertheless, NVA plans for a spring offensive over Easter 1972 were disrupted, and the subsequent conventional attacks, intended to concentrate on targets shown during Tet to have been vulnerable, proved costly, with the NVA losing some 50,000 troops.

Nixon's determination to bomb the North Vietnamese into negotiations, prompted by the Easter offensive, began in May with a resumption of B-52 raids. The airmen found that the air defense system had largely abandoned its reliance on SAMs and instead deployed MiGs, which proved highly effective, with 18 bombers shot down by air-to-air missiles by 1 July at a cost of 24 MiGs. NSA's solution was TEABALL, a U-2 SIGINT mission that monitored the three MiG airfields to provide advance warning of takeoffs so F-4 fighters could be vectored to intercept them before they could reach the bombers. TEABALL was a success, and by mid-October, when the air raids were suspended for truce negotiations, 19 MiGs had been shot down at the cost of 5 Phantoms.

When the truce talks collapsed in mid-December, with Le Duc Tho walking out, the B-52 air raids were resumed with even greater intensity and went unchallenged by MiGs. Just one bomber was lost, to a missile, and the negotiations in Paris were soon resumed, establishing a ceasefire that became effective in February 1973, a date by which all American personnel were to have left the country. The NSA abided by the timetable, withdrawing both ASA and AFSS personnel to other sites in Southeast Asia and donating a further 10 EC-47 aircraft to the SSTB.

The Vietnam War forced the intelligence community to relearn many of the lessons forgotten from World War II and the conflict in Korea, but the overall impact of SIGINT on combat operations was significant, with an estimated 55 percent of target data being supplied from this single source. Similarly, after the Tet Offensive in 1968, SIGINT provided the overwhelming majority of reliable, actionable intelligence to unit commanders at a tactical level.

The North Vietnamese Army's own signals-intelligence organization remained largely unknown until Christmas 1969, when the U.S. First Infantry Division stumbled over a special unit in Binh Duong province and captured 12 technicians, out of the 18 that staffed it, and some 2,000 documents. Analysis by the NSA concluded that the NVA was heavily reliant on communications intelligence that was collected and studied by a dedicated group, with a strength estimated at between 4,000 and 5,000, who concentrated on traffic analysis and U.S. Air Force voice channels and were sufficiently skilled to predict B-52 air raids, albeit with some covert assistance from the People's Republic of China. At that time, communications security was a widely abused concept, but this experience was highlighted by a Pentagon investigation, codenamed PURPLE DRAGON, which tried to tighten up dangerously indiscreet radio procedures and led to the President Johnson's National Security Decision Directive 298, which gave the NSA a global COMSEC responsibility. *See also* CHINESE SIGNALS INTELLIGENCE; DeSOTO; PHOU PHA THI.

VORTEX. The U.S. **National Reconnaissance Office** codename for a signals-intelligence satellite originally codenamed **CHALET**, developed to replace the **CANYON** system. The first VORTEX was launched on 31 January 1984 from Cape Canaveral, Florida, but failed to reach its designated geostationary orbit and was followed by insertions on 2 September 1988 and 10 May 1989. *See also* AIRBORNE COLLECTION PLATFORMS; ZIRCON.

V TRAFFIC. Within the British Security Service, **MI5**, the **VENONA** material was simply referred to as V traffic during the **Cold War** so as to protect the nature of the cryptographic source.

VULTURE. The **Bletchley Park** codename for a large network, initially detected on the Russian Front in June 1941, VULTURE grew into a Wehrmacht military headquarters communications circuit that, by the beginning of 1943, extended to all fronts and was found to carry exceptionally valuable messages of strategic importance. VULTURE was the first German army traffic to be covered by **interception** in any volume and appeared to have reached a peak in December 1941, when the Eastern Front became relatively static.

During the invasion of France, the Wehrmacht had relied upon miles of landlines laid immediately behind the advancing frontline, but evidently in Russia, with a front extending over 1,000 miles, landlines proved impractical. The need for the ground forces to maintain constant contact with the Luftwaffe and with a forward headquarters meant a heavy reliance on encrypted wireless signals, which were available for interception.

W

WALKER, JOHN. A U.S. Navy warrant officer who had joined up in May 1955, John Walker eventually confessed that he had spied for the KGB from December 1967 until his arrest in May 1985. He served on two ballistic-missile nuclear submarines (the USS *Arthur Jackson* and USS *Simon Bolivar*), surface vessels, and at shore establishments, where he had almost continuous access to and responsibility for the security of crypto equipment, including the key cards used to alter the daily settings on the 37 Series, 14 Series, and 7 Series cipher machines and the KY-8 voice scrambler. Having served on the USS *Razorback* as a radio operator, Walker had attended a school to learn repair skills for mending faulty cryptographic equipment, and he was posted as the crypto custodian to a combat supply ship, USS *Niagara Falls*, for duty in Vietnam.

After his offer to spy was accepted, Walker was promised a salary of $4,000 a month, given a Minox camera, a schedule of dead drops, and a rotor reader for a KL-47. At the time, Walker's navy salary was $725 a month. He would not hold another personal meeting with a Soviet contact for 10 years, but over the 17 years of his espionage, he made some 30 drops. When questioned after his arrest, he remarked that he had always been curious that the KGB had expressed no interest in the KG-14 cipher machine, which he interpreted to mean that the Soviets already had another source for that material.

Anticipating his retirement in 1976, and a new career as a private detective, Walker recruited his son Michael, who, at the time of his arrest, was a petty officer serving on the carrier USS *Nimitz* and was found to have 15 pounds of classified material in his locker.

In addition, John Walker had recruited his brother, Lieutenant Commander Arthur Walker, and another navy friend, Jerry Whitworth. In October 1985, father and son pleaded guilty and received two life terms plus 10 years' imprisonment and 25 years' imprisonment, respectively, in return for John Walker's testimony against Whitworth, who surrendered to the Federal Bureau of Investigation in June 1985.

The damage assessment took five years to analyze the scale of the loss, and eventually the Stilwell Commission concluded that more than a million classified U.S. Navy messages had been intercepted and decrypted by the Soviets, many of them in real time. Also compromised was the top-secret **SOSUS** underwater acoustic monitoring network. In return for his information, Walker had received sufficient cash to buy two yachts, a houseboat, a plane, and a home in Norfolk's upscale Algonquin Apartments.

WAR OFFICE Y GROUP (WOYG). Created in March 1926 as the Wireless Section, Eastern Command Signal Company, the War Office Y Group later became known as the Special Y Group and finally the War Office Y Group, designated **MI8**. Initially consisting of just five wireless operators, who were transferred to Chelsea Barracks during the May 1926 general strike to maintain emergency communications for the government, the unit was based at **Fort Bridgewoods**, Chatham. The first intercept was traffic exchanged between Paris and Beirut, and in April 1928, the staff expanded by the recruitment of a further five operators to cover transmissions from the Italian legation in Beijing. Their intercepts were delivered to the **Government Code and Cypher School** at Broadway Buildings once a week by the Post Office. When Japan invaded Manchuria in 1931, the Tokyo traffic became a priority, and three remote intercept recorders were established in London, and the number of operators at Chatham was increased to nine, with a waiting list of 50 ex-servicemen ready for call-up if necessary. More modern equipment was installed, including a Marconi UG6A undulator, and almost all the station's resources were devoted to high-speed Japanese traffic.

Between 1934 and 1935, Whitehall seemed to lose interest in Chatham, and when the commanding officer died suddenly, he was not replaced for months. However, his successor took over in January 1935, and in October, the Italian campaign in Abyssinia provided new channels to monitor. As well as covering Rome, Italian stations in Cairo, Somaliland, and Shanghai, the League of Nations transmissions from Bern and Prangins were also watched, and an initial study was made of German military traffic. A review of Chatham's activities in 1935 showed work in January on a Japanese radio-telephone link between Rome and Japan; some Soviet OGPU channels were found in April; in May, 17 Japanese high-speed circuits were found, the "Bern List" frequencies were tapped, and some Polish channels were monitored following the death in May of Marshal Jozef Pilsudski. In July, an experiment was conducted with the BBC to test the quality of recording the HPB station in Bern, and, in October, work was completed on a **General Post Office (GPO)** line to Cairo for the collection of Middle East traffic.

During 1936, the Tokyo broadcast remained a priority, along with some 200 identified Italian military circuits, all connected with the occupation of Abyssinia, and, in the month of December, some 16,000 individual messages were intercepted, of which 2,500 had been encrypted. This volume of activity prompted an expansion at Fort Bridgewoods, where two vacant casemates were converted into set rooms, one for 6 high-speed sets, the other for 18 hand-speed positions. Accordingly, the operator staff was increased to 15 in January 1937, but the outbreak of the Sino-Japanese War increased Tokyo's traffic by 400 percent. Combined with pressure from signals broadcast in the Spanish Civil War, work on the Italian channels in Abyssinia was to be suspended temporarily and a focus placed on Italian military traffic from Salamanca.

By October 1938, the staff reached 82, working 14 sets around the clock, concentrating on Italian military communications with Salamanca and, during the Munich Crisis, traffic exchanged between Moscow and Berlin. Then, during the first nine months of 1939, Chatham concentrated on the Japanese high-speed links to Europe and extended its priority cover to the German channel between Rocky Point on Long Island and Nauen; Rio de Janeiro to Tokyo and Europe; Buenos Aires to Tokyo, London, Europe, and the United States; Santiago to Rio and Tokyo; Shanghai to Japan and Europe; and Chunking to Europe and the United States. The Italian traffic in Spain was eventually abandoned in June 1939 when the last remnants of the Italian general staff reached Naples aboard the SS *Sardegna*, but the invasion of Albania on Good Friday had led to demands to cover new channels from Rome to Tirana and Durazzo. However, once war was declared, Germany became the top priority, and in the absence of any Air Ministry **interception** facilities until early 1941, the army found itself intercepting Luftwaffe signals for the Royal Air Force.

A severe shortage of trained personnel in the first few months of the war led the WOYG to acquire 120 GPO sorting clerks and telegraphists, who underwent training at Chatham, followed by a course at the No. 6 Intelligence School, **Beaumanor Hall**, near Loughborough, before being posted at one of the outstations at Morley Downs, near Bishops Waltham in Hampshire, with 18 sets concentrating on medium-frequency traffic. This site, which became operational in April 1943, was distinctive because of its vast 350-foot masts, which carried the antenna array. Another satellite, manned entirely by civilians, was established at Kedleston Hall, Derbyshire.

WOYG was also handicapped by a shortage of suitable equipment, and the first American National HRO shortwave receivers did not arrive until February 1941, as part of Lend-Lease. Later, the Hallicrafters SX-28 and finally the RCA AR-88 were made available to supplement the British manufactured DST 100, which went into production in July 1940. Finally, the very advanced American HRO was converted at the Signals Research and Development Establishment into a dedicated intercept set, designated the R206, and became operational at **Chicksands Priory** in October 1943; although, it did not become widely available until the following year.

WOYG's other stations were located at **Sandridge**, **Harpenden**, and **Denmark Hill**. *See also* BRITISH SIGNALS INTELLIGENCE.

WARRANTLESS WIRETAPS. The media term for the **National Security Agency (NSA)** Terrorism Surveillance Program (TSP), which, soon after the Al-Qaida attacks on New York and Washington, DC, freed the organization from some of the restrictions imposed by the 1978 **Foreign Intelligence Surveillance Act (FISA)**. Under the FISA terms, **interception** of telephone calls could only be conducted within the domestic United States where there was probable cause to believe that the target was the agent of a foreign power. At the suggestion of Vice President Dick Cheney, the NSA director, General Michael Hayden, sought presidential approval for a TSP that would sanction the interception of overseas calls to or from a domestic landline or cell-phone number where there was reasonable suspicion that either one of the parties was a terrorist.

In effect, the TSP obviated the requirement to apply to the Foreign Intelligence Surveillance Court for a warrant and permitted the NSA to monitor communications within the territorial United States, subject to certain conditions, one being that the TSP could only be extended beyond 45 days after the Director of Central Intelligence (DCI) had conducted a comprehensive review assessing the current program's intelligence value. Initially, the chairmen and minority party leaders of the House and Senate attended a series of 12 TSP briefings at the vice president's office in the White House, and thereafter the entire committee membership was indoctrinated into the program. The sole dissenter was Senator John D. Rockefeller, the Democrat vice chairman of the Senate Intelligence Committee, who wrote privately to President George W. Bush expressing his reservations.

In 2004, consideration was given to amending FISA to give the TSP permanent legitimacy, but the DCI advised that to hold public hearings on the issue would tip off the targets to NSA's activities and capabilities. Thereafter, when the TSP became public, in a newspaper report written by Jim Risen in the *New York Times* in December 2005, it was characterized by opponents as "warrantless wiretaps" and an infringement of the right to privacy, but in 2009, Congress granted a retrospective immunity to the U.S. carriers who had cooperated with the NSA to implement the TSP. *See also* UNITED STATES SIGNALS INTELLIGENCE.

WEHRKREIS. The German high-frequency radio network across Europe in **World War II**, known as the Wehrkreis, consisted of four major channels covering six southern wireless stations at Prague, Brno, Vienna, Munich, Salzburg, and Graz; six in the west, at Munster, Kassel, Wiesbaden, Nurnberg,

Dresden, and Stuttgart; four in the east, at Königsberg, Danzig, Warsaw, and Breslau; and four in the north, at Hamburg, Hanover, Stettin, and Posen. In addition, in January 1944, two additional channels were established as "Berlin A" and "Berlin E." The traffic generated on 26 high frequencies and 4 medium frequencies, was broadcast 24 hours a day, requiring a minimum of 30 intercept positions to monitor the signals.

WEISBAND, WILLIAM. Born in Odessa in 1908, although his passport showed him to have been born in Alexandria, Egypt, Volodya Weisband had been taken by his parents, Isadore and Sarah, to the United States in 1924 (with his elder brothers, Mark and Harold), where he had become a naturalized citizen in 1938. Isadore had left Odessa first, reaching Egypt via Turkey, with his wife following five years later with Volodya. A jeweler by trade, Isadore eventually owned three shops in New York City, while his wife had run a restaurant in Alexandria, Virginia, when the family had been reunited.

Fluent in Russian, Volodya adopted the name William and attended the American University in Washington, DC, and also worked at the Statler Hotel in New York. He had been drafted into the U.S. Army in 1942 and had served with the **Signal Security Agency** in North Africa and Italy and briefly had acted as a liaison officer with the Red Navy in Naples. All his posts gave him access to highly classified information, and he served as a cryptographic control officer and, at the end of service in Europe, as a liaison officer with the Free French.

Upon his return to the United States, he had been posted to **Arlington Hall** as an interpreter, working in the Russian Section, where he had constant contact with Meredith Gardner and the other **Armed Forces Security Agency (AFSA)** cryptographers attempting to solve the Soviet ciphers. His colleagues recalled that he had taken particular interest in the progress made on the list of atomic scientists that had so excited the **BRIDE** analysts and demonstrated **VENONA**'s counterintelligence potential. Indeed, on one occasion, Meredith Gardner recalls him looking over his shoulder as he worked on the names. Although Weisband had no direct official contact with the team handling VENONA, his sideline of selling jewelry to the Arlington Hall staff had given him considerable freedom of movement, visiting restricted areas, gossiping with the typists, and showing them his wares, which included several good pieces from Harry Winston, inherited from his brother Harold, who had run the family business after Mark's death in the war.

The gregarious Weisband had been popular with the Arlington Hall staff, and he is recalled as the owner of an unusual electrical item, a wire recorder, which in 1948, in the days before tape recorders, was quite a novelty. On one occasion, he took the recorder to a party and recorded the voices of some of the AFSA staff, which was then considered slightly odd behavior. In 1948, he married Mabel Woody, a raven-haired beauty from Tennessee, 20 years his junior, who worked at Arlington Hall as a secretary, and they had four daughters and one son.

Weisband's role as a Soviet spy was not directly disclosed by a VENONA text; although, a message from Pavel Klarin in New York dated 23 June 1943, about the travel movements of LINK, are thought to be the only reference to him; although, it was not fully translated until 1979:

> The last 4 weeks LINK [William Weisband] has spent [1 group unrecovered] school [4 groups unrecovered] ARLINGTON, VIRGINIA where he underwent a course of instruction in Italian. At the moment he has been given leave until the 27th of June [12 groups unrecovered] the ISLAND [England]. By his calculations he will leave the COUNTRY [America] in the first half of July and will arrive on the ISLAND at the end of July. We are setting up the following arrangements with LINK: The password for [1 group unrecovered] contact in any place in Russian or in English: our man: "Hullo Bill. Greetings from Gregori." He: "[7 groups unrecovered] on the West Coast." Starting from 24 July LINK will wait for our man in LONDON on Sundays on the [1 group unrecovered] at the entrance to Leicester galleries, Leicester Square and on Wednesdays [2 groups unrecovered] on the [1 group unrecovered] east corner of Orchard and Wigmore [43 groups unrecoverable] Rush the telephone number and your instructions before [1 group unrecovered] June.

Coincidentally, Weisband was known to have studied Italian at Arlington Hall in June 1944, before leaving by ship for London on 17 July, where he arrived 10 days later. That, combined with Weisband's first name, and his knowledge of Russian, is strong circumstantial evidence that LINK was indeed Weisband, and he was suspended from the AFSA on grounds of his suspected disloyalty.

Although Weisband had worked at Arlington Hall from 1944, he was disconnected by the NKVD as a security precaution in 1945 following the defection of Elizabeth Bentley and was not reactivated until February 1948, when he warned his contact that American cryptographers were reading Soviet traffic. This led to "**Black Friday**," a period when the Soviets began to change their cipher procedures, closing the AFSA's access to the target communications.

Weisband, who died of a heart attack at the wheel of his car in Washington, DC, in May 1967, was never charged with espionage, even though a former girlfriend in New York volunteered the information to the Federal Bureau of Investigation (FBI) that he had used her to empty dead letter drops for him. Instead, he was convicted of contempt in November 1950 and sentenced to a year's imprisonment for having failed to turn up for a grand jury inquiry in his

Communist Party of the United States of America membership, but the clue that led the FBI to him came not from VENONA but from Jones Orin York, a Soviet spy who did appear in several VENONA texts, codenamed NEEDLE.

WILLIAMS, GARETH. In August 2010, the decomposing naked body of Gareth Williams, a 31-year-old **Government Communications Headquarters** cryptographer from Anglesey in Wales, was found in a padlocked holdall in the bath of his top-floor apartment in Alderney Street, London. A police investigation concluded that Williams, who was unmarried and had recently returned from a vacation in the United States, had died from unknown causes but most likely asphyxiation and dehydration. On temporary secondment to the British Secret Intelligence Service, his disappearance was reported by colleagues, and the nature of his work prompted widespread media speculation about the strange circumstances of his death. A description of two visitors to Williams's apartment was circulated by the police, but the suspects were never traced, and nobody was ever charged with a crime.

WIRELESS EXPERIMENTAL STATION (WES). The British wartime regional cryptographic organization in India, the WES was first accommodated in Ramjas College, formerly part of the University of Delhi. Headed by Colonel Patrick Marr-Johnson, the WES was divided into five sections: A, Administration; B, Analysis; C, Codebreaking and Translation; D, Traffic Analysis; and E, Radio Interception. Later, WES moved to Anand Parbat, outside Delhi, and, in July 1943, was linked by radio to **Bletchley Park** as part of the global **Special Communications Unit** network managed by 56, 57, and 153 Special Wireless Sections and 166 Signals Wing, RAF.

The origins of the WES dated back to the end of **World War I**, when Russian military traffic was monitored by a Royal Signals Wireless Company from Cherat, on India's Northwest Frontier. By 1939, the station had moved to Abbottabad, a hill station near Rawalpindi, concentrating on a Russian network controlled from Tashkent. In addition, the WES intercepted German and Japanese diplomatic circuits but was unable to break any high-grade Japanese Army systems until the spring of 1943 when the water transport code was cracked. Then in August, an address code was read, which assisted in the identification of individual units and the development of a comprehensive enemy order of battle, and, in December, the Imperial Army's administrative code succumbed, followed in February 1944 by the capture of the code itself. High-grade army decrypts, codenamed MONGOOSE and LEOPARD, provided valuable insights into Japanese intentions and played a significant role in defeating the Japanese plan to seize Chittagong and invade Bengal in February 1944.

The WES controlled a series of regional intercept sites, located at Yelahanka, near Bangalore, at the 200 Special Wireless Section headquarters at Barrackpore, Dusi, and, from December 1942, a **direction-finding** station at Tezpur on the Chinese border, with forward positions at Akyab, Comilla, Cox's Bazaar, Monywa, and Imphal run by 367 and 368 Wireless Units, RAF. *See also* BRITISH SIGNALS INTELLIGENCE; GOVERNMENT COMMUNICATIONS HEADQUARTERS (GCHQ); LAUNGYAUNG; MI8; SPECIAL WIRELESS GROUP (SWG).

WORLD WAR I. All the modern techniques of wireless **interception**, **traffic analysis**, **direction finding**, and cryptography were developed and refined during World War I when vast resources were devoted to obtaining tactical and strategic advantages over the Central Powers, particularly by the War Office's **MI1(b)** and the British Admiralty, which established **Room 40** in August 1914 to solve enemy ciphers. A civilian, **Alfred Ewing**, was recruited to head a group of amateur linguists and cryptanalysts who studied material culled from airwaves and underwater cables. Exploiting codebooks recovered from shipping, **submarines**, and **zeppelins**, the cryptanalysts produced actionable intelligence that influenced surface and submarine naval operations and, in January 1917, provided the key to the **Zimmermann Telegram**. Although less successful, German codebreakers working for the **Admiralstab** were able to solve some of the Royal Navy's low-grade cipher systems. Following the British example, France and the United States developed similar organizations.

By the end of the war, the two principal British cryptographic organizations were regularly reading the confidential diplomatic communications of numerous neutral countries, and monitoring the enemy's wireless and cable traffic. Further success, achieved during the Treaty of Versailles negotiations, ensured that the British government would retain its SIGINT capability into the peace. *See also* ADMIRALTY Y SERVICE; CODE TALKERS; GOVERNMENT COMMUNICATIONS SECURITY BUREAU (GCSB); OFFICIAL HISTORIES; RADIO FINGERPRINTING (RFP); WIRELESS EXPERIMENTAL STATION (WES).

WORLD WAR II. Although there was every reason to believe in September 1939 that modern cryptographic techniques, and in particular the widespread use of mechanical encryption devices, would make cryptanalysis largely redundant in the imminent conflict, unexpected success, especially with **Enigma** traffic, ensured that the skill of the codebreaker had a very significant impact on all the protagonists. German cryptanalysts also gained significant tactical advantages from **traffic analysis**, preferring to concentrate resources on hand ciphers on the mistaken assumption that Allied machine systems were as secure as they believed their own to be. In a conflict that proved far more mobile than the largely static **World War I**, with almost universal reliance on wireless communications, all sides developed specialist agencies to collect and analyze signals. By far the largest commitment to SIGINT was made by the Allies, and certain campaigns, the conflict in the Atlantic, the battle of El Alamein, and the Normandy landings on D-Day were influenced significantly by cryptographers based at **Bletchley Park**, **Arlington Hall**, and their subordinate organizations in the Middle East, India, Ceylon, and Singapore.

The Germans, confident of the integrity of the Enigma and **Geheimschreiber** machines and the **one-time pads** employed by the Foreign Ministry, were reluctant to rely on landlines in the countries under their occupation, giving the Allies an opportunity to intercept an escalating volume of signals. Although several thousand linguists, analysts, and codebreakers worked on the enemy's encrypted traffic, relatively few knew the actual scale of the operation, which distributed summaries designated **ULTRA** secret to a strictly limited number of indoctrinated recipients. Apart from Prime Minister Winston Churchill, only a couple of senior Cabinet ministers were included on the circulation list for the decrypts. One of the last in the war in Europe heralded the Nazi surrender:

> To call-sign JDM from call-sign JFX Frequency 2319kc/s
> Time ended 0357 G.M.T.
> "Originator signed JODL COL-GENERAL for action. GRAND ADMIRAL DOENITZ GENERAL FIELD MARSHAL KEITEL. Act of military surrender. Para. One. We the undersigned acting by authority of the German High Command hereby surrender unconditionally to the Supreme Commander Allied Expeditionary Force and simultaneously to the Soviet High Command all forces on land, sea and in the air, who are at this date under German control. Para. Two. The German High command will at once issue orders to all German Military, Naval and Air authorities and to all forces under German control to cease active operations at 2301 hours central European time on 8 May and to remain in the positions occupied at that time. No ship, vessel or aircraft to be scuttled or damage done to their hull machinery or equipment. Para. Three. The German High Command will at once issue to the appropriate German commanders and ensure the carrying out of any further orders issued by the Supreme Commanders Allied Expeditionary Force and by the Soviet High Command. Para. Four. This act of military surrender is without prejudice and will be superseded by a general instrument of surrender by the United Nations applicable to Germany and the German armed forces as a whole. Para. Five. In the event of the German High Command or any of the forces under their control failing to act in accordance with this act of surrender the Supreme Commander Allied Expeditionary Force and the Soviet Command will take such punitive or other action as they deem

appropriate. Signed at Rheims France at 0241 hours on the seventh of May 1945. On behalf of the German High command signed J O D L. In the presence of and on behalf of Supreme Commander A.E.F. signed W.B. S M I T H. On behalf of the Soviet high command signed S U S L O P A R O F F. S E V E D. F R E N C H."

Following the German defeat, Allied signals-intelligence operations continued against the Japanese and, when Tokyo surrendered, the resources developed during wartime were concentrated against the Soviet target and Jewish terrorists in Palestine. *See also* ARLINGTON HALL; COMBINED BUREAU MIDDLE EAST (CBME); OP-20-G; VENONA; WIRELESS EXPERIMENTAL STATION (WES).

WYMONDHAM. A **World War II** signals-intercept station run by the **War Office Y Group**, the Wymondham site in Norfolk contained 17 intercept positions.

X

XINJIANG. Following the closure of the **National Security Agency (NSA)** intercept sites at Astara, Behshahr, Kabkan, Khorramabad, and Shirabad in northern Iran in 1979 and the Soviet invasion of Afghanistan in December the same year, the Chinese authorities reached an agreement with the NSA and the **Bundesnachrichtendienst** to establish a pair of jointly run signals-intelligence stations at Korla and Qitai in the Pamir mountains of Xinjiang province to monitor Soviet communications.

Y

YARDLEY, HERBERT O. A highly controversial but brilliant American cryptanalyst, Herbert Yardley headed the War Department's Cipher Bureau and then, at the end of **World War I**, the U.S. State Department's **Black Chamber** in New York until it was closed in October 1929 by Secretary of State Henry Stimson. Famously, he is alleged to have remarked in justification, "Gentlemen do not read each other's mail." Two years later, in 1931, Yardley achieved worldwide notoriety when he published *The American Black Chamber*, an account of his organization's success in solving the ciphers of 19 foreign governments, including the Japanese and Chinese, and the encrypted communications of the 1921 Washington Naval Conference, which settled the 5:5:3 formula for the tonnage of British, American, and Japanese battleships. By intercepting and reading Tokyo's diplomatic cables, the American side had anticipated every one of their opponent's maneuvers, but it was not until Yardley sold the Japanese his cryptographic system in 1930 that they realized their vulnerability. Yardley's indiscretion caused embarrassment to the U.S. administration and outrage in the Diet.

Born in April 1889 in Worthington, Indiana, Yardley was taught telegraphic codes by his father, a railroad station-master. During **World War I**, as a reserve officer, he served in France with the Signal Corps and in May 1916 wrote a 100-page monograph, *A Solution to American Diplomatic Codes*, which drew attention to the flaws in the U.S. government's cipher systems.

In 1938, he would be recruited by the Chinese to work on Japanese ciphers, an act driven by financial worries but one that would always leave him open to criticism. His version of his contract in China, *The Chinese Black Chamber: An Adventure in Espionage*, was eventually declassified and published in 1983.

In June 1941, the **Examination Unit** of the Canadian National Research Council employed him to exploit enemy broadcasts that had been monitored by the Royal Canadian Signals Corps at Rockcliffe Barracks in Ottawa. Under his supervision, the Examination Unit concentrated on Japanese transmissions, but in January 1941, following intense pressure from London and Washington, he was replaced by Oliver Strachey, who had broken the

Abwehr's hand ciphers at **Bletchley Park**. His attempt to return to the cryptographic field in 1942 was rebuffed, and instead he was given an innocuous appointment in the U.S. Office of Price Administration.

Yardley's plans for a sequel, provisionally entitled *Japanese Diplomatic Secrets, 1921–22*, were abandoned when the federal authorities threatened him with dire consequences if he released further classified information, and this episode is the first occasion on which an American government is known to have intervened to ban a book. The manuscript, which bore the optional but provocative titles of *The Listening Post*, *Diplomatic Eavesdropping*, *Embassy Keyholes*, *Embassy Eavesdropping, and Diplomatic Keyholes*, had been circulated by his literary agent.

Attacked for having worked for the Chinese and for the indiscretions contained in his first book, few knew of his deal with the Japanese, and he was buried in August 1958 a national hero in Arlington National Cemetery. The only leak occurred soon after his death when the journalist Ladislas Farago alleged in *The Broken Seal* that Yardley had indeed sold out to the Japanese, but he had been misinformed about the date and stated incorrectly that the sale had been made in 1928 instead of 1930.

Although Yardley was to write several books, including some guides to playing poker, his *Education of a Poker Player* being a bestseller in 1957, his most sensitive book was only declassified in 1979.

Y COMMITTEE. An interdepartmental body created to coordinate British wireless resources in 1941, the Y Committee was intended as a forum in which the **Government Code and Cypher School** could identify **Bletchley Park**'s priorities for the **Admiralty Y Service**, the **War Office Y Group**, and the Air Ministry's **AI-4**. While the Secret Intelligence Service (SIS) and the **Radio Security Service (RSS)** were represented on the Y Committee, **MI5** was not, as the organization was not recognized by the other parties as a significant consumer of signals intelligence and, therefore, made no submissions when the issue of RSS's future was deliberated. In that example, control over RSS passed to SIS, and then after the war to **Government Communications Headquarters**, despite MI5's protests that SIS's member of the committee, Richard Gambier-Parry, was a technician and not an intelligence officer.

The problem of interagency liaison on Y topics was exacerbated by the intelligence consumers' lack of technical knowledge and the compartmentalization that meant some very senior officers, including General Courtenay Fladgate, who was in overall command of Army Signals, had not been indoctrinated into Y needs and capabilities. *See also* BRITISH SIGNALS INTELLIGENCE.

YOUNG, COURTENAY. The son of a British diplomat, Sir George Young, who served in the Naval Intelligence Division in 1915, Courtenay Young was **MI5**'s only Japanese-speaking officer. In 1947, he was posted to the **Australian Security Intelligence Organisation (ASIO)** in Sydney as the security liaison officer (SLO), responsible for indoctrinating ASIO into leads from what was then termed "the source" but was actually VENONA. He would later be transferred as SLO to Malaya.

Z

Z. During the early part of **World War II**, Z was adopted by the British Security Service, **MI5**, as indicating a cryptographic source. Within the category were numerous individual channels, such as **ZPT**.

ZENIT. During the **Cold War**, KGB and **GRU** *rezidenturas* accommodated in diplomatic premises overseas invariably contained a signals-intelligence unit codenamed ZENIT, the Russian word for "zenith" or "pinnacle." The ZENIT post was usually packed with electronic monitoring equipment designed to intercept radio traffic generated by the local hostile counterintelligence apparatus, and before initiating any operation, operators scanned the airwaves to detect signs of compromise and any indication of unusual surveillance activity in the neighborhood that would jeopardize plans. *See also* IMPULSE.

ZEPPELIN. Deployed primarily as reconnaissance platforms, the Imperial German Navy communicated with its fleet of zeppelins based at Emden, Cuxhaven, and Wilhelmshaven by wireless, using transmitters at Nordholz, Borkum, and Sylt. The traffic was routinely monitored by Marconi direction-finding stations in France and the Admiralty's 19 "directionals," enabling Allied cryptanalysts to plot the movements of individual airships to within one and a half degrees and to accumulate sufficient experience to predict the purpose of particular missions. Study of zeppelin operations in the last week of May 1916 was correctly interpreted as an indication that the German High Seas Fleet was about to put to sea, and by midday on 30 May, Admiral Jellicoe's Grand Fleet in Scapa Flow had been alerted to the enemy's intentions, and the subsequent confrontation at the battle of Jutland resulted in the loss of three British battle cruisers, three armored cruisers, and eight destroyers, while the Germans suffered one battleship sunk, one battle cruiser, four light cruisers, and five destroyers lost, with severe damage sustained by several other important warships. Although both sides claimed a victory, the **Admiralstab** would not allow the High Seas Fleet to venture out of port again. *See also* AIRBORNE COLLECTION; MI1(e); WORLD WAR I.

ZIMMERMANN TELEGRAM. On 16 January 1917, the Kaiser's foreign minister, Arthur Zimmermann, sent a secret two-part telegram to his ambassador in Washington, DC, Count Johann Bernstorff, and, in the absence of any German transatlantic cables, it was delivered to the U.S. embassy in Berlin for transmission on a State Department channel, the American cable via Copenhagen and London. The U.S. administration had given its permission for the Germans to use the American communications system in the mistaken belief that the German traffic was supporting peace negotiations.

The cipher employed was a new one, designated 7500, which had replaced the German foreign ministry's 13040, which had been in use since 1907. A copy of Bernstorff's 7500 codebook had been delivered to him by a submarine, the *U-55*, in November 1916. The first part of the message amounted to 850 groups, and the second was 150 groups, and both were decrypted upon delivery on 1 January by the embassy's staff of seven cipher clerks, who then, as instructed in part one, reciphered the entire text in the 13040 code for onward transmission by Western Union to Mexico, at a cost of $85, where the minister, Heinrich von Eckardt, did not possess a copy of 7500.

The full text announced an intention to engage in unrestricted U-boat warfare on 1 February and directed the ambassador in Mexico City to approach his host government with an offer of support if it attacked the United States to recover "lost territory in Texas, New Mexico and Arizona."

> BERLIN TO WASHINGTON W.158 16 JANUARY 1917
> MOST SECRET FOR YOUR EXCELLENCY'S PERSONAL INFORMATION AND TO BE HANDED TO THE IMPERIAL MINISTER IN [?] MEXICO WITH... BY A SAFE ROUTE. WE PROPOSE TO BEGIN ON 1 FEBRUARY UNRESTRICTED SUBMARINE WARFARE IN DOING SO, HOWEVER, WE SHALL ENDEAVOUR TO KEEP AMERICA NEUTRAL.. [?] IF WE SHOULD NOT [succeed in doing so] WE PROPOSE TO [?Mexico] AN ALLIANCE UPON THE FOLLOWING BASIS:
> [Joint] CONDUCT OF WAR
> [Joint] CONCLUSION OF PEACE
> GENEROUS FINANCIAL SUPPORT AND AN UNDERTAKING ON OUR PART THAT MEXICO IS TO RECONQUER THE LOST TERRITORY IN TEXAS, NEW MEXICO AND ARIZONA. THE SETTLEMENT IN DETAIL IS LEFT TO YOU. YOUR EXCELLENCY SHOULD FOR THE PRESENT INFORM THE PRESIDENT [of Mexico] SECRETLY [that we expect] WAR WITH THE USA [possibly] [...Japan].. [Corrupted sentence meaning "Please tell the President] THAT.. OUR SUBMARINES.. WILL COMPEL ENGLAND TO PEACE WITHIN A FEW MONTHS. ACKNOWLEDGE RECEIPT. ZIMMERMANN

The intercepted text would be decrypted in **Room 40** by the Reverend William Montgomery and Nigel de Grey, who had partially reconstructed the 7500 cipher. The cryptographers' task had been made easier by the acquisition from Western Union of a copy of the cable encrypted in the 13040 cipher and by access to a second telegram from Zimmermann, dated 5 February 1917, to his ambassador in Mexico, which was transmitted on a Swedish channel, to Buenos Aires via Madrid. Previously, the Germans had been allowed to send telegrams through Stockholm, but the cable had been routed conveniently through London, and when the practice was discovered, there had been a formal diplomatic protest, which had terminated it. Although the Swedes gave an undertaking not to assist any further German communications, they simply switched the traffic to Latin America, a route that became known as "the Swedish roundabout." The second telegram, encrypted in the compromised 13040 cipher, confirmed:

> Provided there is no danger of secret being betrayed to USA you are desired without further delay to broach the question of an alliance to the President. The definite conclusion of an alliance, however, is dependent on the outbreak of war between Germany and the USA. The President might even now, on his own account, sound out Japan. If the President declines for fear of subsequent revenge [by the USA] you are empowered to offer him a definite alliance after conclusion of peace provided Mexico succeeds in drawing Japan into the alliance.

The original telegram was finally reconstructed by Room 40 on 19 February and delivered to the U.S. ambassador, Walter Hines Page, in London on 23 February by the director of naval intelligence, Admiral Reginald Hall, together with a key to the cipher so the American authorities could authenticate it by comparing it with Western Union's version. Secretary of State Robert Lansing showed the offending document to an outraged President Woodrow Wilson on 27 February, and it was then handed over to a news agency the following day. Surprisingly, when challenged on 3 March, Count Bernstorff confirmed the authenticity of the telegram, and as a direct consequence, President Wilson told Congress on 6 April 1917 that America's neutrality would cease.

The fact that the British had intercepted and read a communication from the U.S. legation in Copenhagen addressed to the State Department in Washington, DC, was of some sensitivity, so a cover story was fabricated to assuage American sensibilities and create the pretence that the same message had been duplicated on other routes, being the Swedish transatlantic cable from Stockholm and the German radio link between Nauen and Sayville, Long Island. To conceal the role played by Room 40, a tale was circulated that the Admiralty had recovered a copy of the 13040 codebook from a captured German agent, Wilhelm Wassmuss, who had been detained at the German consulate in Bushehr earlier in the war.

Embarrassed, Zimmermann expressed his complete confidence in the integrity of his "absolutely secret code" and described the incident as an "act of treachery—one may assume it to have been such," which "appears to have been perpetrated on U.S. territory," but his opinion changed after Bernstorff returned from Berlin, having been withdrawn from Washington on 9 February. "Various indications suggest the treachery was committed in Mexico," he claimed on 29 March, prompting Eckardt to protest that only he and his trusted secretary Magnus had handled the telegram and that it was much more likely that the leak had occurred in Washington. Dr. Otto Göppert, the expert called in to investigate, agreed, concluding that the "betrayal had resulted from American not British action . . . possibly by the treachery or carelessness of some member of the chancery staff in Washington." *See also* GERMAN SIGNALS INTELLIGENCE; UNITED STATES SIGNALS INTELLIGENCE; WORLD WAR I.

ZINNIA SANURA. The American codename for an over-the-horizon radar system.

ZIRCON. Intended for launch in 1988, the ZIRCON signals-intelligence satellite project was scrapped by the British government on grounds of its estimated £500 million cost, calculating that the investment would be better spent in supporting the American **MAGNUM** system. Details of ZIRCON were disclosed in January 1987 in the *New Statesman* magazine, which resulted in police Special Branch raids conducted to seize evidence of a breach of the Official Secrets Act. No prosecution followed, and a television documentary broadcast in September 1988 was based on material gathered from unclassified sources.

ZPT. The British intelligence designation during **World War II** for encrypted wireless communications employed by the Kriegsmarine. *See also* BRITISH SIGNALS INTELLIGENCE.

ZYMOTIC. Allied codeword employed during **World War II** to indicate **special intelligence** provided by Melbourne. *See also* ULTRA.

ZZZZZ. The priority of ULTRA summaries distributed during **World War II** was indicated by up to five "Zs," with just one showing a low priority and ZZZZZ reflecting the top.

Appendix A
VENONA Glossary

When the VENONA decrypts were first declassified and released in July 1995, some of the texts were redacted, and many of the individual spies mentioned only by their codenames were left unidentified, either because the American, Australian, British, Canadian, and Swedish counterintelligence authorities had failed to find their true names or that political expediency had dictated discretion. In the years that followed, GCHQ and the NSA agreed to declassify further information by restoring some redactions, and more analytical research was undertaken by independent scholars, who sometimes were able to succeed where the professional mole hunters had failed. One of the most significant breakthroughs was confirmation of two major British spies, codenamed INTELLIGENTSIA and NOBILITY, both well-connected members of the GRU's mysterious X GROUP. Another achievement was the contribution of a former KGB officer, Alexander Vassiliev, who was able to discreetly copy some files from his organization's Moscow archive and end speculation, for example, about the atomic spy PERS, who turned out to be Russell McNutt, a Manhattan Project engineer.

This glossary is a comprehensive listing of codenames that appear in the NKVD and GRU communications exchanged between Moscow and *rezidenturas* in London, Canberra, Mexico City, New York, Stockholm, and Washington, DC, between 1940 and 1948, with the very latest identifications.

Codename	*Traffic*	*Identification*
19	NKVD Washington	Laurance Duggan
ABRAM	NKVD New York	Jack Soble
ACHIEVEMENT	NKVD Mexico	Unidentified
ACORN	NKVD Washington	William H. Taylor, International Monetary Fund
ACQUAINTANCE	NKVD Mexico	Unidentified
ADA	NKVD Mexico	Kitty Harris
ADAM	NKVD	Rebecca Getzoff
AILERON	NKVD New York	George Silverman
AKIM	NKVD New York	Sergei Lukianov
AL	NKVD New York	Albert D. Slack
ALBERT	NKVD New York	Iskhak Akhmerov, rezident
ALEKSEI	NKVD New York	Anatoli Yakovlev
ALES	NKVD New York	Alger Hiss
AMOR	NKVD Mexico	Margarita Nelken de Paul Mansberger
AMT	NKVD New York	John Abt, CPUSA lawyer
ANITA	NKVD Mexico	Unidentified
ANT	NKVD New York	Kristel Heineman
ANTENNA	NKVD New York	Julius Rosenberg
ANTON	NKVD New York	Leonid R. Kvasnikov
ANTON	NKVD Mexico	Francisco Anton
APTEKAR	GRU London	Nikolai V. Aptekar
ARENA	NKVD New York	Mary Price, China Aid Council
ARNO	NKVD New York	Harry Gold, later GOOSE
ARSENIJ	NKVD New York	Andrei Shevchenko
ARTEK	NKVD New York	Leonid D. Abramov
ARTEM	NKVD New York	Unidentified
ARTHUR	NKVD New York	Unidentified, inside OSS
ARTISTE	NKVD Canberra	Herbert W. Tattersell
ATATMAN	NKVD New York	Bill Gebert
ATTILA	NKVD New York	Major Linn Farish
AUSTRALIAN, THE	Naval GRU	Francia Mitynia, alias Edna Patterson
AUTHOR	NKVD New York	Vladimir B. Morkovin
BARCh	GRU London	Semyon Kremer
BARON	GRU London	Karel Sedlacek
BASQUE	NKVD Mexico	Unidentified
BASS	NKVD New York	Michael J. Burd, banker
BAUER	GRU London	Lieutenant Hein
BEETLE	NKVD Mexico City	Unidentified
BEK	NKVD New York	Sergei Kurnakov
BEN	NKVD Canberra	Detective Sgt Alf Hughes
BENITO	NKVD Mexico	Benito Rodriguez Gutierrez
BERG	NKVD New York	Alexander Koral
BERG	NKVD Washington	Unidentified, later SENOR
BLACK	NKVD New York	Tasso Lessing ("Tom") Black
BOAR	NKVD New York	Winston Churchill
BOB	NKVD London	Boris Krötenschield, alias Boris Krotov
BOB	NKVD New York	Charles Flato
BORIS	NKVD New York	Aleksandr P. Saprygin

Codename	Traffic	Identification
BOSS	NKVD Mexico	Vasili P. Yakubovsky, chargé d'affaires
BREDAN	GRU	Philip Keeney
BRION	GRU London	Major Boris Shvetsov
BUMBLEBEE	NKVD New York	Walter Lippmann
BUMBLEBEE	NKVD New York	David Greenglass, later CALIBER
CALISTRATUS	NKVD New York	Alexander Fomin
CAPITALIST	NKVD New York	W. Averell Harriman
CAUTIOUS	NKVD New York	Julius Joseph, OSS
CAVALRYMAN	NKVD New York	Sergei N. Kurnikov
CHAP	NKVD New York	Zelmond Franklin
CHAR	NKVD New York	Charles Flato
CHARON	NKVD San Francisco	Grigori Kheifets
CHESS KNIGHT	NKVD Mexico	David Siqueiros, artist
CHESTER	NKVD New York	Unidentified in Los Alamos
CLAUDE	NKVD Canberra	Walter Clayton
CLEVER GIRL	NKVD New York	Elizabeth Bentley
COLLEAGUE	NKVD New York	Carl Marzani, OSS
CONSTRUCTOR	NKVD New York	Abe Brothman
COOKIE	NKVD Mexico	Unidentified
CORK	NKVD New York	William Pinsly
CRUCIAN	NKVD New York	Anton Ivancic
CZECH	NKVD New York	Robert O. Menaker, later BOB
DAEDALUS	GRU London	Pierre Cot
DAN	GRU London	Unidentified
DAShA	NKVD New York	Maria Wicher
DICK	NKVD New York	Bernard Schuster
DICK	GRU London	Unidentified rezidentura member
DINAH	NKVD New York	Gertrude Kahn
DIR	NKVD New York	Mary Price
DON	NKVD Mexico	Vicente Uribe Galdeano
DONALD	NKVD New York	W. Ludwig Ullman
DORA	NKVD New York	Helen Silvermaster
DOUGLAS	NKVD New York	Joseph Katz
DROP	NKVD New York	Philip E. Moseley, OSS
DRUG	NKVD Stockholm	Edgar Klaus
ECHO	NKVD New York	Bernard Schuster
EDGAR	GRU Stockholm	Knut Alvar Nyblad
EFIM	NKVD Canberra	Semen Makarov
ELSA	NKVD Mexico	Unidentified
ELZA	NKVD New York	Helen Akhmerova, née Lowry
ERIC	NKVD New York	Leona Franey
ERIE	NKVD New York	Paul Nahin
EXPERT	NKVD New York	Abe Brothman
FERRO	NKVD Canberra	Ric Throssell
FIR	NKVD New York	Helen Scott Keenan, OSS
FIRTREE	NKVD New York	Helen Scott Keenan
FRANK	NKVD New York	Laurence Duggan, later SHERWOOD and PRINCE

Codename	*Traffic*	*Identification*
FRED	NKVD New York	Fred Rose
FREGATO	NKVD Mexico	Redacted
FROST	NKVD New York	Boris Morros
GARD	NKVD New York	Unidentified in OSS
GAVR	NKVD Washington	Identity Redacted, BSC
GIFT	NKVD New York	Grigori Kasparov
GLASSER	NKVD New York	Harold Glasser
GLYCERINE	NKVD Mexico	Unidentified
GNAT	NKVD New York	Victor Kravchenko
GNOME	NKVD Mexico	Ramon Mercader, assassin
GNOME	NKVD New York	William Perl
GOOSE	NKVD New York	Harry Gold
GRANDFATHER	NKVD	Evgeni Kiselev
GRANDSON	NKVD New York	Member of the rezidentura
GREY-HEAD	NKVD Mexico	Edward P. Fitzgerald
GRINGO	NKVD Mexico	Mary Anne Lenson, née Diehl, alias Mary Grohol
GUSTAV	GRU New York	George Gorchoff
HARE	NKVD	Maurice Halperin, OSS
HARRY	NKVD Mexico	Jacob Epstein
HELMSMAN	NKVD New York	Earl Browder
HEN-HARRIER	NKVD New York	Cordell Hull
HENRY	NKVD New York	Henry Malisoff, later TALENT
HICKS	NKVD London	Guy Burgess
HOMER	NKVD New York	Donald Maclean
HUGHES	NKVD New York	Al Sarant
HURON	NKVD New York	Byron Darling, later ERNEST
IDE	NKVD New York	Samuel Krafsur
INFORMER	NKVD New York	Joseph Katz, later X
INTELLIGENSIA	GRU London	Prof. J. B. S. Haldane
INVENTION	NKVD Mexico	Senator Paloma
IRIS	GRU London	Unidentified
IVANOV	NKVD New York	John Scott, OSS
IVY	NKVD New York	Emma Joseph
IZRA	NKVD	Donald Wheeler, OSS
JACK	NKVD London	Unidentified
JEANNE	NKVD New York	Christina Krotkova
JEROME	GRU London	André Labarthe
JOHN	GRU London	Leon Beurton
JOHNSON	NKVD London	Anthony Blunt
JOSE	NKVD Mexico	Unidentified
JUAN	NKVD Mexico	Juan Gaytan
JULIA	NKVD New York	Olga Khlopkova
JURIST	NKVD New York	Harry Dexter White
KANT	NKVD New York	Henry Magdoff
KAPITAN	NKVD Washington	President Roosevelt
KARL	GRU Stockholm	Allan Emanuel Nyblad
KEENEY	NKVD New York	Philip Olin Keeney
KhAZAR	NKVD New York	Unidentified in OSS

Codename	*Traffic*	*Identification*
KIN	NKVD New York	Enos Wicher
KINSMAN	NKVD New York	Unidentified, previously SOLID
KLARK	GRU London	Unidentified rezidentura member
KLAVA	NKVD Mexico	Caridad Mercader
KOCH	NKVD	Duncan C. Lee, OSS
KOLO	NKVD New York	Sava S. Kosanovic
KOLTsOV	NKVD New York	Unidentified
KRAMER	NKVD New York	Charles Kramer
KRITIK	NKVD Stockholm	Unidentified Danish source
KURD	NKVD Mexico	Marcel Gautier
LACQUER	NKVD Mexico	Unidentified
LAREDO	NKVD Mexico	Unidentified
LATA	NKVD Mexico	Dr. Esther Chapa
LAWYER	NKVD New York	Harry Dexter White
LENA	GRU Stockholm	Unidentified Norwegian
LENS	NKVD New York	Michael Sidorovich
LESLEY	NKVD New York	Leontina Cohen
LIBERAL	NKVD New York	Julius Rosenberg
LIGHT	NKVD New York	Aleksandr Raev
LINK	NKVD New York	William Weisband
LION	NKVD New York	Sergei Kudriatsev, rezident in Ottawa
LIZA	NKVD New York	Martha Dodd
LOU	NKVD New York	Marion Berdecio, née Davis
LOUIS	NKVD New York	Alfred Stern
LUKA	NKVD	Pavel Klarin
LYDIA	NKVD Mexico	Lydia Altschuler
MARGARITA	NKVD New York	Identity redacted
MARK	GRU London	Unidentified rezidentura member
MARQUIS	GRU New York	Joseph Bernstein
MARS	NKVD New York	Vasili Sukhomlin
MARTHA	GRU London	Alta Lecoutre
MASHA	NKVD New York	Eufrosina Dvoichenko-Markov
MASTERCRAFTSMAN	NKVD Canberra	Dr. Wilbur Christiansen
MAXSIM	NKVD New York	Vasili Zubilin
MAXWELL	Naval GRU	Jack Fahy
MAY	NKVD New York	Stepan Apresyan, rezident
MAYOR	NKVD New York	Iskhak Akhmerov
METER	NKVD New York	Joel Barr
MIJE	NKVD Mexico	Antonio Mije Garcia
MIRANDA	NKVD New York	Helen Koral, later ART
MLAD	NKVD New York	Theodore Hall
MOLIERE	NKVD Mexico	Pavel Melkishev, alias Pavel Mikhailov
MYRNA	NKVD New York	Elizabeth Bentley
NABOB	NKVD New York	Henry Morgenthau
NAZAR	NKVD New York	Stepan N. Shundenko
NEEDLE	NKVD New York	Jones Orin York
NICK	NKVD New York	Amadeo Sabatini
NIK	GRU London	Unidentified rezidentura member
NIL	NKVD New York	Nathan Sussman

Codename	*Traffic*	*Identification*
NOBILITY	GRU London	The Hon. Ivor Montagu
NONA	NKVD Mexico	Ruth Wilson, married to Jacob Epstein
Okh	NKVD New York	Ing. Adolfo Orive de Alba
OLA	NKVD New York	Sara Veksler
OLD	NKVD New York	Saville S. Sax
OLIVER	NKVD Mexico	Antonio Gomez Deans
OLS	NKVD New York	Christina Krotkova
OSIPOV	NKVD New York	Prince Nikolai Orlov
OSWARD	NKVD New York	Benjamin Gerig, State Dept.
PAGE	NKVD New York	Lauchlin Currie
PAL	NKVD New York	Nathan Gregory Silvermaster
PALM	NKVD Canberra	Boris Lvovich
PANCAKE	NKVD New York	I. F. Stone
PATRIOT	NKVD Mexico	Unidentified source
PAVLOV	NKVD Mexico	Lavrenti Beria
PEAK	NKVD New York	Frank Coe
PEDRO	NKVD Mexico	Jesus Hernandez Tomas
PER	GRU Stockholm	Erik Bjärtnäs
PETER	NKVD New York	Tasso Lessing ("Tom") Black
PHLOX	NKVD New York	Rose Olson
PILOT	NKVD New York	W. Ludwig Ullman
PLATO	NKVD New York	Charles Plato
PLUMB	NKVD New York	Charles Kramer, also LOT and MOLE
POET	NKVD Stockholm	Hella Wuolijoki
POLO	NKVD New York	William Ullman
PRINCE	NKVD New York	Laurence Duggan
PUSH-BUTTON	NKVD Mexico	Nancy Bell
QUANTUM	NKVD New York	Prof. Boris Podolsky
RABIN	NKVD Mexico	Unidentified
RACHEL	NKVD New York	Unidentified in OSS
RADIO ANNOUNCER	NKVD	General Donovan
RAMSAY	NKVD New York	Clarence Hiskey
RAPID	NKVD Mexico	Unidentified
RAS	GRU London	General de Gaulle
RAY	NKVD New York	Frank Oppenheimer, later CARL
RECTOR	NKVD New York	George Zlatovski
REDAKTOR	NKVD Mexico	Konstantin A. Umansky
RELAY	NKVD New York	Philip Morrison, later SERB
REMBRANDT	NKVD Mexico	José Sancha Padros
RESERVIST	GRU London	Kenneth Post
REST	NKVD New York	Klaus Fuchs, later CHARLES
RICHARD	NKVD New York	Harry Dexter White
ROBERT	NKVD New York	Nathan Gregory Silvermaster
ROMAN	NKVD New York	Dr. Robert Soblen
ROSA	NKVD London	Unidentified
ROUBLE	NKVD New York	Harold Glasser
S-1	NKVD New York	Herman R. Jacobson
SACHS	NKVD New York	Solomon Adler, State Dept.

Codename	*Traffic*	*Identification*
SATYR	NKVD New York	Sylvia Caldwell
SCORPION	NKVD Canberra	Prof. Sergei J. Paramonov
SEDA	NKVD Mexico	Elena Vasquez Gomez
SEMEN	NKVD Mexico	Pavel Prosin
SENATOR	NKVD Stockholm	Georg Branting
SENOR	NKVD New York	Unidentified
SERES	NKVD New York	Dr. Ivan Subasic
SERGEI	NKVD New York	Vladimir Pravdin
Sh	NKVD New York	Vicente Lombardo
SIMA	NKVD New York	Judith Coplon
SISTER	NKVD Canberra	Frances Bernie
SLANG	NKVD New York	Jane Foster, OSS
SMART	NKVD New York	Elliot Goldberg
SNOW LEOPARD	NKVD Mexico	Dr. Javier Rojo Gomez
SOLID	NKVD New York	Dr. Reed
SON	NKVD New York	Rudy Baker
SOUND	NKVD New York	Jacob Golos
STAKHANOVITE	NKVD Mexico	Unidentified illegal
STAMP	NKVD New York	Republic Aviation Corp. source
STANLEY	NKVD London	Kim Philby
STENLI	GRU London	Unidentified
STEPAN	NKVD New York	Pavel Fedosimov
STOCK	NKVD New York	Mikhail Shalyapin
STORM RAIDER	NKVD New York	Victor Perlo
SUKhUMSKIY	NKVD New York	Boris M. Boguslavsky
SULLEN	NKVD New York	Viktor Kirilov
SWING	NKVD New York	Raymond Gram Swing
TECHNICIAN	NKVD Canberra	Fedor A. Nosov, TASS
TED	NKVD New York	Edward Fitzgerald
TEXAS	NKVD Mexico	Alejandro Carrillo Marcor
THE PAIR	NKVD Mexico	Nicholas and Maria Fisher
THERAPUTIST	GRU London	Unidentified
THRUSH	NKVD New York	Unidentified, later AKHMED
TINA	NKVD London	Melita Norwood
TOM	NKVD	Leonid Eitingon
TOURIST	NKVD Canberra	James Hill
TRIBUNE	NKVD Mexico	Unidentified
TULIP	NKVD New York	Mark Zborowsky, later KANT
TWAIN	NKVD New York	Semyon M. Semyonov
UCN/9	NKVD New York	Cedric Belfrage
UNCLE	NKVD New York	Isaac Folkoff
URAL	NKVD New York	Nikolai Karpekov
VADIM	NKVD London	Anatoli Gorsky
VAGUE	NKVD New York	James Miller
VARDO	NKVD New York	Elizaveta Zarubina
VEGA	NKVD Mexico	Dmitri Potemkin
VEKSEL	NKVD New York	Unidentified
VICTOR	NKVD New York	Pavel Fitin

Codename	*Traffic*	*Identification*
VIKTOR	NKVD Mexico	Unidentified
VLADIMIR	NKVD Mexico	Unidentified
VLADISLAV	NKVD New York	Lieutenant Nikolai G. Redin
VOGEL	NKVD New York	Russell McNutt, later PERS
VOLUNTEER	NKVD New York	Morris Cohen
VOLVOK	NKVD New York	Andrei A. Orlov
WASP	NKVD New York	Ruth Greenglass
WOLF	NKVD Mexico	Rosendo Gomez Lorenzo
WRITER	GRU London	Unidentified
X	NKVD New York	Joseph Katz
YURI	NKVD Mexico	Lev Tarasov
ZAPATA	NKVD Mexico	Brig.-Gen. Roberto Calvo-Ramirez
ZERO	NKVD New York	Leona Franey
ZhANNA	NKVD New York	Christina Krotkova, later JEANNE and OLS
ZhAROV	NKVD New York	Nikolai E. Ezhov
ZhENYA	NKVD New York	Sonia Gold (née Steinman)
ZhOLUD	NKVD New York	Bill Gold
ZORA	NKVD New York	Flora Wovschin

Appendix B
Agreement between British Government Code and Cipher School and U.S. War Department in regard to Certain "Special Intelligence"

U.S.: SECRET **MAY 17, 1943**

BRITISH: MOST SECRET

Agreement between British Government Code and Cipher School and U.S. War Department concerning cooperation in matters relating to:

U.S.	British
Special Intelligence A	Special Intelligence
Special Intelligence B	Y Intelligence
TA Intelligence	Y Inference

A distinction is made in nomenclature and procedure in handling intelligence derived from the solution of enemy high grade and that obtained from low grade codes and ciphers. The preservation of secrecy in regard to either category is a matter of great concern to both countries and if the highest degree of security is to be maintained, it is essential that the same methods should be pursued by both countries at every level and in every area concerned, since a leakage at any one point would jeopardize intelligence from these sources not in one area only but in all theaters of war and for all services.

This agreement is limited to the traffic specifically designated herein. It does not cover traffic emanating from non-service enemy or neutral sources. These subjects will be covered by future negotiations between Director, G.C.C.S and A.C. of S., G-2, War Department.

1. Both the U.S. and British agree to exchange completely all information concerning the detection, identification and interception of signals from, and the solution of codes and ciphers used by, the Military and Air forces of the Axis powers, including secret services (Abwehr).
2. The U.S. will assume as a main responsibility the reading of Japanese Military and Air codes and ciphers.
3. The British will assume as a main responsibility the reading of German and Italian Military and Air codes and ciphers.

4. Both countries agree that special security regulations shall apply to Intelligence obtained from decoding telegrams in enemy high grade codes and ciphers.
5. Both countries agree to use their most secure codes and ciphers for transmission of the decodes of enemy signals and transmission of technical cryptanalytic data.
6. British or U.S. Commanders-in-Chief, Military or Air, will receive all Special Intelligence necessary to them for the conduct of their operations from either British or U.S. centers as may be mutually agreed. Liaison officers will be appointed as desired for facilitating this. They will be given full access to all decodes.
7. The distribution of intelligence from the sources in question will be governed by the fundamental principle that distribution will be restricted to the minimum and will therefore be confined solely to those who require to receive the intelligence for the proper discharge of their duties.
8. All recipients of Special Intelligence A, whether British or American officers, shall be bound by the same regulations, the regulations (Appendix B) now in force in the theaters of war where British forces are operating to be accepted at the present time. If at a later date either country wishes to modify them in the light of further experience then this may be done by mutual agreement.
9. The extension to officers of a knowledge of the existence of such intelligence shall be confined to as limited a number as possible and restricted to the levels of command in conformity with the above mentioned regulations. Great stress is laid on the principle that Special Intelligence A should not be intermingled in reports with general intelligence from other sources. If, however, it becomes imperative to do so, the whole must be treated as Special Intelligence A and given the same strictly limited distribution. Under no circumstances is it permissible to pass Special Intelligence A in a code or cipher which can be read by other than the authorized recipients.
10. Although Special Intelligence B is not subject to the same stringent regulations as Special Intelligence A, since the two are closely connected, it is essential to maintain a high degree of secrecy in the handling of Special Intelligence B also. In any action taken upon such intelligence and in any documents or telegrams based upon it, it is essential that its origin be disguised and that the codes or ciphers used for its dissemination be absolutely secure.
11. All intelligence available from decodes shall be made available to Liaison officers, and if they deem necessary it will be exchanged between London and Washington. These Liaison officers will be specially appointed and given full facilities for this purpose.

12. British and U.S. will notify one another without delay, giving full particulars, when either has information from any source indicating the compromise of any code or cipher used by the other. Action on such information will be most carefully considered in order not to compromise the source and if possible mutual agreement such action will be sought.
13. Cooperation between and coordination of U.S. Signal Intelligence Service and British "Y" Service must take place at all levels, technical information being exchanged mutually at the same level and each country to agree not to lower the classification of such information or the intelligence derived from it below that level without mutual agreement.
14. Each country shall inform the other of the employment and scope in each joint theater of war of their Signal Intelligence (Y) units in the field.
15. This agreement or the appendices thereto may be supplemented or modified from time to time governing any special feature for which either party wishes to make special provision.
16. Definitions:
    a. Y Service or Signal Intelligence Service. The British, U.S. Army, and U.S. Navy services concerned with intercepting, decoding, interpreting, classifying and dissemination of enemy (and neutral) communications, and the use of D/F and other specialized apparatus for establishing locations and identities of enemy transmitters.
    b. Special Intelligence A. Certain ciphers are placed in a special category, owing to their importance and difficulty of solution. The intelligence derived from these ciphers is known as Special Intelligence A. Such material is treated with most stringent security measures. Special Intelligence A is confined to a very strictly limited number of the most highly placed officers and is mainly of strategical importance.
    c. Special Intelligence B. Intelligence derived from the solution of lower grade ciphers. Such ciphers may under certain circumstances be upgraded to the "Special A" class.
    d. The dissemination of Special Intelligence B is wider though always treated as British Most Secret—U.S. Secret. Special Intelligence B may be used tactically.

APPENDIX A

Special Provisions Regarding Work on German Machine Ciphers

Since it is believed unnecessary and impracticable to duplicate work on German machine ciphers and in view of the large number of personnel required and the unavoidable extra risk to the security of the source involved, agreement which follows has been arrived at. This agreement provides that:

(a) All desired intelligence from this source will be made available to the War Department in Washington.

(b) U.S. personnel will obtain experience by engaging in the solution of this type of cipher in Great Britain.

(c) Research into new methods of attack will made in Washington.

(d) Transmission of Intelligence to Commander-in-Chief in the field will be accomplished by special routes and staffs who will maintain a watch over the use of the intelligence to guard against compromise of the source.</nl>

1. U. S. liaison officers will be appointed at G. C. & C. S. to examine messages and summaries and select those desired for transmittal to Washington for G-2 or Theater Commanders. All decoded material will be made available to those officers. Decodes giving information regarding Order of Battle will be handled as at present, i.e., through U. S. liaison officers in War Office and Air Ministry, respectively.
2. Decodes or summaries to be passed to Washington through existing British channels.
3. U. S. party to effect independent solution of keys will be established in Great Britain, but so coordinated by mutual agreement to avoid duplication. This party will cooperate with the British in regard to tasks and will be given every assistance for instruction of personnel. They will be furnished British machines. Decodes from this section will be passed to Bletchly Park for emendation, translation and distribution, but U. S. party will conduct complete processing, including emendation and translation to such an extent as they desire.
4. Formulas will be supplied by Great Britain for use on machines now at Arlington Hall.
5. U. S. to undertake research for finding a new method for solution and to be rendered every assistance by the British for this purpose.
6. In conformity with British policy, U. S. personnel engaged in solution work in Great Britain will not be transferred elsewhere except for very urgent reasons.
7. Special Intelligence from this source will be passed to Commanders-in-Chief in the field through the Special British units provided for this purpose. The officer in command of these units will have direct access to the Commander-in-Chief and advise as necessary on the security aspect of handling and using this intelligence. Where an American officer is Commander-in-Chief, an American officer, properly trained and indoctrinated at Bletchly Park, will be attached to the unit to advise and

act as liaison officer to overcome difficulties that may arise in regard to differences in language.

8. The Director of the G. C. & C. S. will have the final decision when matters of security are involved in intelligence items (gossip) and as to what is passed to Commanders-in-Chief in the field.

APPENDIX B

British Security Regulations for Special Intelligence CO-ORDINATION OF ROUTING, SECURITY AND USE OF SPECIAL INTELLIGENCE

Part I

1. SPECIAL INTELLIGENCE is the agreed name for the highly secret information obtained by cryptographic means from enemy high grade ciphers.
2. Lower grade cryptographic material classed in general as "Y" Intelligence, is not included in the definition of SPECIAL INTELLIGENCE.

PART I TO BE DESTROYED BY FIRE WHEN READ

Part II

1. All SPECIAL INTELLIGENCE emanating from the United Kingdom and transmitted to Commands abroad will receive the prefix 'Ultra'.
2. All SPECIAL INTELLIGENCE emanating from centres other than the United Kingdom and transmitted either to the United Kingdom or to another command abroad, is to receive a prefix specially allotted to each producing centre as follows:

DELHI	prefix	SIRDAR
WASHINGION	prefix)	
	)	
MELBOURNE	)	ZYMOTIC
	)	
KILINDINI	)	
MIDDLE EAST		"SWELL

3. SPECIAL INTELLIGENCE produced by U.S.A. centres either in U.S.A. or elsewhere if transmitted over British routes either to the United Kingdom or to British Commands overseas, is to receive the prefix of the Command or centre through which it is distributed.

4. Where it is necessary for SPECIAL INTELLIGENCE to be transmitted between Commands or centres other than the United Kingdom, special routes and ciphers are to be arranged and approved by London.

Part III

ULTRA (see Para. 1 above) information can be regarded as reliable and action can be taken on it, but experience has shown that the following security regulations are vital to the preservation of this source. The Commander-in-Chief is held personally responsible for ensuring that they are scrupulously adhered to:

1. The utmost secrecy is to be used in dealing with ULTRA information. Attention is called to the £act that if' from any document that might fall into the hands of the enemy or from any message that the enemy might intercept, from any word that might be revealed by a prisoner of war, or from any ill-considered action based upon it, the enemy were to suspect the existence of the ULTRA source, that source would probably forever be lost to our cause.
2. This loss would vitally affect operations on all fronts, not only the particular front on which the source had been compromised.
3. Commanding Officers of those Commands authorized to receive ULTRA information, i.e., normally only General and Air Officers commanding Armies and Air Forces, are to be instructed that ULTRA messages are for them, their personal representative, and their Senior Intelligence and Operations Staff Officer only, and are not to be seen by, read to, or discussed with any other person. ULTRA messages are to be destroyed by fire immediately action has been taken on them. No records of Intelligence based on ULTRA information may be kept, except at the H.Q. of the Commander-in-Chief.
4. When ULTRA information is to be used by the Commander of an Army or an Air Force as a basis for action to be taken by a subordinate command, the information must be translated, when passed to the subordinate command, into terms of an operational order, so worded that if captured or intercepted by the enemy the origin of the information could not be traced back to the ULTRA source, e.g., orders must never contain the precise time, date or place of an enemy operation revealed by ULTRA. Such orders based on ULTRA information if transmitted by W/T must be encoded only in authorized ciphers. Under no circumstances whatever is it permissible to transmit ULTRA information as such to lower formations.

5. In general, if any action is to be taken based upon ULTRA information, the local Commander is to ensure that such action cannot be traced back by the enemy to the reception of ULTRA Intelligence alone. A momentary tactical advantage is not sufficient ground for taking any risk of compromising the source. No action may be taken against specific sea or land targets revealed by ULTRA unless appropriate air or land reconnaissance has also been undertaken. Names of enemy ships revealed by ULTRA source may never be quoted.
6. The utmost care is to be taken in briefing pilots for an operation based on ULTRA information that only such details are given them as might have been obtained by other means, such as air reconnaissance, and only such as are essential to the success of the operations.
7. No reference to ULTRA information is to be made in any summary whatsoever, however limited the circulation. No discussion of it is permissible except between the senior officers who are immediately concerned with the action to be taken upon it.
8. If it is necessary to ask questions, or make comments on ULTRA material, whether on matters of Intelligence, Operations, Routing or Security, such messages are to be transmitted only over the special channel and in the special ciphers provided for ULTRA traffic.
9. Recipients of ULTRA may not under any circumstances carry on their persons outside their Headquarters, ULTRA messages which have been delivered to them.

1st March 1943

GEO. V. STRONG
Major General
A.C. of S., G-2

15 JUN 1943
Approved for the U.S. War Department.
By order of the Secretary of War.

JOSAPH P. McNARLEY,
Lieut. General, U.S. Army,
Deputy Chief of Staff.

Appendix C
British-U.S. Communication Intelligence Agreement

BRITISH-U.S. COMMUNICATION INTELLIGENCE AGREEMENT

5 March 1946

OUTLINE OF BRITISH-U.S. COMMUNICATION INTELLIGENCE AGREEMENT

1. Parties to the Agreement
2. Scope of the Agreement
3. Extent of the Agreement—Products
4. Extent of the Agreement—Methods and Techniques
5. Third Parties to the Agreement
6. The Dominions
7. Channels between U. S. and British Empire Agencies
8. Dissemination and Security
9. Dissemination and Security—Commercial
10. Previous Agreements
11. Amendment and Termination of the Agreement
12. Activation and Implementation of the Agreement

BRITISH-U.S. COMMUNICATION INTELLIGENCE AGREEMENT

1. Parties to the Agreement

The following agreement is made between the State-Army-Navy Communication Intelligence Board (STANCIB) (representing the U.S. State, Navy, and War Departments and all other U.S. Communication Intelligence[1] au-

1 Throughout this agreement Communication Intelligence is understood to comprise all processes involved in the collection, production, and dissemination of information derived from the communications of other nations.

thorities which may function) and the London Signal Intelligence (SIGINT) Board (representing the Foreign Office, Admiralty, War Office, Air Ministry, and all other British Empire[2] Communication Intelligence authorities which may function).

2. Scope of the Agreement

The agreement governs the relationship of the above-mentioned parties in Communication Intelligence matters only. However, the exchange of as much collateral material as is applicable for technical purposes and is not prejudicial to national interests will be effected between the Communication Intelligence agencies in both countries.

3. Extent of the Agreement—Products
   (a) The parties agree to the exchange of the products of the following operations relating to the foreign communications:[3]
      (1) collection of traffic
      (2) acquisition of communication documents and equipment
      (3) traffic analysis
      (4) cryptanalysis
      (5) decryption and translation
      (6) acquisition of information regarding communication organizations, practices, procedures, and equipment.
   (b) Such exchange will be unrestricted on all work undertaken except when specifically excluded from the agreement at the request of either party and with the agreement of the other. It is the intention of each party to limit such exceptions to the absolute minimum and to exercise no restrictions other than those reported and mutually agreed upon.
4. Extent of the Agreement—Methods and Techniques
   (a) The parties agree to the exchange of information regarding methods and techniques involved in the operations outlined in paragraph 3(a).
   (b) Such exchange will be unrestricted on all work undertaken, except that upon notification of the other party information may be withheld by either party when its special interests so require. Such notification will include a description of the information being withheld, sufficient in the opinion of the withholding party, to convey its significance. It

[2]For the purposes of this agreement British Empire is understood to mean all British territory, other than the Dominions.

[3]Throughout this agreement foreign communications are understood to mean all communications of the government or of any military, air, or naval force, faction, party, department, agency, or bureau of a foreign country, or any person or persons acting or purporting to act therefor, and shall include diplomatic communications of a foreign country which may contain information of a military, political, or economic value. Foreign country as used herein is understood to include any country, whether or not its government is recognized by the U.S. or the British Empire, excluding only the U.S., the British Commonwealth of Nations, and the British Empire.

is the intention of each party to limit such exceptions to the absolute minimum.

5. Third Parties to the Agreement

Both parties will regard this agreement as precluding action with third parties[4] on any subject appertaining Communication Intelligence except in accordance with the following understanding:

(a) It will be contrary to this agreement to reveal its existence to any third party whatever.
(b) Each party will seek the agreement of the other to any action with third parties, and will take no such action until its advisability is agreed upon.
(c) The agreement of the other having been obtained, it will be left to the party concerned to carry out the agreed action in the most appropriate way, without obligation to disclose precisely the channels through which action is taken.
(d) Each party will ensure that the results of any such action are made available to the other.

6. The Dominions

(a) While the Dominions are not parties to this agreement, they will not be regarded as third parties.
(b) The London SIGINT Board will, however, keep the U.S. informed of any arrangements or proposed arrangements with any Dominion agencies.
(c) STANCIB will make no arrangements with any Dominion agency other than Canadian except through, or with the approval of, the London SIGINT Board.
(d) As regards Canada, STACIB will complete no arrangements with any agency therein without first obtaining the views of the London SIGINT Board.
(e) It will be conditional upon any Dominion agencies with whom collaboration takes place that they abide by the terms of paragraphs 5, 8, and 9 of this agreement and to the arrangements laid down in paragraph 7.

7. Channels Between U.S. and British Empire Agencies

(a) STANCIB will make no arrangements in the sphere of Communication Intelligence with any British Empire agency except through, or with the prior approval of, the London SIGINT Board.
(b) The London SIGINT Board will make no arrangement in the sphere of Communication Intelligence with any U.S. agency except though, or with the prior approval of, STANCIB.

---

[4]Throughout this agreement third parties are understood to mean all individuals or authorities other than those of the United States, the British Empire, and the British Dominions.</notes>

8. Dissemination and Security

Communication Intelligence and Secret or above technical matters connected therewith will be disseminated in accordance with identical security regulations to be drawn up and kept under review by STANCIB and the London SIGINT Board in collaboration. Within the terms of these regulations dissemination by either party will be made to U.S. recipients only as approved by STANCIB; to British Empire recipients and to Dominion recipients other Canadian only as approved by the London SIGNT Board; to Canadian recipients only as approved by either STANCIB or the London SIGINT Board; and to third party recipients only as jointly approved by STANCIB and the London SIGINT Board.

9. Dissemination and Security—Commercial

STANCIB and the London SIGINT Board will ensure that without prior notification and consent of the other party in each instance no dissemination of information derived from Communication Intelligence is made to any individual or agency, governmental or otherwise, that will exploit it for commercial purposes.

10. Previous Agreements

This agreement supersedes all previous agreements between British and U.S. authorities in the Communication Intelligence field.

11. Amendment and Termination of the Agreement

This agreement may be amended or terminated completely or in part at any time by mutual agreement. It may be terminated completely and at any time on notice by either party, should either consider its interests best served by such action.

12. Activation and Implementation of Agreement

This agreement becomes effective by signature of duly authorized representatives of the London SIGINT Board and STANCIB. Thereafter, its implementation will be arranged between the Communication Intelligence authorities concerned, subject to the approval of the London SIGINT Board and STANCIB.

For and in behalf of the London Signal Intelligence Board:	For and in behalf of the State-Army-Navy Communication Intelligence Board:
Patrick Marr-Johnson Colonel, British Army General Staff	Hoyt S. Vandenberg Lieutenant General, GSC Senior Member

5 March 1946

Bibliography

CONTENTS

INTRODUCTION

Because of its sensitive nature, relatively little has been published about signals intelligence, but the books on this narrow topic fall into four distinct categories. Firstly, there are the titles that concentrate on the general subject of SIGINT, and Jeffrey T. Richelson has led the field in studying the organizations and the physical locations of the Cold War collection facilities. His work has included *The Ties That Bind*, with the Australian researcher Desmond Ball, published in 1985, *Soviet Ears in the Ether* in 1984, and *Foreign Intelligence Organizations* in 1988 and *America's Space Sentinel* in 1999. The second distinct group consists of those histories of Bletchley Park, which were pioneered in classified section accounts that were never intended for publication. Among those who have written about "BP" are Michael Smith (*Station X*), Ted Enever (*Britain's Best Kept Secret*), Ralph Erskine (*Action This Day*), John Jackson (*The Official History of British Sigint*), Lindsey Butters (*Bletchley Park: Home of Station X*), Richard J. Aldrich (*GCHQ*), and Paul Gannon (*Colossus: Bletchley Park's Greatest Secret*).

Although there were a few references to Allied wartime cryptographic successes in 1968, the deluge began in 1974 with Fred Winterbotham's *The Ultra Secret*, which had been a bid to preempt Anthony Cave Brown, who had learned about Bletchley Park while researching *Bodyguard of Lies*. Thereafter, several of those who had worked at Station X, including Ralph Bennett (*Ultra in the West*), Peter Calvcoressi (*Top Secret Ultra*), John Croft (*Reminiscences of GCHQ and GCB*), Mavis Baty (*Dilly: The Man Who Broke Enigmas*), Marion Hill (*Bletchley Park People*), Gwen Watkins and Asa Briggs (*Cracking the Luftwaffe Codes*), Alan Stripp (*Codebreaker in the Far East*), Harry Hinsley (*Codebreakers: The Inside Story of Bletchley Park*), David Hunt (*A Don at War*), John Cairncross (*The Enigma Spy*), and Gordon Welchman (*The Hut Six Story*), published accounts of their work, having been freed by Foreign Secretary Dr. David Owen in 1977 from some of their undertakings of secrecy. While restrictions remained on the technical details of precisely *how* Enigma had been broken, personnel were advised that they could discuss publicly the impact of their work.

Thirdly, there are the personal memoirs of those who participated in what has been termed the "secret wireless war," and finally there are the books that record particular incidents, such as the construction during the Cold War of the Berlin tunnel in 1954, the 1964 Gulf of Tonkin incident, the attack in 1967 on the USS *Liberty*, the seizure of the USS *Pueblo* in 1968, the loss of Korean Airlines Flight KAL 007 in August 1983, and the individual cases of security lapses, among them Jack Dunlap, Robert Lipka, Bernon Mitchell, William Martin, Victor Hamilton, Christopher Boyce, John Walker, Geoffrey Prime, Ronald Pelton, James Hall III, Robert Hanssen, Aldrich Ames, Ana Montes, and Katharine Gun. From the Soviet perspective of course, the defections of Igor Gouzenko, Vladimir and Evdokia Petrov, and Victor Sheymov represent equally egregious breaches.

The fact that warfare had extended into the ether remained a closely guarded secret until 1920 when Ferdinand Tuohy made some references to SIGINT in his postwar history of the Intelligence Corps, *The Secret Corps*, published in 1920. Tuohy concentrated on interception and direction finding in the context of monitoring the movements of zeppelins and the Kaiser's navy, remarking that "hundreds of ships were located yearly, and thousands of code groups intercepted—all to be classified and studied and restudied day after day as each fresh wireless development arose. . . . Our operators, after years of listening, were able to recognize the principal German operators by their method of transmitting no matter what call-signs they used." The following year, Filson Young, who had served as a junior officer on Admiral David Beatty's flagship at Jutland, handling his communications, wrote *With the Battle Cruisers* and revealed that the enemy's wireless traffic had been intercepted and decrypted, adding that "how it came to be decoded and deciphered is a matter that is, I am informed, even now undesirable to explain."

> The great point was that we read clearly all they said, whether it was in code or cipher. To decide from the mass of material thus obtained an intelligent knowledge of the doings of the German fleet required another staff of experts to coordinate, study and digest—and this is where we lost some of the great advantages which this system should have given us. Obviously it should have been entirely in naval control; and it was not. It was the old story of personal jealousy interfering with public interest. A piece of grit got into the machine here with sad consequences. The result of analysis of the material by non-naval minds, and not the material itself, was given to the Intelligence Division.

No further book raised the sensitive topic until 1927, when Winston Churchill's *The World Crisis* revealed that the German High Seas Fleet had been tracked through the use of "cipher and signal-books of the German navy" removed from the *Magdeburg*. His was indeed a sensational disclosure, all the more so because when Sir Alfred Ewing had touched on his wartime cryptanalytic experiences in the Admiralty during a lecture to the Edinburgh Philosophical Institution in December 1927, he had received a severe official warning about his indiscretion. On that occasion he had observed that

> besides intercepting naval signals, the cryptographers dealt successfully with each cipher to Germany's agents in Madrid, North and South America, Constantinople, Athens, Sofia and other places. One group of deciphered messages threw useful light in advance on the Easter Rising in Ireland, and another on the German intrigues in Persia. Among the many political messages read by the staff was the notorious Zimmermann Telegram.

Ewing's reference to the Zimmermann episode was not entirely new, for in October 1925, the former U.S. ambassador in London, Dr. Walter Page, had credited Admiral Sir Reginald Hall as having been responsible for delivering the decrypted version to him in 1917. Somewhat reluctantly, Hall, a Member of Parliament since 1919, had on that occasion confirmed his role in the affair but thereafter had remained silent on the issue, noting that "it is a tradition of our service never to talk of our work." When, in 1930, Sir George Aston proposed to include references to Room 40 in his *Secret Service*, he was dissuaded from doing so, and his publication, in the view of the Admiralty, made "articles on Naval War Secrets unsaleable for the future."

The Royal Navy's tradition of professional silence on intelligence topics would be undermined in 1931 by two journalists, Hector Bywater of the *Daily Telegrap*h and H. C. Ferraby, the *Daily Express*'s naval correspondent, when they wrote *Strange Intelligence: Memoirs of Naval Secret Service*, in which the two authors claimed, in a chapter entitled "The Men Who Heard the U-boats Talk," that German submarines had been tracked by the Admiralty's Intelligence Division. "The sources of the information were many and varied. The most valuable of all were the wireless direction stations round our coast," but they refrained from going further, asserting that some of the techniques were "still too confidential to be divulged even after this lapse of time." Significantly, the book's second edition contained a foreword written by Hall, who had been refused permission to release his own memoirs (which had been ghosted by Ralph Strauss) and who had employed Bywater during the war. According to Nobby Clarke, Hugh Cleland Foy had been nothing more than "a drunken typist who knew nothing."

Any controversy over *Strange Intelligence* was to be overshadowed by the astonishing *The American Black Chamber* in which Herbert Yardley reopened the Zimmerman affair by stating "this telegram was deciphered by the British Cryptographic Bureau early in 1917, just before we entered the war," and in a chapter entitled "The British Cipher Bureau" Yardley stated that Admiral Hall "because of the information he obtained from the messages that his enormous bureau deciphered, stood next to Lloyd George in power." Yardley recalled having received from Hall "several copies of a certain neutral government's diplomatic codes and a copy of the German naval code in two volumes" and having liaised with Colonel Arthur French of MI1(b) in 1918. He also mentioned having met two other British cryptographers, Captain Brooke-Hunt and Captain Hitchings, "who, according to his superiors, was worth four divisions to the British Army." Yardley concluded that he "did not wonder that England was a great power, for she read practically every code telegram that passed over her cables." Yardley's indiscretions led to Congress passing a law to prevent the unauthorized disclosure of official information, known as the Yardley Act, "For the Protection of Government Records," imposing a penalty of up to 10 years' imprisonment and a fine of $10,000 upon conviction.

Although Yardley narrowly escaped prosecution, because the law was not retrospective, his book encouraged Hector Bywater the following year to publish *Their Secret Purposes: Dramas and Mysteries of the Naval War*, which contained a chapter "Listening-In: Intelligence by Wireless" in which the author asserted that the U.S. State Department's now notorious Black Chamber had a "counterpart at the British Admiralty." Further details of Hall's wartime activities were revealed in the same year by another of his former staff, Hugh Cleland Hoy, whose *40 O.B.* was prefaced by a contribution from Sir Basil Thomson, formerly the director of intelligence at the Home Office. In giving a comprehensive account of wartime naval-intelligence operations, Hoy returned to the Zimmermann episode and said the telegram had been "picked up in the ordinary way in code by the listening stations on the east coast and flashed immediately to the Admiralty along with other messages to be deciphered."

The fact that no effort had been made to suppress *40 O.B.* may have encouraged Walter Gill two years later to release *War, Wireless and Wangles*, arguably the first open-source book to give a detailed account of signals-intelligence operations conducted in World War I. Gill had been commissioned into what he described as "Royal Engineers Wireless Intelligence" to build intercept stations at Alexandria and across the Near East, while insisting that "no secret is being given away in stating that it is possible to intercept a wireless message, and in some cases to decipher it." He recalled the construction of a 160-foot antenna atop the Great Pyramid in 1916 and revealed that the ciphers employed by German troops in Palestine had been broken but that a cryptographic center in Cairo had been able to assist Admiral Hall's experts in London by providing the clear text of re-ciphered enemy signals.

Gill's very entertaining memoirs, illustrated with his own cartoons, were significant because he referred to the British wireless-intercept capability in the Middle East, which until 1923 had concentrated on Turkish traffic and then had been absorbed into No. 2 Wireless Company in Sarafand, commanded by Colonel J. P. G. Worlledge. Upon the outbreak of World War II, the SIGINT experience of both Gill and Worlledge would be called up by the Radio Security Service.

Nothing further on this subject would be published for a further five years, until the release of *The Man of Room 40*, a biography of Sir Alfred Ewing written by his son. Published just before the outbreak of World War II, it is not surprising that it would be another 16 years before the topic of SIGINT was raised again in print, this time by Admiral Sir William James in *The Eyes of the Navy*, his biography of "Blinker" Hall who had died in October 1943. James had served as Hall's gunnery officer aboard HMS *Natal*, his commander on HMS *Queen Mary*, and his deputy in the Naval Intelligence Division, taking over the running of Room 40 in 1917 after Ewing's departure. Accordingly, James was an authoritative figure and would be cited two years later when the distinguished American historian Barbara Tuchman published her detailed analysis *The Zimmermann Telegram*.

Perhaps equally notorious as an enemy message that was decrypted during World War I was the German intelligence signal referring to H-21, a spy later identified by the French Deuxième Bureau as Mata Hari. Although the mythology surrounding her is probably unjustified, her 1930 biographer, Thomas Coulson, ensured that her role as a femme fatale would endure, even if the decrypts presented at her trial, which sealed her fate, would not. Thus, what is arguably the conflict's best-known case of espionage, with her name becoming synonymous with the seductive skills of women spies, was also an early example of signals intelligence being disclosed in a courtroom.

There can be no doubt that the authorities exercised considerable influence during the era following World War II to prevent public discussion of SIGINT; although, no such sanctions were available to prevent the American journalists David Wise and Thomas Ross from making passing references to the National Security Agency (NSA) in *The U-2 Affair* in 1962, *The Invisible Government* in 1964, and *The Espionage Establishment* in 1967. Certainly David Kahn was persuaded to delete every mention of Government Communications Headquarters and its predecessor, the Government Code and Cipher School, from *The Codebreakers*, and Prime Minister Harold Wilson reacted with fury in February 1967 when the veteran *Daily Express* defense correspondent Chapman Pincher had splashed that the "Post Office security department" was being supplied with copies of private overseas cables by the carriers. "It has not been possible to establish when this routine cable probe was introduced but I understand it has been in operation for several years." His information had come from Robert Lawson, a clerk previously employed by Commercial Cables and Western Union, and the story seemed to be a very blatant breach of D-Notice No. 2, issued in October 1961, which required silence on

> the various methods used in the interception of foreign communications for secret intelligence purposes. In this connection the Committee requests you will not refer to the fact that on occasions it is necessary in the interest of defense for the Services to intercept such communications.

Pincher argued that he had received consent to publish by the D-Notice secretary, Colonel Sammy Lohan, an opinion shared by a government panel that reviewed the incident subsequently, but Wilson remained unconvinced and issued a highly contentious White Paper criticizing Pincher and his newspaper. Lohan was sacked, and the issue of D-Notices, rather than the actual subject of the scoop, the interception of communications, became the major political issue of the day. Although many had known about the existence of the British system of self-censorship, subscribed to by both Fleet Street and London's book publishing industry, few had ever seen the texts of the D-Notices, which were themselves considered highly confidential. Accordingly, whereas Americans taking advantage of the First Amendment of the U.S. Constitution could expect to enjoy the protection of the U.S. Supreme Court, any British writer would know their material was likely to be the subject of a ban, if not criminal proceedings. In these circumstances, it is hardly surprising that there was considerable reticence when the ULTRA secret was first disclosed or that it took so long for the century's other great SIGINT secret, the VENONA project, to be revealed.

VENONA's very existence remained a closely guarded secret until 1980, for although Kim Philby published his memoirs, *My Silent War*, in 1968, and made many references to a joint Anglo-American counterespionage operation that led to the arrest of Klaus Fuchs and was in pursuit of a leak codenamed HOMER from inside the British embassy in wartime Washington, DC, he made no mention of BRIDE. This was a strange omission, but very likely he wanted to avoid drawing attention to the material that he must have feared, even from the safety of Moscow, could contain compelling proof not just of his own duplicity but also hot leads to the treachery of so many others. In fact, it was not until 1980 that the *Newsweek* journalist David C. Martin in *Wilderness of Mirrors* gave an account of James Angleton's career and that of Bill Harvey and described a massive cryptological undertaking that had originated with the capture of a Soviet codebook in Finland, only to be betrayed by a spy inside the Armed Forces Security Agency, William Weisband. Martin admits that Angleton had encouraged his "first thoughts of writing a book about Harvey," but the two men fell out as Martin became increasingly critical of the Central Intelligence Agency's (CIA) counterintelligence culture in which supposition and conjecture became blurred with reality.

Despite Martin's revelations, the VENONA cat remained firmly within the bag, as is demonstrated by James Bamford's *The Puzzle Palace*, the very first detailed history of the National Security Agency (NSA), published in 1982, omitting any reference to BRIDE or VENONA. The person who really pursued VENONA, and many other SIGINT topics besides, was Peter Wright, a former Marconi technician whose *Spycatcher* became a cause célèbre in 1986, mainly because of the British government's attempts to suppress it. Although Wright and his coauthor, Paul Greengrass, relied on the MI5 officer's memory, he revealed more about VENONA than any person hitherto, and he may have had an impact on Senator Daniel Patrick Moynihan, who pressed the Bill Clinton White House to declassify the decrypts.

The relative dearth of information was, of course, quite deliberate, for almost all those who had been indoctrinated into the operation hoped that at some future moment a breakthrough would be achieved that would reveal more of the unrecovered groups. To their amazement, the principal enemy, Soviet totalitarianism, simply collapsed, and that astonishing development more than any other allowed the Cold War to be brought to a close and its greatest secret revealed.

The release of VENONA in 1995 generated a small industry devoted to putting the decrypts into their proper context. The NSA analyst responsible for the declassification process, Robert Louis Benson, coauthored a volume of explanation to accompany the releases (*Venona: Soviet Espionage and the American Response, 1939–1957*), and a series of monographs followed, concentrating on particular aspects of the Soviet messages. Then the historians piled in, with John Earl Haynes and Harvey Klehr (*Decoding Soviet Espionage in America*), Allen Weinstein (*The Haunted Wood*), Alexander Vassiliev (*Spies*), Hayden Peake (*OSS and the Venona Decrypts*), Eric Breindel and Herbert Romerstein (*The Venona Secrets*), Joseph Albridge and Marcia Kunstel (*Bombshell*), and Marcia and Thomas Mitchell (*The Spy Who Seduced America*) in the forefront.

After the VENONA revelations, James Bamford, himself a former U.S. Naval Security Group member, returned to write more about the NSA in *Body of Secrets*, while Matthew M. Aid, who once worked at Chicksands Priory as a Russian intercept operator, contributed *The Secret Sentry*. More recently, Richard Aldrich has published *GCHQ*, the most up-to-date account of the organization, even if he was unable to include the BRUSA and UKUSA documents, which were declassified a few weeks too late for his first edition.

Naturally, the closer one comes to considering the signals-intelligence issues of the present day, the more sensitive they become, as President George W. Bush concluded when he opted to use his executive powers to extend the NSA's reach instead of taking the more public route of amending the 1979 Foreign Intelligence Surveillance Act, and the factual literature tails off, leaving plenty of polemic material, ill-informed speculation, and some mischief-making. While extracting authentic information is difficult, Nicky Hager succeeded in 1996, through dogged research, to produce *Secret Power*, a study of New Zealand's Government Communications Security Bureau, and Desmond Ball accomplished much the same with *A Base for Debate*, concentrating on the American satellite ground station at Nurrungar, just as he had examined Pine Gap eight years earlier in *A Suitable Piece of Real Estate*. However, such detailed studies of SIGINT are rare indeed.

REFERENCE WORKS

Adams, Jefferson. *Historical Dictionary of German Intelligence*. Lanham, MD: Scarecrow, 2009.

Blackstock, Paul, and Frank Schaf Jr. *Intelligence, Espionage, Counterespionage, and Covert Operations: A Guide to Information Sources*. Detroit, MI: Gale Research, 1978.

Constantinides, George. *Intelligence and Espionage: An Analytic Bibliography* Boulder, CO: Westview, 1983.

Kahama, Ephraim. *Historical Dictionary of Israeli Intelligence*. Lanham, MD: Scarecrow, 2006.

Kross, Peter. *The Encyclopedia of World War II Spies*. Fort Lee, VA. Barricade, 2001.

Minnick, Wendell L. *Spies and Provocateurs*. London: McFarland, 1992.

O'Toole, G. L. A. *The Encyclopedia of American Intelligence and Espionage*. New York: Facts on File, 1988.

Parrish, Michael. *Soviet Security and Intelligence Organizations, 1917–1990*. Westport, CT: Greenwood, 1992.

Polmar, Norman, and Thomas Allen. *Spy Book: The Encyclopedia of Espionage*. New York: Random House, 1997.

Pringle, Robert W. *Historical Dictionary of Russian/Soviet Intelligence*. Lanham, MD: Scarecrow, 2006.
Sexton, Donal J. *Signals Intelligence in World War II: A Reader's Guide*. Portsmouth, NH: Greenwood, 1996.
Trenear-Harvey, Glenmore. *Historical Dictionary of Air Intelligence*. Lanham, MD: Scarecrow, 2008.
Turner, Michael A. *Historical Dictionary of United States Intelligence*. Lanham, MD: Scarecrow, 2006.
West, Nigel. *Historical Dictionary of British Intelligence*. Lanham MD: Scarecrow, 2005.
———. *Historical Dictionary of Cold War Counterintelligence*. Lanham, MD: Scarecrow, 2007.
———. *Historical Dictionary of International Intelligence*. Lanham, MD: Scarecrow, 2006.
———. *Historical Dictionary of World War II Intelligence*. Lanham, MD: Scarecrow, 2008.

SIGNALS INTELLIGENCE

Aid, Matthew M. *The Secret Sentry*. London: Bloomsbury, 2009.
Aldrich, Richard. *GCHQ*. London: HarperCollins, 2010.
Arnold, David Christopher. *Spying from Space*. College Station: Texas A&M University Press, 2005.
Ball, Desmond. *A Base for Debate: The US Satellite Station at Nurrungar*. Sydney, Australia: Unwin Hyman, 1988.
———. *Soviet Ears in the Ether*. Canberra: Australian National University, 1984.
———. *A Suitable Piece of Real Estate*. Sydney, Australia: Hale and Iremonger, 1980.
Bamford, James. *Body of Secrets*. New York: Doubleday, 2001.
———. *The Puzzle Palace*. Boston: Houghton Mifflin, 1982.
Brownell, George. *The Origin and Development of the National Security Agency*. Laguna Hills, CA: Aegean Park, 1981.
Burrows, William E. *Deep Black*. New York: Random House, 1996.
Carroll, John M. *Secrets of Electronic Espionage*. New York: Dutton, 1966.
Clark, Phillip. "Aspects of the Soviet Photo-reconnaissance Satellite Programme." *Journal of the British Interplanetary Society* 36 (April 1983): 169–84.
Day, Dwayne, John Logsdon, and Brian Latell, eds. *Eye in the Sky*. Washington, DC: Smithsonian Institution, 1998.
Freedman, Lawrence. *The Official History of the Falklands Campaign*. London: Routledge, 2006.
Frost, Mike. *Spyworld: Inside the Canadian and American Intelligence Establishments*. Toronto: Doubleday Canada, 1994.
Hager, Nicky. *Secret Power: New Zealand's Role in the International Spy Network*. Nelson, New Zealand: Craig Heaton, 1996.
Heaps, Leo. *Operation Morning Light*. London: Paddington, 1978.

Hochman, Sandra. *Satellite Spies*. With Sybil Wong. New York: Bobbs-Merrill, 1976.
Hough, Harold. *Satellite Surveillance*. Port Townsend, WA: Loompanics Unlimited, 1991.
Kane, Jock. *GCHQ: The Negative Asset* (Unpublished, 1983).
Killian, James. *Sputnik, Scientists, and Eisenhower*. Cambridge, MA: MIT Press, 1982.
Klass, Philip. *Secret Sentries in Space*. New York: Random House, 1971.
Lanning, Hugh, and Richard Norton-Taylor. *A Conflict of Loyalties: GCHQ, 1984–1991*. Cheltenham, UK: New Clarion, 1991.
McMullen, David. *Chinook! The Special Forces Flight in War and Peace*. London: Simon and Schuster, 1998.
Miller, Jay. *Lockheed SR-71*. Arlington, TX: Aerojax, 1985.
Peebles, Curtis. *The Corona Project*. Annapolis, MD: Naval Institute, 1997.
——. *Guardians*. Novato, CA: Presidio, 1987.
Penkovsky, Oleg. *The Penkovsky Papers*. Garden City, NJ: Doubleday, 1965.
Petrov, Vladimir, and Evdokia Petrov. *Empire of Fear*. New York: Praeger, 1956.
Richelson, Jeffrey. *America's Space Sentinels*. Lawrence: University Press of Kansas, 1999.
——. *Foreign Intelligence Organizations*. Cambridge, MA: Ballinger, 1988.
Sheymov, Victor. *Tower of Secrets*. Annapolis, MD: Naval Institute, 1993.
Taubman, Philip. *Secret Empire*. New York: Simon and Schuster, 2003.
Taylor, John W. R., and David Monday. *Spies in the Sky*. New York: Charles Scribner, 1972.
Thorp, David J. *The Silent Listener: British Electronic Surveillance*. Stroud, Gloucestershire, UK: Spellmount, 2011.
Wiley, Richard G. *The Interception of Radar Signals*. Norwood, MA: Artech House, 1985.

EUROPE

Andrew, Christopher. *Secret Service*. London: Heinemann, 1985.
Borovik, Genrihk. *The Philby Files*. London: London: Little, Brown, 1994.
Costello, John. *Mask of Treachery*. London: Collins, 1988.
Damaskin, Igor. *Kitty Harris: The Spy with Seventeen Names*. London: St. Ermin's, 2001.
Foote, Alexander. *Handbook for Spies*. London: Museum, 1964.
Heiskanen, Raimo. *Stella Polaris*. Helsinki: Otava, 1994.
Knightley, Philip. *Philby: KGB Masterspy*. London: Jonathan Cape, 1997.
Koehler, John O. *Stasi: The Untold Story of the East German Secret Police*. Boulder, CO: Westview, 1999.
Macrakis, Christie. *Seduced by Secrets: Inside the Stasi's Spy-Tech World*. Cambridge: Cambridge University Press, 2008.

Mitchell, Marcia, and Thomas Mitchell. *The Spy Who Tried to Stop a War*. New York: Polipoint, 2008.
Sudoplatov, Pavel. *Special Tasks*. Boston: Little, Brown, 1994.
Werner, Ruth. *Sonia's Report*. London: Chatto and Windus, 1991.
Whitehouse, Arch. *Espionage and Counterespionage*. New York: Doubleday, 1964.
Wright, Peter. *Spycatcher*. New York: Viking, 1987.

UNITED STATES

Allen, Thomas, and Norman Polmar. *Merchants of Treason: America's Secrets for Sale*. New York: Delacorte, 1988.
Anonymous. *British Security Coordination*. London: St. Ermin's, 1998.
Barron, John. *Breaking the Ring: The Bizarre Case of the Walker Family Spy Ring*. Boston: Houghton Mifflin, 1987.
Batvinis, Ray. *The Origins of FBI Counterintelligence*. Lawrence: University Press of Kansas, 2007.
Bernikow, Louise. *Abel*. New York: Trident, 1970.
Blum, Howard. *I Pledge Allegiance . . . : The True Story of the Walkers: An American Spy Family*. New York: Simon and Schuster, 1987.
Borne, John E. *The USS Liberty: Dissenting History vs. Official History*. New York: Reconsideration, 1995.
Brandt, Ed. *The Last Voyage of the USS* Pueblo. New York: Norton, 1969.
Burrows, William E. *Deep Black*. New York: Random House, 1986.
Carmichael, Scott W. *The Believer*. Annapolis, MD: Naval Institute, 2007.
Clark, Robert M. *The Technical Collection of Intelligence*. Washington, DC: CQ, 2009.
Crawford, Don. Pueblo *Intrigue: A Journey of Faith*. Wheaton, IL: Tyndale House, 1969.
Cristol, A. Jay. *The Liberty Incident: The Israeli Attack on the U.S. Navy Spy Ship*. Washington, DC: Brassey's, 2002.
Dallin, Alexander. *Black Box: KAL 007 and the Superpowers*. Berkeley: University of California Press, 1985.
Dallin, David. *Soviet Espionage*. New Haven, CT: Yale University Press, 1955.
Damashkin, Igor. *The Spy with Seventeen Names*. London: St. Ermin's, 2004.
Deriabin, Piotr. *The Secret World*. New York: Doubleday, 1959.
Earley, Pete. *Family of Spies: Inside the John Walker Spy Ring*. New York: Bantam, 1988.
Ennes, James. *Assault on the* Liberty*: The True Story of the Israeli Attack on an American Intelligence Ship*. New York: Random House, 1979.
Gallery, Rear Admiral (Ret.) Daniel V. *The* Pueblo *Incident*. Garden City, NY: Doubleday, 1970.

Granatstein, J. L., and David Stafford. *Spy Wars: Espionage and Canada from Gousenko to Glasnost*. Toronto: Key Porter Books, 1990.

Haynes, John Earl, Harvey Klehr, and Alexander Vassiliev. *Spies: The Rise and Fall of the KGB in America*. New Haven, CT: Yale University Press, 2009.

Hendricks, Steve. *A Kidnapping in Milan*. New York: Norton. 2010.

Herrington, Stuart A. *Traitors among Us*. Novato, CA. Presidio, 1999.

Hunter, Robert W. *Spy Hunter: Inside the FBI Investigation of the Walker Espionage Case*. Annapolis, MD: Naval Institute, 1999.

Hyde, H. Montgomery. *Room 3603*. New York: Farrar, Straus, 1962.

Johnson, Thomas R. *American Cryptology during the Cold War, 1945–1989*. Washington, DC: National Security Agency, declassified 2006.

Kahn, David. *The Codebreakers*. New York: Scribner, 1996.

Kneece, Jack. *Family Treason: The Walker Spy Case*. Briarcliff Manor, NY: Stein and Day, 1986.

Lerner, Mitchell, B. *The* Pueblo *Incident: A Spy Ship and the Failure of American Foreign Policy*. Lawrence: University Press of Kansas, 2002.

Liston, Robert A. *The* Pueblo *Surrender: A Covert Action by the National Security Agency*. New York: Evans, 1988.

Massing, Hede. *This Deception*. London: Sloane and Pearce, 1951.

Olmsted, Kathryn. *Red Spy Queen*. Chapel Hill: University of North Carolina Press, 2002.

Pearson, Anthony. *Conspiracy of Silence: The Attack on the U.S.S.* Liberty. London: Quartet Books, 1968.

Pearson, David. *KAL 007: The Cover-Up*. NY: Summit Books, 1987.

Richelson, Jeffrey T. *America's Space Sentinels*. Lawrence: University Press of Kansas, 1999.

Richelson, Jeffrey T., and Desmond Ball. *The Ties That Bind*. Sydney: Allen and Unwin, 1985.

Schlossberg, Bert. *Rescue 007: The Untold Story of KAL 007 and Its Survivors*. South Windsor, CT: International Committee for the Rescue of KAL 007 Survivors, 2000.

Sontag, Sherry, and Christopher Drew. *Blind Man's Bluff*. Annapolis, MD: Naval Institute, 1999.

Tully, Andrew. *The Super Spies*. New York: William Morrow, 1969.

Vyborny, Lee, and Don Davis. *Dark Waters: An Insider's Account of the NR1, the Cold War's Undercover Nuclear Sub*. New York: Penguin Putnam, 2003

Walker, John. *My Life as a Spy*. New York: Prometheus Books, 2008.

Walker, Laura. *Daughter of Deceit: The Human Drama behind the Walker Spy Case*. With Jerry Horner. Dallas: Word, 1988.

Weinstein, Allen, and Alexander Vassiliev. *The Haunted Wood*. New York: Random House, 1999.

Woodward, Bob. *Veil: The Secret Wars of the CIA, 1981–1987*. New York: Simon and Schuster, 1987.

WORLD WAR I SIGNALS INTELLIGENCE

Aston, Sir George. *Secret Service*. London: Faber and Faber, 1930.

Baker, W. J. *A History of the Marconi Company*. New York: St. Martin's: 1971.

Beesly, Patrick. *Room 40*. New York: Harcourt Brace Javanovich, 1982.

Bontaz, Heinz. *Die Deutsche Marine-Funkaufklärung*. Darmstadt, 1970.

Churchill, Winston. *The World Crisis, 1911–1918*. London: Thornton Butterworth, 1932.

Coulson, Major Thomas. *Mata Hari: Courtesan and Spy*. New York: Harper, 1930.

Ewing, Alfred. *The Man of Room 40*. London: Hutchinson, 1940.

Ferris, Paul, ed. *The British Army and Signals Intelligence during the First World War*. Stroud, Gloucestershire, UK: Allan Sutton, 1992.

Fisher, Admiral of the Fleet Lord. *Memories*. London: Hodder and Stoughton, 1919.

Foss, Kenelm, ed. *The Best of A. J. Alan*. London: Richards,1954.

Freeman, Peter. "MI1(b) and the Origins of British Diplomatic Cryptanalysis." *Intelligence and National Security* 22, no. 2 (April 2007): 206–28.

Friedman, William F. *Solving German Codes in World War I*. Laguna Hills, CA: Aegean Park, 1977.

Friedman, William F., and Charles J. Mendelsohn. *The Zimmermann Telegram of January 16, 1917 and Its Cryptographic Background*. Laguna Hills, CA: Aegean Park, 1994.

Gannon, Paul. *Inside Room 40*. London: Ian Allan, 2011.

Gill, E. B. *War, Wireless and Wangles*. Oxford: Basil Blackwell, 1934.

Hay, Alice Ivy. *Valiant for Truth: Malcolm Hay of Seaton*. London: Neville Spearman, 1971.

Hay, Malcolm. *Wounded and a Prisoner of War, by an Exchange Officer*. New York: G. H. Doran, 1917.

Howe, Russell Warren. *Mata Hari: The True Story*. New York: Dodd Mead, 1986.

Huebsch, Edward. *The Last Summer of Mata Hari*. New York: Crown, 1979.

James, William. *The Eyes of the Navy*. London: Metheun, 1955.

Jolly, W. P. *Marconi*. New York: Stein and Day 1972.

Keay, Julia. *The Spy Who Never Was: The Life and Loves of Mata Hari*. London: Michael Joseph, 1987.

Millar, Ronald. *Mata Hari*. Geneva: Heron Books, 1970.

Morgan, Janet. *The Secrets of the Rue St. Roch*. London: Allen Lane, 2004.

Newman, Bernard. *Inquest of Mata Hari*. London: Robert Hale, 1956.

Nickles, David P. *Under the Wire: How the Telegraph Changed Diplomacy*. Cambridge, MA: Harvard University Press: 2003.

Occleshaw, Michael. *Armour against Fate: British Military Intelligence in the First World War*. London: Columbus Books, 1989.

Ostrovsky, Erika. *Eye of Dawn: The Rise and Fall of Mata Hari*. New York: Dorset, 1978.

Ramsay, David. *Blinker Hall, Spymaster*. Stroud, Gloucestershire, UK: Sutton, 2004.

Seligman, Matthew. *Spies in Uniform*. Oxford: Oxford University Press, 2006.

Singer, Kurt. *Mata Hari*. London: Tandem, 1967.

Tuchman, Barbara. *The Zimmermann Telegram*. New York: Dell, 1963.
Tuohy, Ferdinad. *The Secret Corps*. London: John Murray, 1920.
Waagenar, Sam. *Mata Hari: A Biography*. New York: AppletonCentury, 1965.
Wheelwright, Julie. *The Fatal Lover: Mata Hari and the Myth of Women in Espionage*. London: Collins and Brown, 1992.
Young, Filson. *With the Battle Cruisers*. Annapolis, MD: Naval Institute,1998.

WORLD WAR II SIGNALS INTELLIGENCE

Accoce, Pierre, and Pierre Quet. *The Lucy Ring*. London: W. H. Allen, 1967.
Aaseng, Nathan. *Navajo Code Talkers: America's Secret Weapon in World War II*. New York: Walker, 2000.
Baty, Mavis. *Dilly: The Man Who Broke Enigmas*. London: Biteback, 2009.
Beesly, Patrick. *Very Special Intelligence*. London: Hamish Hamilton, 1977.
Bennett, Ralph. *Ultra and Mediterranean Strategy*. New York: William Morrow, 1980.
———. *Ultra in the West*. London: Hutchinson, 1979.
Birch, Frank, and John Jackson, eds. *The Official History of British SIGINT, 1914–1945*. Vol. 1, part 1, of *The Bletchley Archive*. Milton Keynes, UK: Military Press, 2004.
Bixler, Margaret T. *Winds of Freedom: The Story of the Navajo Code Talkers of World War II*. Darien, CT: Two Bytes, 1995.
Boyd, Carl. *Hitler's Japanese Confidant*. Lawrence: University Press of Kansas, 1993.
Butters, Lindsey. *Bletchley Park: Home of Station X*. Hampshire, UK: Pitkin Unichrome, 2000.
Cairncross, John. *The Enigma Spy*. London: Random Century, 1997.
Calvocoressi, Peter. *Top Secret Ultra*. London: Cassell, 1980.
Clark, Ronald. *The Man Who Broke Purple*. Boston: Little, Brown, 1977.
Clayton, Aileen. *The Enemy Is Listening*. London: Hutchinson, 1980.
Croft, John. "Reminiscences of GCHQ and GCB 1942–45." *Intelligence and National Security* 13, no. 4 (Winter 1998): 133–43.
Denniston, A. G. "The Government Code and Cypher School between the Wars." *Intelligence and National Security* 1, no. 1 (1986): 48–70.
Dingman, Roger. *Deciphering the Rising Sun*. Annapolis, MD: Naval Institute, 2009.
Drea, Edward J. *MacArthur's Ultra: Codebreaking and the War against Japan 1942–45*. Lawrence: University Press of Kansas, 1986.
Ellingworth, Marshall. *The History of the War Office Y Group*. Unpublished, 1948.
Enever, Ted. *Britain's Best Kept Secret: ULTRA's Base at Bletchley Park*. Gloucestershire, UK: Alan Sutton, 1994.
Erskine, Ralph. "Churchill and the Start of the Ultra-Magic Deals." *International Journal of Intelligence and Counterintelligence* 10, no. 1 (1997): 57–74.
Ferris, John. "The Road to Bletchley Park. The British Experience with Signals Intelligence, 1892–1945." *Intelligence and National Security* 17, no. 1 (2002): 53–84.

Filby, P. William. "Bletchley Park and Berkeley Street." *Intelligence and National Security* 3, no. 2 (1988): 272–84.

———. "Floradora and a Unique Break into One-Time Pad Ciphers." *Intelligence and National Security* 10, no. 3 (1995): 408–22.

Fitzgerald, Penelope. *The Knox Brothers*. London: Macmillan, 1977.

Gannon, Paul. *Colossus: Bletchley Park's Greatest Secret*. London: Atlantic Books, 2006.

Hill, Marion. *Bletchley Park People: Churchill's Geese That Never Cackled*. Phoenix Mill, UK: Sutton, 2004.

Hinsley, F. H., and Alan Stripp, eds. *Codebreakers: The Inside Story of Bletchley Park*. Oxford: Oxford University Press, 1993.

Hinsley, Harry. *British Intelligence in the Second World War*. With E. E. Thomas, C. F. G. Ranson, and R. C. Knight. London: HMSO, 1979–88.

Hodges, Andrew. *Alan Turing: The Enigma*. London: Hutchinson, 1983.

Holmes, W. J. *Double-Edged Secrets: US Naval Intelligence Operations in the Pacific during World War II*. Annapolis, MD: Naval Institute, 1979.

Jackson, John. *Code Wars*. Barnsley, UK: Pen and Sword Books, 2011.

Kahn, David. *Seizing the Enigma*. Boston: Houghton Mifflin, 1991.

Lewin, Ronald. *American MAGIC*. New York: Farrar Strauss, 1982.

———. *ULTRA Goes to War*. London: Hutchinson, 1978.

Macksey, Kenneth. *The Searchers*. London: Cassell, 2003.

Page, Gwendoline. *They Listened in Secret*. Wymondham, UK: George Reeve, 2003.

Parker, Frederick D. *A Priceless Advantage: US Navy Communications Intelligence and the Battles of Coral Sea, Midway, and the Aleutians*. Washington, DC: National Security Agency, 2006.

Parrish, Thomas. *The Ultra Americans*. New York: Stein and Day, 1985.

Pearson, Joss. *Cribs for Victory*. Clifton, UK: Polperro Heritage Press, 2011.

Pearson, Michael. *Voices of the Code Breakers*. Newton Abbot, UK: David and Charles, 2007.

Prados, John. *Combined Fleet Decoded: The Secret History of American Intelligence and the Japanese Navy in World War II*. New York: Random House, 1995.

Skillen, Hugh. *Spies of the Airwaves*. Skillen, 1989.

Smith, Bradley, F. *The ULTRA-MAGIC Deals*. Shrewsbury, UK: Airlife, 1993.

Smith, Michael. *Station X*. London: Channel 4 Books, 1998.

Smith Michael, and Ralph Erskine, eds. *Action This Day: Bletchley Park from the Breaking of the Enigma Code to the Birth of the Modern Computer*. London: Bantam Books, 2000.

Spector, Ronald. *Listening to the Enemy*. Wilmington, DE: Scholarly Resources, 1988.

Stripp, Alan. *Codebreaker in the Far East*. London: Frank Cass, 1989.

Thomas, Andy. "British Signals Intelligence after the Second World War." *Intelligence and National Security* 3, no. 4 (1988): 103–10.

Watkins, Gwen. *Cracking the Luftwaffe Codes: The Secrets of Bletchley Park*. London: Greenhill Books, 2006.

Welchman, Gordon. *The Hut Six Story*. New York: McGraw-Hill, 1982.

Winterbotham, Frederick. *The Ultra Secret*. London: Weidenfeld and Nicolson, 1974.

COLD WAR SIGNALS INTELLIGENCE

Aid, Matthew. "The National Security Agency and the Cold War." *Intelligence and National Security* 16, no. 1 (2001): 27–66.
Aid, Matthew, and Cees Wiebes. "Introduction on the Importance of Signals Intelligence in the Cold War." *Intelligence and National Security* 16, no. 1 (2001): 1–26.
Aid, Matthew M., and Cees Wiebes, eds. *Secrets of Signals Intelligence during the Cold War and Beyond*. London: Frank Cass, 2001.
Aldrich, Richard. "GCHQ Sigint in the Early Cold War 1945–70." *Intelligence and National Security* 16, no. 1 (2001): 67–96.
Aronsen, Lawrence. "Seeing Red: U.S. Air Force Assessments of the Soviet Union, 1945–1949." *Intelligence and National Security* 16, no. 2 (2001): 103–32.
Boyd, J. A., D. B. Harris, D. D. King Jr., and H. W. Welch, eds. *Electronic Countermeasures*. Los Altos, CA: Peninsula, 1961.
Faligot, Roger. "France, Sigint and the Cold War." *Intelligence and National Security* 16, no. 1 (2001): 177–208.
Jacobsen, Alf. "Scandinavia, Sigint and the Cold War." *Intelligence and National Security* 16, no. 1 (2001): 209–42.
Lashmar, Paul. *Spy Flights of the Cold War*. Stroud, Gloucestershire, UK: Sutton, 1996.
Lindgren, David. *Trust but Verify: Imagery Analysis in the Cold War*. Annapolis, MD: Naval Institute, 2000.
Modin, Yuri. *My Five Cambridge Friends*. London: Headline, 1994.
Page, Bruce, Phillip Knightley, and David Leitch. *The Philby Conspiracy*. New York: Doubleday, 1968.
Penrose, Barry, and Simon Freeman. *Conspiracy of Silence*. London: Grafton, 1986.
Philby, Kim. *My Silent War*. London: McGibbon and Kee, 1968.
Philby, Rufina. *The Private Life of Kim Philby*. London: St. Ermin's, 1999.
Prime, Rhona. *Time of Trial: The Personal Story behind the Cheltenham Spy Scandal*. With Jean Watson. London: Hodder and Stoughton, 1984.
Ring, Jim. *We Come Unseen: The Untold Story of Britain's Cold War Submariners*. London: John Murray, 2001.
Seale, Patrick, and Maureen McConville. *Philby: The Long Road to Moscow*. London: Hamish Hamilton 1973.
Stafford, Davd. *Spies beneath Berlin*. London: John Murray, 2002.
Wiebes, Cees. "Dutch Sigint during the Cold War, 1945–94." *Intelligence and National Security* 16, no. 1 (2001): 243–84.
Urban, Mark. *UK Eyes Alpha: The Inside Story of British Intelligence*. London: Faber and Faber, 1996.

POST–COLD WAR SIGNALS INTELLIGENCE

Corera, Gordon. *The Art of Betrayal*. London: Weidenfeld and Nicolson, 2011.
Rudner, Martin. "Canada's Communications Security Establishment from Cold War to Globalization." *Intelligence and National Security* 16, no. 1 (2001): 97–128.

Schmit-Eenboom, Erich. "The Bundesnachrichtendienst, the Bundeswehr and Sigint in the Cold War and After." *Intelligence and National Security* 16, no. 1 (2001): 129–76.

Woodward, Bob. *Obama's Wars*. New York: Simon and Schuster, 2010.

VENONA

Adamson, Iain, *The Great Detective*. London: Frederick Muller, 1966.

Agrell, Wilhelm, and Bo Huldt, eds. *Clio Goes Spying: Eight Essays on the History of Intelligence*. Lund, Sweden: Department of History, University of Lund, 1983.

Albright, Joseph, and Kunstel, Marcia. *Bombshell*. New York: Random House, 1997.

Belfrage, Cedric. *The American Inquisition*. New York: Bobbs-Merill, 1973.

——. *The Frightened Giant*. London: Secker and Warburg, 1957.

——. *Something to Guard*. New York: Columbia University Press, 1978.

Benson, Robert Louis, and Michael Warner. *Venona: Soviet Espionage and the American Response 1939–1957*. Washington, DC: National Security Agency, 1996.

Bentley, Elizabeth. *Out of Bondage*. New York: Devin-Adair, 1951.

Bly, Herman O. *Communism: The Cold War and the FBI Connection*. New York: Huntingdon House, 1998.

Bohm, David. *Unfolding Meaning*. Abingdon, UK: Routledge and Kegan Paul, 1985.

Breindel, Eric, and Herbert Romerstein. *The Venona Secret*. New York: HarperCollins, 1999.

Burt, Leonard. *Commander Burt of Scotland Yard*. London: Heinemann, 1959.

Carpozi, George. *Red Spies in Washington*. New York: Trident, 1965.

Clark, Ronald. *J. B. S.* London: Quality Books, 1968.

Clubb, O. Edmund. *The Witness and I*. New York: Columbia University Press, 1974.

Dallin, David. *Soviet Espionage*. New Haven, CT: Yale University Press, 1955.

Donovan, James B. *Strangers on a Bridge*. London: Atheneum, 1964.

Gamow, George. *My World Line*. New York: Viking, 1970.

Haldane, Charlotte. *Truth Will Out*. London: Vanguard, 1950.

Haldane, J. B. S. *The Inequality of Man, and Other Essays*. London: Chatto and Windus, 1932.

Haynes, John Earl, and Harvey Klehr. *Venona: Decoding Soviet Espionage in America*. New Haven, CT: Yale University Press, 1999.

Hiss, Tony. *Laughing Last: Alger Hiss*. Boston: Houghton Mifflin, 1957.

——. *The View from Alger's Window: A Son's Memoir*. New York: Knopf, 1999.

Huss, Pierre J., and George Carpozi. *Red Spies in the UN*. New York: Coward-McCann, 1965.

Jordan, George Racey. *From Major Jordan's Diaries*. New York: Harcourt, Brace, 1952.

Kalugin, Oleg. *The First Directorate*. New York: St. Martin's, 1994.

Kessler, Lauren. *Clever Girl*. New York: HarperCollins, 2003.

Kirschner, Don S. *Cold War Exile*. Columbia: University of Missouri Press, 1995.

Klehr, Harvey, and John Haynes. *The Secret World of American Communism*. New Haven, CT: Yale University Press, 1995.
Klehr, Harvey, and Ronald Radosh. *The Amerasia Spy Case*. Chapel Hill: University of North Carolina Press, 1996.
Knightley, Philip. *The Hack's Progress*. London: Bloomsbury, 1997.
Kravchenko, Viktor. *I Chose Freedom*. London: Robert Hale, 1947.
——. *I Chose Justice*. London: Scribner, 1950.
Lamphere, Robert. *The CIA-KGB War*. New York: Random House, 1986.
Martin, David C. *Wilderness of Mirrors*. New York: Harper and Row, 1980.
Modin, Yuri. *My Five Cambridge Friends*. London: Hodder Headline, 1995.
Montagu, Ivor. *The Youngest Son*. London: Lawrence and Wishart, 1970.
Moorhead, Alan. *The Traitors*. New York: Harper and Row, 1952.
Morros, Boris. *My Ten Years as a Counterspy*. London: Werner Laurie, 1959.
Moynihan, Daniel Patrick. *Secrecy*. New Haven, CT: Yale University Press, 1998.
Pale, Erkii. *Totuus Stella Polariksesta*. Helsinki: Ahtokari, 1994.
——. *Suomen Radiotiedustelu 1927–1944*. Helsinki: Hakapaino Oy, 1998.
Peake, Hayden. "OSS and the Venona Decrypts." *Intelligence and National Security* 12, no. 3 (1997): 14–34.
Peierls, Rudolf. *Bird of Passage*. Princeton, NJ: Princeton University Press, 1985.
Philby, Kim. *My Silent War*. London: McGibbon and Kee, 1968.
Radosh, Ronald, and Joyce Milton. *The Rosenberg File*. New York: Holt, Rinehart and Winston, 1983.
Rees, David. *Harry Dexter White*. New York: Coward, McCann and Geoghegan, 1973.
Report of the Royal Commission. Ottawa, Canada, 1946.
Report of the Royal Commission. Canberra, Australia, 1956.
Sandilands, Roger J. *The Life and Political Economy of Lauchlin Currie: New Dealer, Presidential Adviser, and Development Economist*. Durham, NC: Duke University Press, 1990.
Smyth, Henry D. *Atomic Energy for Military Purposes*. Princeton, NJ: Princeton University Press, 1945.
Stephenson, William. *British Security Coordination 1941–45*. London: St. Ermin's, 1998.
Stone, I. F. *The Hidden History of the Korean War 1950–51*. London: Little, Brown, 1952.
——. *The War Years, 1939–45*. London: Little, Brown, 1988.
Whitaker, Paul, and Louis Kruh. "From Bletchley Park to Berchtesgaden." *Cryptologia* 11, no. 3 (1987): 129–41.

WEBSITES

Air Force Security Service http://unitpages.military.com/unitpages/unit.do?id=712251
Air Intelligence Agency www.afisr.af.mil

Army Security Agency www.nasaa-home.org
Bletchley Park www.bletchleypark.org.uk
Bundesnachrichtendienst www.bnd.de
Communications Security Establishment www.cse-cst.gc.ca
Danish Defense Intelligence http://fe-ddis.dk/eng/Pages/default.aspx
www.thehollandbureau.com/tag/fortress-langeland/
Defence Signals Directorate www.dsd.gov.au
Dutch Militaire Inlichtingendienst www.defensie.nl
Federation of American Scientists www.fas.org
Government Communications Headquarters www.gchq.gov.uk
Government Communications Security Bureau www.gcsb.gov.nz
Hanslope Park www.hmgcc.gov.uk
National Reconnaissance Office www.nro.gov
National Security Agency www.nsa.gov
Naval Security Group www.navycthistory.com
RAF Menwith Hill www. raf.mod.uk/organisation/rafmenwithhill.cfm
Russian FAPSI www.globalsecurity.org/intell/world/russia/fapsi-ops.htm
Russian Federal Protective Service www.fso.gov.ru
Swedish Försvarets Radioanstalt www.fra.se/snabblankar/english.10.html
UK SIGINT sites www.secret-bases.co.uk

Index

About the Author

Nigel West is a military historian specializing in intelligence and security issues, and while still at university, he worked as a researcher for two authors: Ronald Seth, who had been parachuted into Silesia by SOE, and Richard Deacon, a former wartime Naval Intelligence officer and latterly the foreign editor of *The Sunday Times*. He later joined BBC TV's General Features department to work on the *SPY!* and *ESCAPE* series.

His first book, coauthored with Richard Deacon in 1980 and published by BBC Publications, was the first book of the *SPY!* series and was followed by other nonfiction: *British Security Service Operations, 1909–45* (Bodley Head, 1981), *A Matter of Trust: MI5, 1945–72* (Weidenfeld and Nicolson, 1982), *MI6: British Secret Intelligence Service Operations, 1909–45* (Weidenfeld and Nicolson, 1983), *The Branch: A History of the Metropolitan Police Special Branch* (Secker and Warburg, 1983), *Unreliable Witness: Espionage Myths of the Second World War* (Weidenfeld and Nicolson, 1984), *GARBO* (coauthored with Juan Pujol, Weidenfeld and Nicolson, 1985), *GCHQ: The Secret Wireless War* (Weidenfeld and Nicolson, 1986), *Molehunt* (Weidenfeld and Nicolson, 1987), *The Friends: Britain's Postwar Secret Intelligence Operations* (Weidenfeld and Nicolson, 1988), *Games of Intelligence* (Weidenfeld and Nicolson, 1989), *Seven Spies Who Changed the World* (Secker and Warburg, 1991), *Secret War: The Story of SOE* (Hodder and Stoughton, 1992), *The Faber Book of Espionage* (Faber and Faber, 1993), *The Illegals* (Hodder and Stoughton, 1993), *The Faber Book of Treachery* (Faber and Faber, 1995), *The Secret War for the Falklands* (Little, Brown, 1997), *Counterfeit Spies* (Little, Brown, 1998), *Crown Jewels* (with Oleg Tsarev, HarperCollins, 1998), *VENONA: The Greatest Secret of the Cold War* (HarperCollins, 1999), *The Third Secret* (HarperCollins, 2000), *Mortal Crimes* (Enigma, 2004), *The Guy Liddell Diaries* (Routledge, 2005), *MASK* (Routledge, 2005), *Historical Dictionary of British Intelligence* (Scarecrow, 2005), *Historical Dictionary of International Intelligence* (Scarecrow, 2006), *On Her Majesty's Secret Service* (Greenhill, 2006), and *Historical Dictionary of Cold War Counterintelligence* (Scarecrow, 2007), *Historical Dictionary of World War II Intelligence* (Scarecrow, 2008), *Historical Dictionary of Sexspionage* (Scarecrow, 2009), *TRIPLEX* (with Oleg Tsarev, Yale University

Press, 2009), *Historical Dictionary of Naval Intelligence* (Scarecrow, 2010), and *SNOW: The Double Life of a World War II Spy* (with Madoc Roberts, Biteback, 2011).

In 1989, he was voted The Experts' Expert by a panel of spy writers selected by the *Observer*. He is currently the European editor of the *International Journal of Intelligence and Counterintelligence* and teaches the history of postwar intelligence at the Centre for Counterintelligence and Security Studies in Alexandria, Virginia. In October 2003, he was awarded the U.S. Association of Former Intelligence Officers first Lifetime Literature Achievement Award.

www.ingramcontent.com/pod-product-compliance
Ingram Content Group UK Ltd.
Pitfield, Milton Keynes, MK11 3LW, UK
UKHW041127071026
15587UKWH00009B/284

9 780810 871878